INTRODUCTION TO

ASTRONOMY

by **Theodore Metochites**

INTRODUCTION TO ASTRONOMY by Theodore Metochites

(Stoicheiosis Astronomike 1.5-30)

Emmanuel A. Paschos
Technische Universität Dortmund, Germany

Christos Simelidis
Aristotle University of Thessaloniki, Greece

World Scientific

NEW JERSEY • LONDON • SINGAPORE • BEIJING • SHANGHAI • HONG KONG • TAIPEI • CHENNAI • TOKYO

Published by

World Scientific Publishing Co. Pte. Ltd.
5 Toh Tuck Link, Singapore 596224
USA office: 27 Warren Street, Suite 401-402, Hackensack, NJ 07601
UK office: 57 Shelton Street, Covent Garden, London WC2H 9HE

Library of Congress Cataloging-in-Publication Data
Names: Paschos, E. A. (Emmanuel A.) | Metochites, Theodoros, 1270–1332. | Simelidis, Christos, 1977–
Title: Introduction to astronomy by Theodore Metochites (Stoicheiosis Astronomike 1.5-30) / Emmanuel Paschos (Technische Universitat Dortmund, Germany), Christos Simelidis (Aristotle University of Thessaloniki, Greece).
Description: New Jersey : World Scientific, 2017. | Includes bibliographical references and index.
Identifiers: LCCN 2017002294 | ISBN 9789813207486
Subjects: LCSH: Astronomy, Medieval. | Metochites, Theodoros, 1270-1332. | Astronomy--Byzantine Empire. | Astronomy--Early works to 1800. | Calendars--History--To 1500.
Classification: LCC QB23 .P36 2017 | DDC 520--dc23
LC record available at https://lccn.loc.gov/2017002294

British Library Cataloguing-in-Publication Data
A catalogue record for this book is available from the British Library.

Cover illustration: The model of the universe with the spherical Earth and the rotation of the Sun through the zodiac. From codex Dionysiou 224 (s. XVI), f. 490r. © Holy Monastery of Dionysiou, Mount Athos.

Back cover: Map of Constantinople (1422) by Cristoforo Buondelmonti. Wikimedia Commons

Desk Editor: Edward C. Yong

Printed in Singapore

PREFACE

This book makes available for the first time a large section of an important text of late Byzantine astronomy. The first five, philosophical, chapters of Theodore Metochites' *Stoicheiosis Astronomike* have been well edited and commented on by Börje Bydén. However, the astronomical bulk of the work remains unpublished and this impedes serious study and full appreciation. Metochites' treatise is significant for the study of medieval astronomy, and especially the reception of Ptolemy in Byzantium.

This work is the product of collaboration between a physicist and a philologist with each contributing in the field of his expertise. Emmanuel Paschos produced the Introduction, the English Translation, and the Analysis, while Christos Simelidis prepared the Textual Introduction and the Critical Edition of the Greek text. Metochites' sentences are often long and repetitive, making the text difficult to translate. Paschos tried to make the translation easier to read by limiting repetitions in order to preserve the content of sentences and of entire paragraphs.

We are grateful to Ioannis Vassis and Dimitrios A. Christidis for reading the Greek text and making valuable suggestions. We would also like to thank Andrew Faulkner, Mark Huggins, Nikolaos Konomis, Dimitri Korobeinikov, Maria Mavroudi, Georgi Parpulov, Filippomaria Pontani, Ioanna Skoura, Panagiotis Sotiroudis, Anne Tihon, and Nigel G. Wilson for their helpful advice and encouragement. We are indebted to Haralampos Kimikoglou for preparing the drawings and to Edward C. Yong for his editorial help in preparing the final camera-ready manuscript.

December 2015

Emmanuel A. Paschos
Department of Physics
Technische Universität Dortmund

Christos Simelidis
Department of Philology
Aristotle University of Thessaloniki

TABLE OF CONTENTS

1. Introduction

The *Stoicheiosis Astronomike* is an extensive treatise on astronomy from the late Byzantine period. It was written by Theodore Metochites (1270–1332), who was a distinguished politician serving in high positions (1305–1328), including the highest posts of Grand Logothete (Chancellor) and *mesazon* or prime minister, under the emperor Andronikos II Palaiologos. During his illustrious career as statesman and scholar, he produced, besides the treatise on astronomy, literary and scholarly works including commentaries on Aristotle, hexameter poems, orations and hagiographical *encomia*.[1] He donated part of his wealth for the restoration and redecoration of the church in the Chora Monastery, where his portrait survives in a mosaic of the narthex. He enriched the library of the monastery and bequeathed it to his student Nikephoros Gregoras.[2] In 1313, at the advanced age of 43, he studied astronomy under Manuel Bryennios and wrote his *Introduction to Astronomy*, which is the subject of this book.

Metochites' astronomical treatise expounds upon the movements of celestial bodies as they were understood in the late 13th and early 14th century. The text covers most topics related to the solar system: the Sun, the Moon and the five planets. It was written at a time in which Byzantium was experiencing a revival of interest in classical studies, especially mathematics and astronomy (Palaiologan Revival). This was a difficult time for Byzantium, when the finances and the dominion of the Empire were decreasing. In spite of these difficulties, there was flourishing astronomical activity as represented by the many surviving texts. The surviving astronomical texts fall into two categories. The first follows the ancient astronomical tradition: the texts and tables include results from ancient Greek and Hellenistic astronomy, presenting and elucidating astronomical results and then applying them to contemporary problems. The second group brought to Byzantium elements from Persian and Arabic astronomy and is more original. Metochites' *Stoicheiosis Astronomike*

[1] For a summary of his life, works and thought, see Bydén (2011).
[2] See Förstel (2011).

follows the former school without recourse to direct observations. Metochites' aim was partly educational and he succeeded in reviving the study of ancient astronomy in Palaiologan Byzantium. According to Pingree, *Stoicheiosis* "raised the level of sophistication in Byzantine astronomy to a height it had not attained for centuries".[3]

In his effort to revive the interest in classical Greek knowledge, Metochites includes five introductory chapters covering ancient Philosophy, Mathematics and Astronomy. These chapters have already been studied and edited by Börje Bydén.[4] Chapter one contains a narrative of how Metochites became involved with the project. The emperor Andronikos II was a cultivated man and asked him to undertake the project in order to improve the level of education and understanding. Metochites searched for a tutor and found Manuel Bryennios with whom he completed his studies.[5] Bryennios had been taught by a man educated in Persia[6] and thus one may suspect a connection with Persian astronomy. However, in *Stoicheiosis*, especially the text we analysed there is no evidence for such a connection. There is historical evidence that Gregory Chioniades visited Persia and brought back texts which he translated into Greek. Some scholars have discussed the interaction between the two schools of astronomy mentioned above and the transmission of knowledge, especially from the Maragha School.[7] This transmission is absent in *Stoicheiosis*, but it appears in other works of Chioniades and Theodore Melitiniotes.[8]

The educational aim of *Stoicheiosis* is evident in many respects. At the very beginning of chapter five it is stated that "in this treatise we set out to present an elementary account of spherical astronomy". Many topics are explained in detail and in the middle of such discussions are interjected comments of philosophical or epistemological nature. For example, in chapters six and seven Metochites describes the revolutions of the two higher spheres (the diurnal rotation and the precession of the equinoxes) in greater detail

[3] Pingree (1964) 137.
[4] Bydén (2003).
[5] See Constantinides (1982) 95-97.
[6] As Metochites himself reveals in one of his poems. See Bydén (2003) 248-50. For the text of this poem (Carmen I), see now Polemis (2015) 5-51, at 26.
[7] Mavroudi (2006).
[8] See, respectively, Paschos – Sotiroudis (1998) and Leurquin (1990).

than in the *Almagest* 1.8. Establishing the precession was a difficult problem and Ptolemy presents in a later chapter (*Alm.* 7.2-3) extensive data to support it. Metochites' description is very clear and he mentions that it is difficult to observe it. This gives him the opportunity to discuss the scientific method that one must follow in order to establish small effects by quoting the *Harmonics* of Ptolemy and Aristotle (7.59-77). This approach is found throughout the *Stoicheiosis*. The educational purpose is also revealed by concluding sentences like: "we elucidated with sufficient clarity the time difference between meridians" (27.96-97) or "the tables for the uniform and anomalous motion of the Sun … are constructed in this manner. They are well understood and have been presented in a way I consider useful" (25.1-4). Then he mentions explicitly the sources for the tables and includes instructions and examples on how to use them.

In Byzantium, Astronomy was part of the *quadrivium* and was taught in combination with mathematics.[9] Our edition and study demonstrates the influence of mathematics on astronomy as well as Metochites' level of mastering the subject and the issues discussed and investigated at that time by himself, his students and perhaps a small part of the public. There are precise physical and geometric statements, as well as numerical tables covered in his book. Finally, the text of the *Stoicheiosis* has a special style. At the beginning of many chapters Metochites includes a brief summary of the previous chapter. In many places the sentences are long and repetitive making the translation difficult. We tried to make the reading of the translation easier by limiting repetitions in order to present the content of sentences and of entire paragraph(s).

Medieval treatises on scientific topics are frequently difficult to appreciate, and sometimes misunderstood, because at that time the concepts, notation and terminology were very different from those we use today. The task of appreciating them becomes even more difficult when their author is not a professional scientist but rather a newcomer, an amateur or, as in our case, a high state official who presents his arguments in a scholastic manner. For these reasons it is necessary for studies of these works to analyse their contents (descriptive and mathematical) in order to see the connection with the past and the influence they exerted in the future. The complete understanding

[9] See Markopoulos (2008) 789.

of Metochites' work requires a critical edition of the Greek text, as well as a translation and exposition of its contents. B. Bydén's analysis of the first five chapters of the treatise is based on an impressive knowledge of the reception of ancient philosophy in Byzantium. However, the bulk of the treatise, which contains the astronomical part of the work, remains unpublished. In this book we start with the 5th chapter of *Stoicheiosis*, which outlines the purpose of the treatise. We include all chapters up to the 30th, an extensive section of the treatise which is self-contained. It covers the revolutions of the upper three spheres: the diurnal rotation, the precession of the equinoxes and the apparent rotation of the Sun on the inclined (ecliptic). Our work stops at the place where the study of the Moon begins. There are several other topics covered in these chapters, such as the Egyptian, Greek and Roman calendars leading to a proposal for changing the initial date for measuring time. It describes the correlation of times among the various geographic zones (ascensions), the obliquity, and contains extensive tables with instructions on how to use them. The format of our study, which offers a textual introduction, text, translation and analysis, makes available for the first time the astronomical content of chapters 6-30.

Metochites' work prepared the ground for further work and we mention a few of the recorded developments. A recurring topic those days were methods for measuring time and the accuracy of the Julian calendar. The interest in these topics is related to the determination of the day for the celebration of Easter (Paschalion). An early proposal for revising the calendar appears in chapters 25 and 26 of *Stoicheiosis*. Metochites proposes a new beginning for the calendar coinciding with the first year of the reign of Andronikos II. In these chapters he determines the day of the year for the new beginning and the position of the Sun on that day. The chapters attracted attention, as indicated by many marginal notes, with the most extensive written by John Chortasmenos (ca. 1370–ca.1436/7). Many copies of the *Stoicheiosis* also contain annotations by Nikephoros Gregoras (1295–1358), who was a pupil of Metochites and became an accomplished astronomer himself. Notable among his contributions is the description of a plan to correct the calculation for the length of the year and thus determine the date for celebrating Easter. In the *Roman History* he mentions the revision, but

does not include details or numerical values.[10] In the surviving manuscripts it is stated that the length of the tropical year estimated by Ptolemy must be corrected. Nine years later Barlaam of Calabria (ca. 1290–1348) revisits the determination of the date of Easter.[11] For the spring equinox he follows the method and values from the *Megiste Syntaxis* (*Almagest*) and then he diverts the discussion to the prediction of full Moons. Much later, Isaac Argyros (ca. 1300–ca. 1375), a student of Gregoras, commenting on Gregoras' discovery states that "the correction required is obtained by subtracting 1/200 of a day from the Ptolemaic figure or about 1/100 of a day from 365 and one-quarter days."[12] This value is very accurate for the length of the tropical year (Argyrus' value of 365.2417 days is close to the present value of 365.2422 days).

Needless to say, that none of the proposals was adopted, because of varying opinions among astronomers and because of the political events that followed. Andronikos II was deposed in 1328 and Gregoras' political and religious views fell into disfavour to the extent that he was imprisoned. The short narrative shows that many astronomers were indebted to the works of Metochites. The *Stoicheiosis* was studied for quite some time, and John Chortasmenos notes, two generations later, that with the *Stoicheiosis* it was then possible to navigate across the sea of Ptolemy's thought.[13]

The detailed presentation of the text, including the numerous tables with instructions on how to use them and the exposition of intricate concepts support the view that Metochites contributed to the revival of astronomy along the lines of the classical Greek tradition. There are changes on various topics and we elaborate on two of them. As mentioned already, in chapters 25 and 26 there is a proposal for selecting a new date for measuring time by determining the date of the year and the position of the Sun on that date. The proposal attracted a lot of attention for a few generations as indicated by the commentaries on the margins. The second change has to do with the catalogues for the positions in longitude of the fixed stars of first and second

---

[10] *History* 8.13; see van Dieten (1979) 68-69.

[11] See Tihon (2011).

[12] Pingree (1964) 138; Tihon (2001) 180.

[13] In the preface he added to the *Stoicheiosis*, edited by Ševčenko (1962) 44-45 (n. 1): "ἀλλὰ τὸ μέγα ἐκεῖνο καὶ ἄπλωτον πέλαγος τῆς Πτολεμαίου διανοίας παρεχομένη περαιοῦσθαι παντάπασιν ἀσφαλῶς, ὥσπερ ἐπὶ τινος σχεδίας φερομένους". Cf. Wilson (1996) 263.

magnitude. Metochites explains at length that the positions change due to the slow precession of the equinoxes at a rate of one degree in one hundred years. We studied the tables in the manuscripts and found the following. In Vat. gr. 182 (V), ff. 249v-250r, the values for the longitudes advanced by approximately 11° from the values of Ptolemy. In Vat. gr. 1365 (C), ff. 241v-242r, the values for the positions of the stars are the same with those given in the *Megiste Syntaxis*. Bydén has given serious arguments, which were confirmed by our study, that V precedes C and was closer to the original text of Metochites. Once we accept this chronological sequence the earlier document includes the correction from the precession of one degree in one hundred years. For unknown reasons the scribe of C decided to return to the values of *Megiste Syntaxis*.[14] In Vat. Gr. 1087 (G) the scribes (which included Gregoras) have only partially completed some of the tables, with the tables of the fixed stars being empty.

Inconsistencies among tables are also present in the various manuscripts of the "Tables of fixed stars in the Persian Syntaxis of George Chrysokokkes", compiled and discussed by Kunitzsch.[15] In this work the tables of type-III are closest to the tables of the *Stoicheiosis*. In table of type III, the values in the various manuscripts are very different, varying from Ptolemy's values +11° 55′ up to Ptolemy's +19° 30′. Regarding the tables there is a recurring pattern of discrepancies that has not been clarified yet. One possibility may be that since the tables were used for the determination of Easter, some scribes for religious reasons decided to return to the values of Ptolemy. In our case, the scribe of G was perhaps planning to correct and complete the tables at an opportune time but he never came to complete this task.

A curious aspect in *Stoicheiosis* is that Metochites gives instructions on how to calculate the position of the Sun and of the stars and gives answers without including numerical calculations. This is most evident in chapters 25 and 26 for the position of the Sun on the sixth of October 1283. The numerical work was added by John Chortasmenos in the margins (scholia). Metochites gives the impression that he provides the theory and expects others, perhaps his assistants, to complete the calculations which were carried out only in a few cases.

---

[14] Cf. p. 28-29, 349.

[15] Kunitzsch (1964) 382.

Among the topics covered in the book are the following:

1. A geometrical description of the coordinates on the celestial globe, including spherical coordinates. Nine celestial spheres are introduced and are used to describe the positions of the stars, the Sun, the Moon, and the planets, as well as their rotations.
2. A brief classification of the geographical zones on the surface of the Earth and their relation to the celestial coordinates.
3. The two dominant astronomical theories of that time, one with eccentrics and the other with epicycles are elaborated and demonstrated to be equivalent.
4. The Sun has the simplest trajectory which is described including uniform and anomalous rotations. Then the model is used to define units of time with which one can define various calendars. In *Stoicheiosis* there is a precise description of the Egyptian calendar used by astronomers. Metochites uses the opportunity to propose a revision of the calendar by selecting the initial time for counting years to be the sixth of October of the first year of the reign of Andronikos II Palaiologos.
5. The models are supported by detailed tables adopted either from the *Almagest* or from the *Handy Tables* accompanied by explanatory chapters on how to use them.
6. For the ascensions of the Sun and the horoscope there are two chapters (27 and 30). They are supported by tables for oblique ascensions of the seven zones and for Byzantium. They are borrowed from the *Handy Tables* with instructions on how to use them. In chapter 28 we include the tables for right ascensions and the oblique ascensions for Meroë and Byzantium. The remaining tables for the six zones from Syene to Borysthenes are included in the Appendix. Finally, it is worth mentioning that Metochites attributes the *Handy Tables* to Ptolemy[16] and mentions Theon only when he presents abridged tables.[17]

[16] *Stoicheiosis* 1.670-71: "οἱ δὲ Πρόχειροι Κανόνες ἐπονήθησαν αὐτῷ τε Πτολεμαίῳ καὶ μεθύστερον Θέωνι, ἐστενωμένῃ παντάπασι παραδόσει" etc.

[17] Contrary to widespread modern opinion, Theon did not introduce any changes in the *Handy Tables*. See Tihon (2000) 357-85.

2. Table of Contents of *Stoicheiosis Astronomike*[18]

Θεοδώρου Μετοχίτου

ἀστρονομικῆς κατ᾽ ἐπιτομὴν στοιχειώσεως

βιβλίον πρῶτον, ἐν κεφαλαίοις ϛα, ἤγουν

First book, in 81 chapters

[18] This Table of Contents follows V for Book One and C for Book Two. Sathas (1872) 111-118 has published the chapter titles of the first book. He used the two Marciani manuscripts, M and S (see the Textual Introduction).

ζʹ. περὶ τῆς δευτέρας φορᾶς τῶν οὐρανίων σωμάτων τῆς ἀπὸ δυσμῶν εἰς ἀνατολάς
7. On the second revolution of the celestial bodies from west to east
ηʹ. ὅτι λοξὸς νοεῖται κύκλος ἐπὶ τῆς οὐρανίας σφαίρας, μέγιστος καὶ αὐτὸς ὢν τῶν ἐν αὐτῇ κύκλων, καὶ εἰς ἴσα δύο ἡμισφαίρια τέμνων αὐτὴν καὶ τὸν ἰσημερινὸν κύκλον
8. The inclined circle of the celestial sphere is a major circle dividing the celestial sphere into two hemispheres and the equator into two semicircles
θʹ. ὅτι περὶ τοὺς τοῦ τοιούτου λοξοῦ κύκλου πόλους ἀπὸ δυσμῶν εἰς ἀνατολὰς πάντες φέρονται οἱ ἀστέρες
9. The motion of all the stars on the inclined [circle] is from west to east
ιʹ. ὅτι τὸ ὑπὸ τὸν ἰσημερινὸν τμῆμα τῆς γῆς ἀοίκητόν ἐστι, καὶ δι᾽ ἥντινα τὴν αἰτίαν
10. The region of the Earth below the [celestial] equator is uninhabited; what is the reason for this.
ιαʹ. ὅτι τὰ παρ᾽ ἑκάτερα τούτου τῆς γῆς τέτταρα τμήματα, τὰ βόρεια καὶ νότια, δύο ἄνω καὶ δύο κάτω, πέφυκεν εἶναι οἰκούμενα ὡς εὔκρατα
11. The four regions of the Earth on each side of the equator, two in the north and two in the south, are habitable because they are temperate
ιβʹ. ὅτι ἡ καθ᾽ ἡμᾶς οἰκουμένη ἐν τῷ ἑνὶ τῶν τεττάρων τμήματι καταλαμβάνεται, τῷ βορείῳ τῷ ἄνω, καὶ πόθεν ἡ ταύτης καταλαμβάνεται ἀρχή, καὶ περὶ τῶν ἐπ᾽ αὐτῆς νοουμένων κύκλων
12. Our ecumene is one of the four upper regions to the north with a description of its beginning and of its parallel circles
ιγʹ. περὶ τῆς τάξεως τῶν οὐρανίων σφαιρῶν, καὶ ὅτι πᾶσαι ὑποκάτω εἰσὶ τῆς ἀπὸ ἀνατολῶν εἰς δυσμὰς κατὰ τὴν πρώτην ὡς εἴρηται φορὰν κινουμένης σφαίρας, καὶ πᾶσαι ἐναντίως αὐτῇ ἀπὸ δυσμῶν εἰς ἀνατολὰς φέρονται
13. The order of the celestial spheres: the first sphere that moves from east to west with all other spheres located below and rotating from west to east

ιδ΄. ὅτι οἱ κύκλοι πάντες, ἐφ' ὧν φέρονται ὅ τε ἥλιος καὶ ἡ σελήνη καὶ οἱ πέντε πλανῆται, οὐχ ὁμόκεντροί εἰσι τῷ παντί, ἀλλ' ἔκκεντροι, ὡς εἶναι τὰ μὲν αὐτῶν ἀπόγεια μέρη, τὰ δὲ περίγεια
14. All the circles which carry the Sun, the Moon and the five planets are not homocentric to the universe, but eccentric, and how their apogees and perigees are arranged
ιε΄. ὅτι τὰ τοιαῦτα ἀπόγεια καὶ περίγεια αὐτῶν οὐκ ἐπὶ τῶν αὐτῶν καταλαμβάνονται μορίων ἀεὶ τοῦ ζωδιακοῦ, ἀλλὰ διόλου μετατιθέμενα ἐν ἄλλοτε ἄλλοις
15. The apogees and perigees of the planets and the Moon are not always at the same place of the ecliptic but they move
ις΄. ὅτι τὸ τῆς ἡλιακῆς σφαίρας μόνης ἀπόγειον καὶ περίγειον ἐπὶ τῶν αὐτῶν μορίων τοῦ ζωδιακοῦ εἰσιν ἀεὶ ἀκίνητα, καὶ ὅτι μόνη ἡ πάροδος τοῦ ἡλίου ἀεὶ ἐπὶ τοῦ μέσου ἐστὶ τοῦ ζωδιακοῦ, αἱ δὲ τῶν ἄλλων πάροδοι παρεκκλίνουσιν ἐφ' ἑκάτερα τοῦ μέσου καὶ βόρεια καὶ νότια
16. Only the apogee and the perigee of the solar sphere are always at the same place on the zodiac and only the trajectory of the Sun is always in the middle of the zodiac. The trajectories of the others deviate from the centre being at some times to the north and at other times to the south
ιζ΄. ὅτι ἐπὶ τοῦ ἡλίου καὶ τῆς σελήνης καὶ τῶν πέντε πλανωμένων ἀστέρων θεωροῦνται ὁμαλαὶ καὶ ἀνώμαλοι κινήσεις, καὶ ὅτι κατὰ δύο τρόπους αἱ ἀνώμαλοι καταλαμβάνονται τῶν ἀστέρων κινήσεις, κατά τε τὸ ἐπὶ ἐκκέντρων, ὡς εἴρηται, φέρεσθαι καὶ κατὰ τὸ ἐπὶ ἐπικύκλων ἰδίᾳ πάλιν αὐτοὺς παροδεύειν τοῦς ἀστέρας
17. On the motions of the Sun, the Moon and the five planets — some of them are considered to be regular and others irregular. The irregular motions of the stars are understood in two ways: they are carried along on the eccentric circles, as we have mentioned already, or they move on their epicycles

ιη΄. ὅτι τῆς σελήνης καὶ τῶν ἄλλων ἀστέρων καὶ κατ' ἀμφοτέρους τοὺς τρόπους κινουμένων ἀνώμαλον κίνησιν, μόνος ὁ ἥλιος καθ' ἕνα τούτων μόνον τρόπον ἀνωμάλως κινεῖται καὶ ἁπλῆν μόνην ἔχει τὴν αἰτίαν τῆς ἀνωμάλου κινήσεως
18. The Moon and the other planets have anomalous motions described by one of the two methods. Only the Sun moves according to one of the models and has a simple explanation
ιθ΄. ὅτι οἱ ἀστρονόμοι αἰγυπτιακοῖς ἔτεσιν ἀπαριθμοῦνται τὰς τῶν ἀστέρων κινήσεις καὶ οὐχ ἑλληνικοῖς εἴτουν ῥωμαϊκοῖς, καὶ τίς ἡ τῶν αἰγυπτίων ἐτῶν πρὸς ταῦτα διαφορά
19. The astronomers compute the motions of the stars according to the Egyptian and not in the Graeco-Roman calendar. Description of the difference between the Egyptian calendar and the others
κ΄. πῶς ἐμεθοδεύθη τοῖς ἀστρονόμοις ἡ τῶν ὁμαλῶν κινήσεων κατὰ χρόνους ὡρισμένους ἀπαρίθμησις, καὶ ὅτι ἀναγκαῖον ταύταις ὑποτεθῆναι τροπικὴν ἀρχήν, ἀφ' ἧς ἔσται ἡ προχώρησις καὶ ἡ ἀπαρίθμησις, ἔτι γε μὴν καὶ χρόνου τινὸς ἀρχήν, ἀφ' οὗ ἔσται ὡσαύτως ὁ ἐπιλογισμὸς τῶν ὁμαλῶν κινήσεων ἡλίου καὶ σελήνης καὶ τῶν εἰρημένων ἄλλων ἀστέρων
20. How did astronomers come to the idea to calculate the uniform movement in certain time periods, and why it is necessary to postulate a beginning, from which we count the subsequent movements. Indeed, we specify a time from which we begin to calculate the uniform rotations of the Sun, the Moon and the other planets
κα΄. κανόνες τῶν ὁμαλῶν τοῦ ἡλίου κινήσεων
21. Tables for the solar mean motion
κβ΄. διασάφησις τῆς ἐκθέσεως τῶν αὐτῶν κανόνων
22. Explanation of the contents of the tables
κγ΄. περὶ τῆς ἀνωμάλου τοῦ ἡλίου κινήσεως καὶ ἔκθεσις κανόνος τῆς τοιαύτης ἀνωμάλου αὐτοῦ κινήσεως
23. On the anomalous motion of the Sun and the corresponding table
κδ΄. διασάφησις τοῦ αὐτοῦ κανόνος τῆς ἀνωμάλου τοῦ ἡλίου κινήσεως
24. Explanation of the table for the anomalous movement of the Sun

κε΄. παράδοσις μετὰ ἐπιλογισμοῦ χρονικῆς ἀρχῆς, ὅθεν δεῖ τὰς ψηφοφορίας τοῦ ἡλίου λογίζεσθαι
25. Presentation, with calculations, of the origin from which we begin measuring the time that will be used in computations
κς΄. παράδοσις ὅπως ἐστὶ καθ’ ὁντιναοῦν δεδομένον χρόνον μετὰ τὴν ὑποτεθεῖσαν ἀρχὴν ψηφοφορεῖν καὶ ἐπιλογίζεσθαι τὰς ὁμαλὰς καὶ ἀνωμάλους κινήσεις καὶ ἐποχὰς τοῦ ἡλίου
26. Description of how we calculate and determine, at a given moment of time, after the assumed starting time, the regular and anomalous movement and longitude of the Sun
κζ΄. περὶ διαφορᾶς ὡρῶν, καὶ καθ’ ὅσους τρόπους αἱ διαφοραὶ θεωροῦνται τῶν ὡρῶν, καὶ ὅτι ἀναγκαία ἡ περὶ τῶν ἀναφορῶν τῶν τοῦ ἰσημερινοῦ τμημάτων εἴτουν μοιρῶν κατάληψις
27. Various definitions of hours with a description of their differences, whose understanding is essential for describing the ascensions at various sections of the equator
κη΄. ἔκθεσις κανόνος ἀναφορῶν ἐπὶ τῆς ὀρθῆς νοουμένης οὐρανίου σφαίρας, ἔτι δὲ ἔκθεσις κανόνων ἀναφορῶν ἐπὶ τῶν ἑπτὰ κλιμάτων καὶ ἐπ’ αὐτοῦ τοῦ διὰ Βυζαντίου, εἴτουν τῆς βασιλίδος πόλεως, παραλλήλου
28. Presentation of the table for the ascensions of the Sun on *sphaera recta*; in addition tables for the ascensions of the Sun on the seven zones and on the parallel of Byzantium - the Queen City
κθ΄. περὶ τῆς τοῦ ἡλίου λοξώσεως ἑκάστοτε ἀπὸ τοῦ ἰσημερινοῦ, καὶ κανὼν τῆς αὐτοῦ λοξώσεως καὶ διασάφησις τοῦ αὐτοῦ κανόνος
29. The obliquity of the Sun with a table and its explanation
λ΄. περὶ τῆς εὑρέσεως τοῦ ὡροσκοποῦντος καὶ μεσουρανοῦντος ἑκάστοτε τμήματος τοῦ ζωδιακοῦ
30. Determination of the horoscope and the culmination for each section of the zodiac
λα΄. περὶ τῆς σελήνης καὶ τῶν ὑποθέσεων τῶν κατὰ ταύτην κινήσεων
31. Model for the rotations of the Moon

μη΄. ἔκθεσις κανόνων παραλλάξεων καθ' ἕκαστον κλίμα, ἀλλὰ δὴ καὶ ἐπὶ τοῦ διὰ Βυζαντίου παραλλήλου
48. Table of the parallax at each zone and for the parallel of Byzantium
μθ΄. διασάφησις τῆς ἐκθέσεως τῶν αὐτῶν κανόνων
49. Explanation of the presentation of the same tables
ν΄. ἔκθεσις κανονίου τοῦ ἐπικαλουμένου διορθώσεως
50. Presentation of the table known as the correction
να΄. διασάφησις τοῦ τοιούτου κανονίου
51. Explanation of this table
νβ΄. παράδοσις πῶς ἐστιν ἀνευρίσκειν ἑκάστοτε τὰς τῆς σελήνης παραλλάξεις
52. Instructions on how to find the parallax of the Moon at any given time
νγ΄. περὶ τῶν ἐκλειπτικῶν σελήνης καὶ ἡλίου ὑποθέσεων, καὶ τίνες ὅροι ἄκροι, ἐκλειπτικοὶ ἐπί τε τῆς σελήνης καὶ τοῦ ἡλίου
53. Models for the eclipses of the Moon and the Sun and a few exceptional cases concerning the eclipses
νδ΄. περὶ τῶν ἐκλειπτικῶν τῆς σελήνης κανόνων, καὶ ὅτι ἀναγκαῖον αὐτοὺς εἶναι δύο, μεγίστου καὶ ἐλαχίστου ἀποστήματος, καὶ ἔκθεσις τῶν τοιούτων ἀμφοτέρων κανόνων
54. On the tables for lunar eclipses and the necessity of two cases for maximal and minimal deviations. Presentation of these two tables.
νε΄. διασάφησις τῆς τοιαύτης ἐκθέσεως τῶν κανόνων
55. Explanation of the tables
νς΄. παράδοσις πῶς ἐστι δι' αὐτῶν ἀνευρίσκειν καὶ ἐπιλογίζεσθαι τὰς τῆς σελήνης ἐκλείψεις
56. How one calculates and finds lunar eclipses
νζ΄. περὶ τῶν ἡλιακῶν ἐκλείψεων καὶ ἔκθεσις κἀνταῦθα δύο ἡλιακῶν ἐκλειπτικῶν κανόνων
57. Solar eclipses and presentation of two tables for them
νη΄. διασάφησις τῆς ἐκθέσεως τῶν τοιούτων ἡλιακῶν ἐκλειπτικῶν κανόνων
58. Explanation to the presentation of these solar eclipse tables

πθ´. περὶ τῶν ἀπλανῶν ἀστέρων καὶ τῶν κατ᾽ αὐτοὺς ὑποθέσεων, καὶ ὅπως εἰς τὰ ἑπόμενα τοῦ παντός καὶ αὐτοὶ κατ᾽ ὀλίγον προχωροῦσι περὶ τοὺς πόλους τοῦ ζωδιακοῦ
89. The model for the fixed stars which explains how they rotate slowly around the poles of the ecliptic
ϛ´. περὶ τῶν κατὰ μῆκος καὶ πλάτος ἐποχῶν τῶν τοιούτων ἀπλανῶν ἀστέρων
90. The longitude and latitude of the fixed stars
ϛα´. ἔκθεσις τῶν τοιούτων κανόνων τῶν ἀπλανῶν ἀστέρων
91. Presentation of these tables for the fixed stars

βιβλίον δεύτερον

Second book

α´. προοίμιον, ἐν ᾧ ὅπως χρεία διασαφήσεως τινῶν κεφαλαίων δι᾽ ὧν ἑκάστοτε ὁ Πτολεμαῖος τοὺς ἀστρονομικοὺς ἐπιλογισμοὺς καὶ τὰς τῶν γραμμικῶν ἀποδείξεων ἐφόδους συμπεραίνεται
1. Preface, where it is necessary to clarify a few chapters related to how Ptolemy deals with problems of astronomical calculations and of geometrical proofs
β´. περὶ πολλαπλασιασμῶν μοιρῶν τε καὶ ἑξηκοστῶν, εἴτουν ὅλων καὶ μορίων
2. The multiplication of degrees and minutes, i.e. entire degrees or fractions thereof
γ´. ἔτι πῶς γίνονται οἱ πολλαπλασιασμοὶ τῶν τε ὅλων καὶ τῶν μορίων
3. How one performs the multiplications for entire degrees and fractions of them
δ´. Περὶ τῶν μερισμῶν τῶν αὐτῶν καὶ ὅπως οἱ μερισμοὶ γίνονται
4. Their divisions and how we carry them out
ε´. πῶς ἐστι καὶ διὰ τίνος μεθόδου ἑκάστου δεδομένου ἀριθμοῦ ἢ χωρίου ἢ ρητοῦ ἢ ἀρρήτου τετραγωνικὴν ἀνευρίσκειν πλευράν
5. A method for finding the square root of a given rational or irrational number or of an area

In several manuscripts, including the most important ones (V and C), *Stoicheiosis Astronomike* is followed by Metochites' *Ἐπιτετμημένον ἐγκώμιον τῆς τοῦ Πτολεμαίου Μαθηματικῆς Συντάξεως* ('Epitomized praise of Ptolemy's Mathematical Syntax'). This was written as a separate work or was intended as an appendix to the *Stoicheiosis*.

3. Textual Introduction

1. *The Manuscripts*

The manuscript tradition of the *Stoicheiosis Astronomike* was first described by Ševčenko (1962: 280-286), who initially undertook a critical edition of the *Stoicheiosis Astronomike* for a second doctorate in Louvain, which he later turned into a broader study on Metochites and Choumnos.[19] More recently, B. Bydén (2003) studied 11 manuscripts for his critical edition of Book 1, ch. 1-5. They transmit Metochites' treatise in full or in part:[20]

V	Vaticanus gr. 182, ff. 1-250, which initially formed a single codex with V^1, written ca. 1317–1331.
V^1	Vaticanus gr. 181, ff. 2-38.
C	Vaticanus gr. 1365, ff. 1-276, written ca. 1326–1332.
M	Marcianus gr. 330 (coll. 915), ff. 16-203, written before the middle of the 14th century.
G	Vaticanus gr. 1087, ff. 3-122, 148-221, written ca. 1326–1334.
Z	Vaticanus gr. 2176, ff. 53-293, written probably soon after 1361.
S	Marcianus gr. 329 (coll. 734), ff. 1-95, from the middle of the 14th century.
E	Scorialensis 242 (Y. I. 3), ff. 1-354, of the 16th century.
Esc	Scorialensis 28 (R. II. 8), ff. 157-261, of the 16th century.
Mon	Monacensis gr. 100, ff. 349-457, dated to 1551.
A	Ambrosianus gr. 1005 (E 1 inf.), ff. 1-237, of the 16th century.

Bydén's conclusions and his *stemma codicum*[21] have been used in this edition:

[19] See Mavroudi (forthcoming). We thank Maria Mavroudi for sharing her article with us prior to publication.

[20] For the possible dates of the manuscripts see Bydén (2003) 385-86. For G see now Menchelli (2013).

[21] In this stemma "sources of corrections are not accounted for" (Bydén 2003: 411).

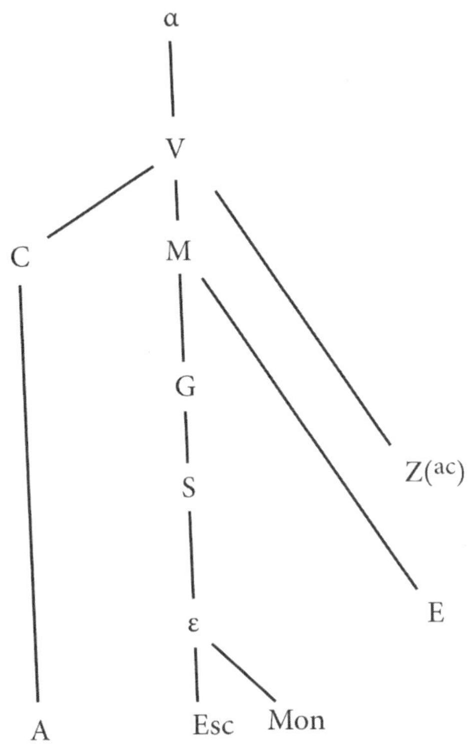

According to Bydén, the ancestor of all extant manuscripts is probably V, dating from Metochites' time and possibly containing corrections by him. The same applies to C, which however appears to be a copy of V. So, it seems that the editor of the treatise needs only concern himself with V and C. These manuscripts have also been read by Nikephoros Gregoras, who sometimes corrects or annotates the text. Gregoras' hand is found in other copies of the text as well and it is evident that he wanted personally to inspect the copies of Metochites' work, especially those produced at the Chora Monastery.[22]

In addition to V and C, we decided to collate also G (a copy of M), because of its close connection with Nikephoros Gregoras, who was one of its scribes and added some marginal notes.[23] But we found nothing of note

22 Cf. Agapitos – Hult – Smith (1996) 10; Bianconi (2005) 426; Förstel (2011).

23 See Ševčenko (1962) 113 (n. 5), 280, 284 (n. 1); Ševčenko (1964) 446; Bianconi (2005) 417; Pontani (2013) and Menchelli (2013).

in the section of the treatise edited here, other than Gregoras' correction on f. 33r (where at **23.40-44** he adds in the lower margin a sentence missed by the main scribe) and the fact that scribes have only partially completed some of the tables and omitted several others, with blank space left to introduce them at a later stage.[24]

Ševčenko (1962: 280-281) and Bydén (2003: 386) list some additional manuscripts, which contain short excerpts from Metochites' treatise: Laurentianus Plut. 28.46, ff. 206-226 (s. XIV); Vaticanus gr. 213, ff. 58-101 and 106-120 (s. XIV); Vaticanus gr. 1059, ff. 136-137 (s. XV); Vaticanus Urbinas gr. 80, ff. 109-110 (s. XV); Cantabrigiensis Gg II, 33 (1463), ff. 57-63 and 203-210 (s. XV/XVI); and Yale University, Beineke MS 424, ff. 103-142 (s. XVI). From this list, we have consulted two manuscripts, which contain excerpts from ch. 5-30: Laur. Plut. 28.46 transmits 1.22-24, 1.46 and 2.6, while Vat. gr. 213 includes parts of ch. 19, 20, 23, 25, 27 and 29, among several others of the first book.

Not surprisingly, Laur. Plut. 28.46 (L) is not significant for the edition of the text. It derives from C, as the following common errors suggest:

> **22.26** πάροδος V : περίοδος C L **24.1** δύο post πρῶτα transp. C L **24.6** μοίρας ante εἰς τὰ ἡγούμενα transp. C L **24.6** τξ deest C (rasura) : om. L **24.37** μέχρι σο V : μέχρις ο C L

It has several additional errors (listed below) and omits the table of ch. 23.

> **23.6** αἱ V C : om. L **23.10-11** ταύτην τὴν V C : ταυ τὴν L **23.19** ἀληθῆ καὶ V C : ἀληθῆ καὶ καὶ L **23.23** τοῦ κόσμου : τοῦ om. L **23.26** ἀσφαλέστατα V C : σαφέστατα L **23.40-44** καὶ ὅση ... θεωρουμένην om. L (homoeotel.) **23.48** οὐ om. L **23.50-53** ἀλλὰ ... προσθετικὴν om. L (homoeotel.) **23.57** ἀπὸ τοῦ om. L **24.1** ὁρᾶν : ὡρᾶν L

---

[24] In ch. 21 the tables of months and days have been left incomplete (f. 32v), while blank space has been left for the introduction of the tables of hours (f. 23ar); f. 38v-41v are also blank with ch. 27.208 – 28.10 missing; after 29.58 a blank page has again been left for the table of the Sun's obliquity.

The scribe of Vat. gr. 213 (Va) has compiled selections from Metochites' treatise (incomplete chapters) and felt free to rephrase the text, as he did, for example, at **20.263-64**, where ἰστέον γὰρ καὶ τοῦτο, ὅτι οἱ ἀστρονόμοι τὰς ἀρχὰς τῶν ἡμερῶν, μᾶλλον was changed to: οἱ γὰρ ἀστρονόμοι τὰς ἀρχὰς τῶν ἡμερῶν, μᾶλλον. No tables have been included in the excerpts from ch. 23 and 29. To give an example of his selective use of the treatise, the first three lines compiled from ch. 19 consist of sentences from lines 21-22, 41-42, 44-47:

> ἰστέον δὴ ὅτι οἱ ἀστρονόμοι τοῖς αἰγυπτιακοῖς χρῶνται ἔτεσιν ἐν πάσαις ταῖς ψηφοφορίαις. καὶ Ῥωμαῖοι μὲν καὶ Ἕλληνες συμπεραίνουσι τὸν ἐτήσιον χρόνον δι' ἡμερῶν τξε καὶ τετάρτου· Αἰγύπτιοι δὲ δι' ἡμερῶν τξε μόνων καὶ τὸ λοιπὸν τέταρτον τῆς ἡμέρας παρεῶσι· καὶ ἀρχὴν τίθενται ἑτέρου ἔτους.

Unlike L, Va has readings that are derived from V, as the following cases suggest:

> **20.246** ὅθεν V Va : ὅπερ C **20.294** ἀπαρτίζεται V Va : ἀπαρτίζονται C **20.302** τῶν εἰρημένων ϛ μηνῶν V Va : τῶν ϛ μηνῶν τῶν εἰρημένων C **25.94** ἀποπλήρωσιν V Va : ἀναπλήρωσιν C

The three manuscripts used for this edition were collated from digital scans of microfilms. Some more information about them ought to be mentioned:

V = Vaticanus gr. 182, ca. 1317-1331 AD

> Paper manuscript, ff. VI + 250, mm. 308x235. It transmits Book 1 of *Stoicheiosis Astronomike*. Chapters 5-30 are on ff. 16v-75r. Vaticanus gr. 181 was initially a second part of the same manuscript, offering *Stoicheiosis*'s second book, written by the same anonymous scribe.[25]

[25] On this scribe see Bianconi (2005) 425-28 and Pérez Martín (2008) 436 (n. 177). For a longer description of these manuscripts, see Mercati – Franchi de' Cavalieri (1923) 207-8.

C = Vaticanus gr. 1365, ca. 1326-1332 AD

Parchment manuscript, ff. 386. Chapters 5-30 are on ff. 19v-77r. The *Stoicheiosis Astronomike* is followed by Metochites' eulogy and extended summary of the *Almagest*. This manuscript was written by the so-called *Metochitesschreiber*, who has now been identified with Michael Klostomalles,[26] an imperial notary who in the first decades of the fourteenth century wrote official documents, copies of Metochites' works, and other significant manuscripts.[27] The copies produced by Klostomalles may appear to offer an authoritative version of Metochites' works, but in the case of the *Stoicheiosis* it is clear that V is superior to C, as is explained below. The codex belonged to John Chortasmenos who wrote notes in the margins and added a new preface to the *Stoicheiosis Astronomike*.[28]

G = Vaticanus gr. 1087, ca. 1326-1334 AD

Western paper manuscript, ff. 320, mm. 315x230.[29] The watermarks suggest that it was produced between the 1320s and the 1330s.[30] It was probably written ca. 1326–1334. G is an astronomical miscellany, which includes, among others, part of the *Stoicheiosis Astronomike* (chapters 5-30 are on ff. 14ar-44v), Metochites' eulogy and extended summary of the *Almagest*, as well as Nikephoros Gregoras' treatise *On the Construction and Origin of the Astrolabe* (for the most part copied by Gregoras himself). G has attracted attention for its remarkable illustrations, a set of constellation pictures with the zodiac, the planisphere and two hemispheres; they are all found in 10 folios with

26 Lamberz (2000) 158-59.

27 See Lamberz (2006) 44-48; Menchelli (2000) 170-75; Wilson (1996) 229, on Codex Crippsianus (Burney 95).

28 The text of this preface has been edited by Ševčenko (1962) 44-45 (n. 1). For Gregoras' and Chortasmenos' hands on f. 1r of C see Bianconi (2005) 407 and 418 (with bibliography). Cf. Hunger (1969) 24-25.

29 According to Pérez Martín (1997) 83 (n. 53), there are exceptions: ff. 34-41: 300x230 mm.; ff. 123-147: 290/5-210 mm.

30 Pérez Martín (1997) 83 (n. 53 and 54) and Menchelli (2013) 22-23.

no text.[31]

Bydén studied these manuscripts *in situ* and the findings of his thorough analysis of their readings (for ch. 1-5 only) have been confirmed by our collation of ch. 5-30. The fewest erroneous readings are found in V, which seems to be the exemplar from which C was copied. As far as C is concerned, Bydén concludes that "for a MS (and a quasi-official 'deluxe' edition to boot: Ševčenko 1962, 282; 1975, 49 n. 220) at such a relatively early stage of the tradition as is implied by the chronology, C^{ac} contained exceptionally many mistakes".[32] The fact that only one of the main manuscripts of the text derives from C may not be coincidental.

Erroneous readings of V^{ac} not found in C are usually easy to explain:

7.53 κατελείφθησαν V^{ac} : κατελήφθησαν C. This would have been an easy mistake in V, corrected in both V and C.
19.8 ὥσπερ V^{ac} : ὡς περὶ C. This is also a trivial mistake of V^{ac}.
25.25 παροίκει V^{ac} : παρήκει C. Another trivial mistake.

Another superior feature of V is that the diagrams of ch. 17 and 18 are properly placed in the body of the main text, while Klostomalles left no space for them in the main text, and the diagrams are found in the margins of C. There they are mixed with later additional diagrams (see ch. 17) and other marginalia. These additional diagrams are not found in G, which only has the diagrams of V.

Later on, in ch. 91, it is V that adds 11° to Ptolemy's values due to the precession of the equinoxes, in accordance with what Metochites himself promises to do in his treatise.[33] Although Ševčenko knew that V had the right

[31] Cf. Weitzmann (1970) 96. The constellation pictures were perhaps originally associated with the *Catasterismi* of Eratosthenes of Cyrene.

[32] Bydén (2003) 400. On the other hand, C has corrections introduced by later learned readers of the text, e.g. **17.182** ζ C^{pc} (manus secunda) : β V C^{ac}; **29.31** τὸ τοῦ Σκορπίου καὶ τὸ Τοξότου C^{pc} : τὸ τοῦ Τοξότου καὶ τὸ Σκορπίου V C^{ac}.

[33] It is impossible to know who is responsible for the change of the values in C, where, however, "a Byzantine reader added the words 'in his [i.e., Ptolemy's] times' (fol. 231v, lower margin) and 'according to assumptions prevalent in the times of Ptolemy' (fol. 233v, margin) next to the tables and discussion of the angular distances of Venus and Mercury from the Sun. The reader

values, his view of the superiority of C led him to conclude that Metochites did not keep his promise and his "tables reproduce Ptolemy exactly".[34]

Generally speaking, Bydén's view (based on his findings for ch. 1-5) that the differences between V and C "are nearly all quite trivial from the point of view of language and content"[35] is also true for our section. *Stoicheiosis Astronomike* is transmitted very well by manuscripts originating in the author's lifetime and bearing authorial revisions. Further scrutiny of additional manuscripts and their variant readings is unlikely to contribute to the establishment and understanding of the text. However, a closer study of all the manuscripts would be important for the reception of Metochites' treatise. As cultural products, they might shed light on intellectual life in Byzantium. Moreover, the examination of their contents and any special characteristics, as well as their circulation and use by Byzantine scholars and readers, could reveal significant developments in the history of astronomy. Indeed, G (with its unique illustrations mentioned above) was the subject of a colloquium organized by a research group on astronomical illustrated manuscripts at the Scuola Normale Superiore, Pisa, on February 8th, 2012.[36]

2. *The Apparatuses*

Given the primary importance of both V and C, and their relatively few differences, the critical apparatus cites all variant readings, including orthographic variants. However, discrepancies in the accentuation of the enclitics are not recorded, because they would have unduly cluttered the apparatus. Some examples of such differences are offered in the section below. In addition, several corrections (of earlier mistakes) that could be traced

was aware that the tables and data had not been made to agree with the new starting point in calculations of 1283, which Metochites had introduced, but were taken over, without change, from Ptolemy's *Almagest* or from Theon of Alexandria's *Handy Tables*" (Ševčenko 1982: 43, n. 179).

34 Ševčenko (1982) 43 adds that "if Metochites, or his secretary, had kept his promise, the difference for each star would have amounted to about twelve degrees, using the Ptolemaic system. In reality it amounted to more and the discrepancy could have been ascertained just by using the astrolabe". See also Ševčenko (1962) 92-93 (n.4). Cf. p. 6 and 349.

35 Bydén (2003) 384.

36 The proceedings have been published in Guidetti – Santoni (2013).

mainly in C are not reported (for examples from ch. 1.5 see Bydén [2003] 401), partly because it is not easy to describe them accurately from reproductions of the original MS. Even with access to the original, it was sometimes not clear to Bydén whether a second hand in C ('possibly the original scribe, but in ink of a distinct tint') 'has actually altered the text or simply reinforced faint script'. A sample of such corrections can be found in Bydén's introduction. G is cited only occasionally, for example, at **23.40-44**, where Nikephoros Gregoras has added in its margin a passage omitted by the scribe because of *homoeoteleuton*; at **27.151-55** to record a scholion in its margin; and at **30.78**, where its text has a necessary improvement adopted in this edition.

The apparatus fontium includes quotations, as well as references to Ptolemy, Theon and other mathematicians.

3. *Punctuation, Accentuation And Word Divisions*

Modern editions of Byzantine texts used to follow the conventions for orthography, punctuation and accentuation established for classical texts. Especially in matters of punctuation, editors have different policies, sometimes influenced by the practices of their native modern languages. Even in the present edition, the punctuation of ch. 5 (notably, in regard to commas) is quite different from that used by Bydén. But in recent years it has been argued that Byzantine authors had a "rhetorical logic"[37] which should be reflected in the manuscripts' punctuation, until now ignored by editors, with only a few exceptions.[38] The matter is more complicated than it looks at first sight and is currently the subject of a scholarly debate.[39] Byzantine authors used punctuation marks as a guide for reading aloud and these marks might not be very helpful to the modern reader, who wants to understand the syntax of usually demanding texts.

The punctuation of the manuscripts, especially when it reflects an author's practice, is definitely related to the author's creative process and

[37] Reinsch (2012) 145.

[38] E.g. Munitiz (1984).

[39] See e.g. Giannouli – Schiffer (2011) and the contributions of Reinsch and Bydén in Bucossi – Kihlman (2012). For earlier scholars who have contributed to this theory or have edited Byzantine texts from autographs, see Reinsch (2002) 140-141 (with bibliography).

is thus able to illuminate a medieval text with a true light; indeed, with its original punctuation "the text will gain a life it never would acquire with mere grammatical punctuation".[40] On the other hand, Byzantine punctuation would make better sense if one reads the text aloud (and this is not the usual practice today), and at the same time it would often make Byzantine texts more difficult to understand for modern readers, who find the conventional punctuation helpful in deciphering the syntax.[41] Readers would need to forget the standard practices and get familiar with the Byzantine ones, in a process that would be confusing. And, most importantly, the Byzantine practices have not yet been sufficiently studied and understood. What is certain is that even contemporary manuscripts differ in their conventions; in fact, there are often inconsistencies even in the same manuscript, although one would expect fewer inconsistences in a manuscript that can be identified as an author's master copy. It has been suggested that differences could be due to scribal "mistakes" or just indicate better and worse choices: a critical process would again be needed to establish the "original" or just "better" Byzantine punctuation.[42]

We look forward to seeing and using new editions of Byzantine texts following Byzantine conventions, but we did not think that an astronomical text would be a suitable text for such a venture. The confusion of modern readers, who in this case would be mostly historians of science familiar with classical texts, is not irrelevant.[43] A scientific text is also less rhetorical and less likely to be performed orally (certainly today, but perhaps even in Byzantine times) than other texts. As a result, in this edition no attempt has been made to represent the punctuation of the manuscripts. However, two examples should be cited, to offer an idea of the situation as far as this particular text is concerned. The first example (**27. 280-86**) shows C's punctuation to be slightly heavier than V's:

---

40 Again in the words of the main advocate of this theory, D. R. Reinsch (2002) 146.

41 Cf. Pérez Martín (2012): "what is at stake here is the legibility of future editions. If the reproduction of a rhythmic punctuation prevails, reading a Byzantine text will be more difficult than now".

42 Cf. Reinsch (2012) 148.

43 See Bydén (2012) 161.

> διὰ τοῦτο καὶ ἐπεὶ πᾶσαν ἡμέραν καὶ νύκτα εἰς ιβ ὥρας, ἃς καὶ καιρικὰς καλοῦσιν, ὁποῖαί ποτ᾽ ἂν ὦσιν, οἱ ἀστρονόμοι διαιροῦσι—καὶ εἰσὶ λοιπὸν αἱ τοιαῦται ὧραι, νῦν μὲν ἰσημεριναί, νῦν δὲ μείζους τῶν ἰσημερινῶν, νῦν δὲ ἐλάττους—, οὐ πᾶσαι ὧραι πάσης ἡμέρας ἰσαρίθμοις χρόνοις, ἤτοι τμήμασι τοῦ ἰσημερινοῦ, συγκαταλογίζονται καὶ συγκαταριθμοῦνται ἐν τῇ περιόδῳ καὶ περιστροφῇ τοῦ παντός, ἀλλὰ διαφόροις καὶ ἀνίσοις.

V διὰ τοῦτο καὶ ἐπεὶ πᾶσαν ἡμέραν καὶ νύκτα εἰς ιβ ὥρας ἃς καὶ καιρικὰς καλοῦσιν ὁποῖαι πότ᾽ ἂν ὦσιν οἱ ἀστρονόμοι διαιροῦσι· καὶ εἰσὶ λοιπὸν αἱ τοιαῦται ὧραι· νῦν μὲν ἰσημεριναί· νῦν δὲ μείζους τῶν ἰσημερινῶν· νῦν δὲ ἐλάττους, οὐ πᾶσαι ὧραι πάσης ἡμέρας ἰσαρίθμοις χρόνοις ἤτοι τμήμασι τοῦ ἰσημερινοῦ, συγκαταλογίζονται καὶ συγκαταριθμοῦνται, ἐν τῇ περιόδῳ καὶ περιστροφῇ τοῦ παντὸς· ἀλλὰ διαφόροις καὶ ἀνίσοις·

C διὰ τοῦτο καὶ ἐπεὶ πᾶσαν ἡμέραν καὶ νύκτα εἰς ιβ ὥρας ἃς καὶ καιρικὰς καλοῦσιν ὁποῖαι πότ᾽ ἂν ὦσιν, οἱ ἀστρονόμοι διαιροῦσι· καὶ εἰσὶ λοιπὸν αἱ τοιαῦται ὧραι, νῦν μὲν, ἰσημεριναί· νῦν δὲ μείζους τῶν ἰσημερινῶν· νῦν δὲ ἐλάττους, οὐ πᾶσαι ὧραι πάσης ἡμέρας ἰσαρίθμοις χρόνοις. ἤτοι τμήμασι τοῦ ἰσημερινοῦ συγκαταλογίζονται καὶ συγκαταριθμοῦνται, ἐν τῇ περιόδῳ καὶ περιστροφῇ τοῦ παντός· ἀλλὰ διαφόροις καὶ ἀνίσοις·

In comparison with the text as edited here, the manuscripts do not use punctuation in relative clauses, while they use it for the prepositional phrase with ἐν. The second example (**6.85-89**) shows V and C in agreement and perhaps indicates that modern practice may sometimes be quite close to the Byzantine one:

> οἱ γὰρ κύκλοι καὶ τὰ αὐτῶν ἡμικύκλια, ἐφ᾽ ὧν δοκοῦσι περιφέρεσθαι οἱ ἀστέρες τῇ πρώτῃ τοῦ οὐρανίου σώματος κινήσει καὶ περιφορᾷ, ἄνισα μὲν (ἄλλα γὰρ ἄλλων καὶ μείζω καὶ ἥττω), ἀλλά γε πάνθ᾽ ὅμοια (παράλληλα γάρ)—περὶ γὰρ τοὺς αὐτούς, ὡς ἔφημεν, περιφέρεται πόλους.

V οἱ γὰρ κύκλοι καὶ τὰ αὐτῶν ἡμικύκλια ἐφ' ὧν δοκοῦσι περιφέρεσθαι οἱ ἀστέρες, τῇ πρώτῃ τοῦ οὐρανίου σώματος κινήσει καὶ περιφορᾷ, ἄνισα μὲν· ἄλλα γὰρ ἄλλων, καὶ μείζω καὶ ἥττω· ἀλλά γε πάνθ' ὅμοια· παράλληλα γάρ· περὶ γὰρ τοὺς αὐτοὺς ὡς ἔφημεν, περιφέρεται πόλους·

C οἱ γὰρ κύκλοι καὶ τὰ αὐτῶν ἡμικύκλια ἐφ' ὧν δοκοῦσι περιφέρεσθαι οἱ ἀστέρες, τῇ πρώτῃ τοῦ οὐρανίου σώματος κινήσει καὶ περιφορᾷ, ἄνισα μὲν· ἄλλα γὰρ ἄλλων, καὶ μείζω καὶ ἥττω· ἀλλά γε πάνθ' ὅμοια· παράλληλα γὰρ. περὶ γὰρ τοὺς αὐτοὺς ὡς ἔφημεν, περιφέρεται πόλους·

As far as accentuation is concerned, it would be easier to adopt any conventions followed by the manuscripts.[44] But again, in the cases of the enclitics, for example, there is no general agreement or consistency in the practices of the two principal manuscripts of this text: e.g. **7.19** αὗται γε εἰσὶν V C (but **7.91** οἷον τέ ἐστιν V C); **7.94** παραστῆσαι τί V : παραστῆσαί τι C; **8.14** μέγιστος ἐστὶ V C, **8.28** μέγιστοι εἰσὶ V C, but **10.44** ἀείκητόν ἐστι V C, **17.146** δήποτέ ἐστιν V C; **12.13** ἐγγύτερος ἐστὶ V : ἐγγύτερός ἐστι C. However, there is one exception: the particle τέ does not lose its accent, when it follows a paroxytone, but is always accentuated, e.g. **5.36** διεξόδους τὲ καὶ κινήσεις V C; **5.42-43** εὐτυχέστατά τε διάξειν τὸν βίον τελευτήσας τὲ εἰς τόπους V C; **5.46** πάντων τὲ τῶν V C; **5.47** δι' ἄλλων τὲ πολλῶν V C; **5.139** ὅπῃ τὲ ἔχοι καὶ V C; **5.321** διαιροῦσι τὲ καὶ V C; **29.53** ἔστι τὲ καὶ V C.[45] This convention is adopted in this edition.

In the manuscripts, numerals are either fully written or expressed by letter signs with no consistency even within the same sentence, for example: **20.169** κε ἔτη ἢ ὀκτωκαίδεκα. In this case I have allowed this diversity in my edition, as it is not confusing and has been followed by editors of similar texts in the past.[46]

---

44 For a thorough examination of the accentuation in some Byzantine manuscripts, see De Groote (2012) lxxiii-xciv.

45 This is reported by Dendrinos (2011) 33 to have also been the policy of Makarios Makres (1382/3 – 7.1.1431).

46 For example, Theon, *in Ptolemaei canones commentarium parvum* 2 ... ἐκ τῶν πέντε κεφαλαίων μοίρας ἢ καὶ ἑξηκοστά. Καὶ ἐὰν μὲν αὗται αἱ συναχθεῖσαι ἐκ τῶν ε κεφαλαίων ... (ed. Tihon 1978: 208.15-209.1).

Byzantine manuscripts often combine words which appear separately in editions of classical texts.[47] Once again, V and C do not show agreement or consistency in this regard. C has the tendency to combine words written separately in V: for example, **23.13** & **27.167** ἐξ ἀνάγκης V : ἐξανάγκης C (although this appears usually as one word in both MSS, for example **9.29** & **30.6** ἐξανάγκης V C); **27.302** πρὸ ὀλίγου V : προολίγου C; **29.41** κατ' ὀλίγον V : κατολίγον C; **24.1** εἴτ' οὖν V : εἴτουν C[pc] (εἴτ' οὖν C[ac]), but only a few lines below (**24.8**) εἴτουν V C. However, there are exceptions: **27.294** τοανάπαλιν V : τὸ ἀνάπαλιν C.

[47] See, e.g., Munitiz (1984) xlviii.

4. Sigla and Abbreviations

V	Vaticanus gr. 182
C	Vaticanus gr. 1365
G	Vaticanus gr. 1087

ac	ante correctionem
add.	addidit, -erunt
Chort.	Ioannes Chortasmenos
corr.	correxit, -erunt
exhib.	exhibe(n)t
Greg.	Nikephoros Gregoras
homoeotel.	homoeoteleuton
mg.	margen (in mg. = in margine)
om.	omisit, -erunt
pc	post correctionem
ras.	rasura
sch.	scholion
sl	supra lineam
transp.	transposuit/-erunt

Works cited in the apparatus fontium

Cleom.
= Cleomedes, ed. R. B. Todd, *Cleomedis Caelestia*, Leipzig 1990.

Eucl., *Elem.*
= Euclid, *Elementa*, ed. E. S. Stamatis (post J. L. Heiberg), *Euclidis elementa*, vol.1, 2nd edn, Leipzig 1969.

Herm., *in Phaedr.*
= Hermias, *In Platonis Phaedrum scholia*, edd. C. M. Lucarini – C. Moreschini Berlin; Boston 2012.

[Pl.], *Epin.*
= *Epinomis*, ed. E. Des Places, Paris 1956.

Porph. *in Harm.*
= Porphyry, *in Ptolemaei Harmonica*, ed. I. Düring, Göteborg 1932.

[Procl.], *Par.Ptol*
= [Proclus], *Paraphrasis in Ptolemaei Tetrabiblon*, ed. L. Allatius, Leiden 1635. This work is transmitted by Vat. gr. 1453 and any readings of this MS reported in the apparatus derive from Bydén (2003).

Ptol. *Alm.*
= Ptolemy, *Syntaxis mathematica*, ed. J. L. Heiberg, *Claudii Ptolemaei opera quae exstant omnia*, vol. 1 (2 parts): *Syntaxis Mathematica*, Leipzig 1898-1903.

Ptol. *Geog.*
= Ptolemy, *Geography*, ed. K. Müller, *Claudii Ptolemaei geographia*, 2 vols. Paris 1883-1901 (for books 1-5); ed. C. Nobbe, *Claudii Ptolemaei geographia*, vol. 2, Leipzig 1845 (for books 6-8).

Ptol. *Harm.*
= Ptolemy, *Harmonics*, ed. I. Düring, *Die harmonielehre des Klaudios Ptolemaios*, Göteborg 1930

Ptol. *Proch. Kan.*
= Ptolemy, *Handy Tables*, ed. A. Tihon, *Ptolemaiou Procheiroi Kanones: Les Tables Faciles de Ptolémée*, Louvain-la-Neuve 2011; ed. N. Halma, Vol. 1: *Commentaire de Théon d'Alexandrie sur le livre III de l'Almageste de Ptolemée. Tables manuelles des mouvemens des astres*, Paris 1822, Vols. 2 and 3: *Tables manuelles astronomiques de Ptolemée et de Théon*, Paris 1823 and 1825.

Sext. *Adv. Math.*
= Sextus Empiricus, *Adversus mathematicos*, ed. J. Mau – H. Mutschmann, *Sexti Empirici opera*, vols. 2 & 3, 2nd edn. Leipzig 1914-1961.

Theon, *PC*
= Theon, *Small Commentary on Ptolemy's Handy Tables*, ed. A. Tihon, *Le 'petit commentaire' de Théon d'Alexandrie aux tables faciles de Ptolémée*, Vatican City 1978.

Theon Sm.
= Theon Smyrnaeus, ed. E. Hiller, *Theonis Smyrnaei philosophi Platonici, Expositio rerum mathematicarum ad legendum Platonem utilium*, Leipzig 1878. (Corrections by F. M. Petrucci, *Teone di Smirne: expositio rerum mathematicarum ad legendum Platonem utilium: introduzione, traduzione, commento*, Sankt Augustin 2012).

Theodos. *Sph.*
= Theodosius, *Sphaerica*, ed. J. L. Heiberg, *Theodosius Tripolites. Sphaerica*, Berlin 1927

5. *STOICHEIOSIS ASTRONOMIKE* 1. 5-30:

TEXT & TRANSLATION

Θεοδώρου Μετοχίτου

Ἀστρονομικῆς κατ' ἐπιτομὴν Στοιχειώσεως

βιβλίον πρῶτον

5 Ἔπαινος ἀστρονομίας, καὶ τίς ὁ σκοπὸς καὶ ἡ πᾶσα πραγματεία ταύτης

Περὶ ταύτης δὴ τῆς ἀστρονομικῆς, εἴτουν σφαιρικῆς, ἐπιστήμης νῦν στοιχειώσασθαι προυθέμεθα, ἥτις δὴ καὶ τὸ καλλιστεῖον ἐξαίρετόν ἐστι τοῦ ὅλου μαθηματικοῦ τῆς φιλοσοφίας τῆς θεωρητικῆς εἴδους, ὃ δὴ καὶ αὐτὸ τιμιώτερον ἀπεδείχθη τῶν ἄλλων τῆς θεωρητικῆς φιλοσοφίας πρώτων εἰδῶν. ἀντερεῖ γάρ, οἶμαι, τῶν ἁπάντων οὐδεὶς ὅτι τοῦτο δὴ τὸ ἀστρονομικὸν τῆς μαθηματικῆς ὅλης ἐπιστήμης ἐστὶ τὸ πρῶτον φύσει καὶ τιμιώτερον καὶ ἀξιολογώτερον, οὐ μόνον διὰ τὴν σεμνότητα καὶ τὸ ἀξίωμα τοῦ ὑποκειμένου δὴ τούτου περὶ ὃ σπουδάζει, καθ' ὃν λόγον, ὡς ἀνωτέρω εἴρηται, ἄλλη ἄλλης τέχνη τέχνης καὶ ἐπιστήμη ἐπιστήμης ἐστὶ προτιμοτέρα, ἀλλ' ὅτι καὶ τοῖς ἄλλοις σχεδὸν εἴδεσι τοῦ μαθηματικοῦ τὴν αἰτίαν αὐτὸ παρέσχε καὶ τὴν γένεσιν.

Καὶ τοῦτο πρόδηλον καὶ ῥᾷστά τις ἂν ἔχοι δεικνύναι καὶ διὰ πολλῶν· οὐ μὴν ἀλλὰ καὶ ὁ θαυμάσιος αὐτὸς Πλάτων ἐν τῇ προειρημένῃ Ἐπινομίδι τοῦτο σαφῶς καὶ λίαν ἐπιμελῶς κατασκευάζει. διόλου μὲν οὖν ἐν τῇ τοιαύτῃ Ἐπινομίδι παρίστησι τὸ μαθηματικὸν τῆς φιλοσοφίας, ὁποῖον ἔχει τὸ ἀξίωμα καὶ ὁπόσην τὴν χρείαν παρέχεται φιλοσοφίᾳ, μᾶλλον δὲ ὅπως τὴν ἀνθρωπίνην κατασκευάζει τελειότητα καὶ τὴν ὅλην ἀγωγὴν τοῦ κόσμου τὲ καὶ τοῦ βίου. καὶ ἁπλῶς τὸν μαθηματικὸν ἐπιστήμονα, ὁποῖον εἶναι καὶ ὅθεν κατεσκευάσθαι ὡς οἷόν τέ ἐστιν

13-23 [Pl.], *Epin.* 977a-b; 978b-79a.

3 εἴδους : εἶδος V[ac] C **6** μαθηματικῆς post ὅλης transp. C

Theodore Metochites

A Brief Introduction to Astronomy

First Book

5 Praise of Astronomy – Its general aim and method

In this book we set out to present an elementary account of astronomy — to be specific, of *spherical* astronomy — which is the jewel in the crown of the entire field of mathematics. Mathematics in turn has been shown to be the more honourable among the basic contemplative philosophies. I believe that no one will dispute that astronomy is by nature the first and most important part of mathematics, not only on account of its dignity and the reputation of the subject that it studies, but, as mentioned earlier, is more important as an art among the arts and as a science among the sciences. It is also important because it has provided the basis for the development of almost all branches of mathematics.

This is evident and easy to demonstrate in many ways, and the admirable Plato himself establishes it clearly and very carefully in the *Epinomis*. In this work, he gives a general presentation of mathematics, of its intrinsic value and its usefulness for philosophy, or rather on how it institutes the perfection of mankind and the whole course of the world and life. There, he indicates and demonstrates the nature of the mathematical expert and the sources of his

ὑποτυπῶν καὶ ἀποδεικνύων, τοῦτον εἶναι, φησί, τὸν τῆς σοφίας ἁπάσης ἄκρον καὶ ἀρχηγὸν καὶ τὸν τῇ ἀληθινῇ ἀνθρωπίνῃ εὐδαιμονίᾳ ξυνόντα. καὶ καθόλου μὲν τοιοῦτός ἐστιν, ὡς εἴρηται, ὁ ἐν τῷ εἰρημένῳ λόγῳ τῇ Ἐπινομίδι σκοπὸς αὐτῷ καὶ ἡ πρόθεσις ἐπιμελῶς κατασκευαζομένη.

Ἰδίᾳ δὲ ἀστρονομία ὅτι σοφώτατον ἁπάντων φησί, καὶ ἀπὸ τῶν κοινῶν ἐννοιῶν καὶ τῶν φαινομένων καὶ ἀπὸ παντοίας ἄλλης ἐπιχειρήσεως, καὶ ὅτι 'σοφώτατον ἀνάγκη τὸν ἀληθῶς ἀστρονόμον εἶναι, μὴ τὸν καθ' Ἡσίοδον ἀστρονομοῦντα,' ἵνα τοῖς ἐκείνου χρήσωμαι ῥήμασι, 'καὶ πάντας τοὺς τοιούτους οἷον δυσμάς τε καὶ ἀνατολὰς ἐπεσκεμμένους, ἀλλὰ τὸν τῶν ὀκτὼ περιόδων' ἱκανὸν ἐπιστήμονα, 'διεξιούσης τὸν αὐτῆς κύκλον ἑκάστης οὕτως ὡς οὐκ ἂν ῥᾳδίως ποτὲ πᾶσα φύσις ἱκανὴ γένηται θεωρῆσαι μὴ θαυμαστῆς μετέχουσα φύσεως.'

Αἵτινες δὴ σεπταί, κατ' ἐκεῖνον ἐρεῖν, περίοδοι ὀκτὼ αὗται καὶ κατὰ μέρος αὐτῷ δηλοποιοῦνται, οὐ μὴν ἀλλὰ καὶ ἡμῖν κατὰ καιρὸν εἰρήσονται· τὸν γάρ τοι οὐρανόν, καὶ τὰς ἐν αὐτῷ ποικίλας τὲ καὶ καλλίστας διακοσμήσεις, καὶ θαυμασίας τῶν ἐν αὐτῷ ἄστρων ἁπάσας διεξόδους τὲ καὶ κινήσεις, εἶναι θέαμα τῷ τῶν ἀξιολόγων ὄντως πραγμάτων θεωρῷ νῷ. καὶ μὴν αὐτὸν οὐρανὸν Πλάτων φησὶν ἐν αὐτῇ τῇ Ἐπινομίδι ἀρχηγὸν εἶναι πάσης σοφίας καὶ διδάσκαλον ἀνθρώποις καὶ 'ἀγαθῶν ἡμῖν αἴτιον ξυμπάντων', καὶ 'τοῦ πολὺ μεγίστου τῆς φρονήσεως αἴτιόν' φησι γεγονέναι, 'ὃν ὁ μὲν εὐδαίμων πρῶτον μὲν ἰδὼν ἐθαύμασεν, ἔπειτα δὲ ἔρωτα ἔσχε τοῦ καταμαθεῖν ὁπόσα θνητῇ φύσει δυνατά, ἡγούμενος ἄρισθ' οὕτως εὐτυχέστατά τε διάξειν τὸν βίον τελευτήσας τὲ εἰς τόπους ἥξειν προσήκοντας ἀρετῇ, μεμυημένος ἀληθῶς καὶ ὄντως μεταλαβὼν φρονήσεως' καὶ 'θεωρὸς τῶν καλλίστων γενόμενος, ὅσα κατ' ὄψιν διατελεῖ.'

Ὅτι δὲ διδάσκαλος πάντων τὲ τῶν ἄλλων ἀγαθῶν ἀνθρώποις αἴτιος οὐρανὸς καὶ τὰ κατ' αὐτὸν θεάματα δι' ἄλλων τὲ πολλῶν

26-31 [Pl.], *Epin.* 990a5-b2 **39-45** [Pl.], *Epin.* 977a1-2; 986c5-d4

20 ἀποδεικνύων : ὑποδεικνύων C **28** δυσμάς τε καὶ ἀνατολὰς : ἀνατολάς τε καὶ δυσμὰς C **29** ἐπεσκεμμένους : ἐπεσκεμμένον [Pl.] **29** post περιόδων habet τὰς ἑπτὰ περιόδους [Pl.] **30** αὐτῆς : αὐτῶν [Pl.] **31** γένηται : γένοιτο [Pl.] **41** ἰδὼν om. [Pl.] **43** post ἀρετῇ habet καὶ [Pl.] **44** post ἀληθῶς habet τε [Pl.]

expertise, he says that he is the master of all wisdom and the one who understands true human happiness. This is the general aim and purpose of the careful discussion in the *Epinomis*.

Moreover he argues especially about astronomy that it is the wisest of all things, both from common concepts and phenomena and from all other endeavours, and that "the true astronomer must be a man of great wisdom: not the one who practices astronomy on the lines of Hesiod and all those who observe settings and risings but one who has sufficient knowledge of the eight orbits, each completing its own circle in a way whose nature is not easy to comprehend without possessing a wonderful intellect".

These venerable periods are according to him eight and are elucidated separately. In due course we shall also speak about them. Thus the heavens with the diverse and most beautiful arrangements of the stars and the wonderful passages and movements are indeed among the most significant sights for the intellect. Plato says in the *Epinomis*, that the sky is the origin of all wisdom, the teacher to mankind, "the cause of all blessings" and "has become the main source for our understanding". When the happy man first sees it he is amazed, and then he is seized with the passion to learn all that a mortal can learn for he believes "he will spend his days best and with most good fortune, and after he dies and reaches the proper abodes of virtue, having been truly initiated and attaining thoughtfulness, he experiences the finest things that a vision can reach".

Plato demonstrated in many other writings that the heavens and all the associated phenomena are the teacher of every good to mankind, and

αὐτὸς ἀποδείκνυσι Πλάτων, καὶ ὅτι καὶ ἀριθμὸν αὐτὸς ἡμῖν δίδωσιν, ὃν εἴπερ ‘ἐκ τῆς ἀνθρωπίνης φύσεως ἐξέλοιμεν’, φησίν, ‘οὐκ ἄν ποτε φρόνιμοι γενοίμεθα.’ ἀριθμὸν γὰρ βούλεται Πλάτων καὶ τὴν περὶ αὐτὸν ἐπιστήμην, καὶ Πυθαγόρας καὶ πάντες μαθηματικοί, ἀρχέγονον αἴτιον εἶναι πάσης μαθηματικῆς ἐπιστήμης συνεισαγόμενον ταῖς ἄλλαις, οὐ τὰς ἄλλας αὐτῷ συνεισάγοντα, συναναιροῦντα καὶ μὴ συναναιρούμενον, καὶ ὅλως πρῶτον κατὰ φύσιν καὶ στοιχεῖον, ὡς αὐτοί φασι, τῆς μαθηματικῆς οὐσίας. εἰ μὴ γὰρ καθ’ ἑαυτὸν νοοῖτο καὶ οὕτως ἐπιστητὸν θεωροῖτο πρῶτον, οὐκ ἂν ἔπειτ’ ἐν τῇ τοῦ πρός τι σχέσει θεωροῖτο καὶ ἁρμονικοὺς λόγους καὶ μουσικὴν οὐσιοῖ. γεωμετρίας δὲ αἱ πᾶσαι πλοκαὶ καὶ τομαὶ καὶ τῶν σχημάτων ἐπωνυμίαι καὶ γωνίαι καὶ πλευραὶ τὸν ἀριθμὸν ἀναγκαστῶς συνεισάγουσι, καὶ οὐκ ἔστιν ἄνευ ἀριθμητικῆς παρωνυμίας τὴν τῶν γεωμετρικῶν ἐπιστήμην συνίστασθαι.

Οὕτως ἀριθμὸς ἀρχέγονον αἴτιον καὶ στοιχεῖον, ὡς ἔφημεν, πάσης μαθηματικῆς ἐπιστήμης. ἀριθμὸν δὲ δίδωσιν, ὥς φησι Πλάτων, ᾗ γεννᾷ οὐρανὸς καὶ αἱ κατ’ αὐτὸν στροφαὶ καὶ περίοδοι καὶ ἡ συνέχεια τῆς αὐτοῦ κινήσεως. ἡ γὰρ συνέχεια αὕτη τῆς αὐτοῦ κινήσεως τοῦτο μὲν τοὺς ἀριθμοὺς ἐγέννησε τῶν περιφορῶν, τοῦτο δὲ τεμνομένη κατὰ τὴν διαιρετὴν φύσιν τοῦ συνεχοῦς—συνεχὲς δὲ πάντως ἡ κίνησις—τὸ διωρισμένον ποσὸν οὐσιοῖ. ἡ γὰρ κίνησις ἐστὶ μὲν ὡς συνεχὴς ἐν διαιρετῷ ποσῷ, ἔστι δὲ σύστοιχος χρόνῳ· ὁ γὰρ χρόνος συμπέφυκεν ἀδιάστατος τῇ κινήσει. καὶ ἔστιν ὁ χρόνος διὰ μὲν τὴν συνέχειαν τῆς κινήσεως ὑπὸ τὸ συνεχὲς ὡσανεὶ ποσόν, καθ’ ἑαυτὸν δὲ ἀριθμός.

Ἀτὰρ δὴ τὸν ἀριθμὸν οὕτω γεννῶν οὐρανὸς τὴν ἀρχέγονον, ὡς ἔφημεν, αἰτίαν καὶ στοιχειώδη τῆς ὅλης μαθηματικῆς οὐσίας καὶ ἐπιστήμης συλλογιστικῇ διανοίᾳ προβάλλεται. καὶ πλεῖστ’ ἂν ἔχοιμεν περὶ τούτων λέγειν, ταῖς τοῦ Πλάτωνος ἀφορμαῖς καὶ αὐτοὶ τἀληθῆ συνεισφέροντες οἴκοθεν καὶ τοῖς ἐκ τῶν ἀρχαίων καὶ μάλιστα τῶν

49-50 [Pl.], Epin. 977c1-3 **62-64** cf. [Pl.], *Epin.* 977a-b; 978b-79a.

49 post ποτέ habet τι [Pl.] **53** αὑτῷ Bydén : αὐτῷ codd. **54** πρῶτον — στοιχεῖον V^{pc} : πρῶτον καὶ στοιχεῖον κατὰ φύσιν V : πρῶτον καὶ κατὰ φύσιν στοιχεῖον C **57** οὐκ ἂν — θεωροῖτο C^{sl} **56-60** ὅτι κατὰ φυσικὴν ἀνάγκην πρῶτον ἐστὶ τῶν τεσσάρων μαθημάτων ἀριθμός sch. in mg. C (Chort.) **71** οὕτω : οὕτως C^{ac} **74** περὶ τούτων post ἀφορμαῖς transp. C

especially that they created for us the concept of number, and "if it were banished from mankind", he says, "we could never become wise at all". Plato, Pythagoras and all mathematicians profess that the [concept] of number and its associated sciences are the original motivation for developing all branches of mathematics which enter other sciences but do not depend on them; numbers can disprove them but cannot be disproved by them, being the natural origin and element, as they say, of the foundations of mathematics. If they were not considered to be building blocks themselves, we would not have considered them to give relations for harmonic fractions and music. [Finally], geometric constructions, cross-sections, as well as the names of figures, angles and sides by necessity introduce numbers and it is impossible to conceive geometry without arithmetic terms.

As we said, the concept of number is the originating cause and element for every branch of mathematics. As Plato says, the numbers are either given or generated by the heavens with their periods, the revolutions and the continuous motion. The continuity of the motion generated the number of revolutions, and did so while being subdivided into parts specified by numerical values, since motion is continuous and every continuous quantity is divisible. Consequently, since motion is continuous, it is also divisible and congruent to time. Time, on the other hand, has no spatial dimension, but flows along with motion. Time, being very similar to motion, is also continuous and expressible with numbers, as if it were a quantity.

Thus, the heavens being responsible for generating the concept of number, they provided through logical reasoning the motivation for the field of mathematics. I could add quite a few things of my own to this motivation of Plato and also employ the sayings of the ancients, especially clever pronouncements of the Pythagoreans, but discussing them here would lead us far away from the topic.

Πυθαγορείων εὐφυῶς προσχρώμενοι νῦν εἰς τὸ προκείμενον, ἀλλ᾽ ἴσως καὶ πόρρω τοῦ καιροῦ γενοίμεθ᾽ ἂν τρίβοντες ἐνταῦθα.

Ὅτι δ᾽ οὖν καὶ ἄλλως ἥδιστον καὶ χαριέστατον καὶ πάνυ τοι θαυμάσιον χρῆμα τοῖς σοφῶς ὁρῶσιν ἡ ἀστρονομικὴ ἐπιστήμη πάντ᾽ ἂν οἶμαι ξυνομολογῆσαι, καὶ οὐδεὶς οὕτω δυστυχὴς ἢ βασκανίας καθάπαξ δοῦλος, ὅς οὐκ ἂν θαυμάσαι δικαίως, εἴ τις ἐν γῇ μένων καὶ τοῖς ὀφθαλμοῖς εἰς ὀλίγον, οὐκ οἶδ᾽ ὁπόσον, χρῆσθαι δυνάμενος ἔπειτ᾽ εἰς οὐρανὸν ἀκριβέστατα καταθρεῖν οἷός τέ ἐστιν, ἢ ὥσπερ ἐκεῖθεν κατάπεμπτος ἥκων δύναται λέγειν περὶ πάντων ἀψευδέστατα, οἷον νῦν μὲν καιρὸν εἶναι προεπιτέλλειν τόνδε τινὰ τῶν ἀστέρων ἡλίου, νῦν δ᾽ ἐπιδύνειν, καὶ τὸ διάστημα τοὐν μέσῳ τοσόνδ᾽ εἶναι, ὥσπερ κατὰ σπιθαμὰς λογισάμενος, καὶ τὸν χρόνον καθ᾽ ὅν, ὥσπερ ἀπὸ συνθήματος καὶ ὡς ἐν στάθμῃ ξυμμετρησάμενος ἔπειτα προαγορεύων, νῦν δὲ σελήνην ἀποτεμνομένην ἡλίου καὶ ἐπιδύνουσαν, νῦν συνοῦσαν ἀδιαστάτως σχεδὸν ὁτῳοῦν ἀστέρι, νῦν ἀφισταμένην ἐπ᾽ ἀμφότερα βόρειά τε καὶ νότια, νῦν ἐκλείπουσαν, καὶ τόνδε τὸν χρόνον καὶ τοσόνδε τὸν χρόνον καὶ τοσόνδε τοῦ κατ᾽ αὐτὴν σώματος, ἤ γε μὴν ὅλον καὶ ἔτι γε μὴν πρός, τὸ ξυμβαῖνον αὐτῇ σκότος μένον, καὶ μὴν ἀστέρας ὑπαύγους ἡλίῳ, νῦν δὲ φεύγοντας τὰς αὐτοῦ δᾳδουχίας, καὶ οὐκ ἐπὶ τῶν αὐτῶν ἀεὶ μέτρων ἀλλ᾽ ἄλλοτ᾽ ἄλλων, καὶ νῦν μὲν μεσουρανοῦντας αὐτῷ δύνοντι ἢ καὶ ἀνατέλλοντι, πάντως πιστῶς ὅσα γε ἐκ τῆς ἐποπτείας τῶν ἄλλων, νῦν δ᾽ ἀκρονύκτους, νῦν δ᾽ ἄλλοτ᾽ ἄλλα τῶν κατ᾽ οὐρανὸν καὶ περὶ τοὺς ἀστέρας θαυμάτων, περὶ ὧν ἔστιν οὕτω προλέγειν καὶ διϊσχυρίζεσθαι ταῖς τῆς ἐπιστήμης ἀσείστοις πίστεσιν, ὡς οὐδὲ περὶ τῶν ἐν χερσὶ πραττομένων ἢ τῇ ὄψει σύνεγγυς ὑποκειμένων.

Ἐμοὶ γοῦν οὐ μόνον θαυμάσιον τοῦτο δοκεῖ καὶ οἷον τοὺς πολλούς, ἅμα δὲ καὶ τοὺς ὑπὲρ τοὺς πολλούς, καταπλήττειν καί, ὡς ἔπος εἰπεῖν, καταναγκάζειν ἐπιθειάζειν, ἀλλὰ καὶ σφόδρα ἥδιστον ἐνταῦθα χρῆσθαι καὶ μεγάλην τινὰ τὴν γλυκυθυμίαν καὶ οἵαν ἄρρητον ἐμποιεῖν τοῖς τὰς ἐν οὐρανῷ τῶν ἀστέρων χορείας παναρμονίους καὶ ὡς ἀληθῶς συμφώνους ἐφορωμένοις, καὶ ὥσπερ πολιτείας

84 ἀψευδέστατα : ἀκριβέστατα C **88** προσαγορεύων V[ac] **91** βόρειά τε καὶ νότια : νότιά τε καὶ βόρεια C

I think everybody will confess anyway that astronomy is a most agreeable, delightful and admirable affair for those who observe wisely. In addition, no one is so enslaved in misfortune or envy that he does not justifiably admire the one who dwelling on earth turns his eyes for some time, and I do not know how long, to the sky to observe accurately how it looks or, as if coming down from there, is able to describe everything accurately. For example, that now is the time when a given star rises before the Sun, or now sets after the Sun, and the intermediate distance is so or so great, as if gauging the distance with his fingers and measuring the time as on a scale. Sometimes the Moon is separated from the Sun and sets before it; now it is in conjunction with a given star almost without intervening distance; at other times it stands apart to the north or to the south; sometimes it is partially or totally eclipsed being dark for that interval of time. Indeed some stars begin to shine to the Sun, at other times fleeing away from its rays, not always at the same but at different distances. Thus there are various formations of the stars in the sky, which are predicted and affirmed by science with unshaken accuracy, which we do not possess for events happening in our vicinity on earth.

I for one not only think that this is wonderful, and can make a great impression not only to the [general] public, but also to exceptional people, and forces one to consider divine influence. Furthermore it brings unspeakable pleasure, delight and makes the motions of the stars harmonious and agreeable

ἀστασιάστους καθιστοροῦσι καὶ πάντοθεν ὁμολογούσας εἰρηνικάς τε καὶ ἀσυγχύτους, καθ' ἅς οἱ ἀστέρες τὸν ἅπαντα αἰῶνα πολιτεύονται τὰς αὐτῶν ἀγωγάς, καὶ νόμοις πολὺ τὸ καθεστὸς ἔχουσι καὶ ἀνάλογον διοικοῦσι τὰ σφέτερα αὐτῶν.

Τοῦτο μέν γε καὶ ἀκίνδυνον παντάπασι καὶ ἀζήμιον τῇ καθ' ἡμᾶς χριστιανικῇ θεοσεβείᾳ καὶ πίστει, τὸ περὶ τῶν κινήσεων τῶν ὁμαλῶν τε καὶ ἀνωμάλων καὶ τῶν πρὸς ἀλλήλους σχηματισμῶν ἡλίου τὲ καὶ σελήνης καὶ τῶν πέντε πλανωμένων, ἔτι δὲ καὶ ἀπλανῶν ἀστέρων, ἀκριβῶς ἐπίστασθαι καὶ καταλογίζεσθαι καὶ λέγειν ἔχειν, περὶ ὧν, ὡς ἔπος εἰπεῖν, ἡ πᾶσα σπουδὴ καὶ πραγματεία Πτολεμαίῳ ἐν τοῖς τρισὶ καὶ δέκα βιβλίοις τῆς Μαθηματικῆς Συντάξεως.

Τί γὰρ ἐν τούτοις τὸ προξενοῦν ζημίαν τῇ πίστει; ἐπεὶ καὶ τί γε ἄλλο τουτὶ ἢ καθὼς καὶ περὶ ἄλλων ὡντινωνοῦν ὄντων θεωρία τις καὶ γνῶσις, ὅπῃ ποτ' ἄρ' ἔχοι, καὶ κατάληψις καὶ πολυπραγμοσύνη φιλοσοφίας; ἢ τί γὰρ μᾶλλον τοῦτο λυμαινόμενον ἡμῖν τῷ θείῳ σεβάσματι ἢ τὸ ἁπλῶς οὕτω καὶ ἰδιωτικῶς, ὥσπερ οἱ πολλοὶ τῶν ἀνθρώπων καὶ ἀμαθέστεροι, τοὺς τῆς σελήνης δρόμους ἐπιλογίζεσθαι καὶ καταριθμεῖν, νῦν μὲν ἀνευρίσκοντας τὴν πανσεληνιακὴν αὐτῆς συζυγίαν, ἣν αὐτοὶ καλοῦσιν ἀπόχυσιν, νῦν δὲ τὴν συνοδικήν, ἣν αὐτοὶ καλοῦσι γένναν, ἢ τὰς τοῦ ἡλίου ὁλοσχερεῖς ἐποχὰς καὶ τόπους ἐν οἷς ἑκάστοτε γίγνεται, ὡς τῷδε μὲν τῷ μηνὶ ἐν τῷδε τῷ ζῳδίῳ λέγειν αὐτὸν εἶναι, τῷδε δὲ ἐν ἄλλῳ τῳ, νῦν μὲν ἐν Κριῷ, νῦν δὲ ἐν Ζυγῷ, ἄλλοτ' ἐν Καρκίνῳ, ἄλλοτ' ἐν Αἰγοκέρωτι, ἄλλοτ' ἐν ἄλλῳ;

Ὥσπερ γὰρ οὐδ' ἡντιναοῦν ὄνησιν πάντως ἢ ζημίαν φέρει, ὅσα γε πρὸς τὸ θεοσεβεῖν καὶ ἀσφαλῶς ἔχειν τῆς πίστεως, τὸ περὶ ταῦτ' ἐνίους πραγματεύεσθαι ἀμαθῶς οὑτωσὶ καὶ ἁπλοϊκῶς, καθάπερ πολλοὺς τῶν ἀνθρώπων ὁρῶμεν, τὸν αὐτὸν τρόπον οὐδὲ τοῖς ἐπιστημονικῶς τὰ τοιαῦτα καὶ ἠκριβωμένως καταμεμαθηκόσι ζημία τις ἢ ὄνησις ἐντεῦθεν πρὸς τὸ καθαρῶς θεοσεβεῖν ἐγγίνεται.

Καὶ τοῦτ' εὔδηλόν τε καὶ παντὶ ξυνιδεῖν· γνῶσις γάρ ἐστιν, ὡς ἔφην, καὶ τοῦτο διὰ φιλοσοφίας ἐπιστημονικὴ περί τινων κτιστῶν καὶ γενητῶν ὄντων καὶ φύσεων ὑπὸ τῆς παντοκρατορικῆς θεϊκῆς

138 γεννητῶν V C^{ac}

as if they are political states without revolutions, being always peaceful and orderly in which the stars arrange eternally their movements by obeying strict laws which govern the relations among them.

This is not dangerous and is also absolutely harmless to our Christian devotion and faith, that is to investigate precisely and calculate the regular and anomalous motions and relative formations of the Sun, the Moon, the five planets and in addition of the fixed stars. And these things are practically the exclusive concern and study of Ptolemy in the thirteen books of his *Mathematical Syntaxis.*

What aspect of this may cause harm to the faith? What else is contained in them besides a particular view and knowledge which in any case belongs to the realm and practice of philosophy? Alternatively, is this more harmful to our divine faith than what many untrained people calculate and tabulate for the orbits of the Moon; now that full Moon is in *syzygy*, which they call "effluence" or now in synodic *syzygy* which they call "birth"; or calculate the longitudes and positions where the Sun is located, by saying that at this month the Sun is on this sign of the zodiac, at another time elsewhere, now it is in the sign of Aries or in Libra and other times in Cancer or in Capricorn or somewhere else.

No benefit or harm to the reverence and certainty of our faith is caused by those who treat these topics with ignorance and naïvety, as we witness many people doing. In the same way, no harm or advantage to pure piety comes from those who study them scientifically and accurately.

It is evident to everyone and easy to see that knowledge is, as we said, the scientific description of things and entities that are constructed, born and created by the Almighty Divine Power as they are or as they behave, obeying divine rules and commands.

δημιουργηθέντων δυνάμεως, ὅπῃ τὲ ἔχοι καὶ ὅπως ταῦτα φέρεται, θεϊκοῖς πάντως ὅροις πειθαρχοῦντα καὶ προστάγμασι.

Καὶ τοῦτο μὲν πρόδηλον ὡς οὕτως ἔχει· ἀλλ᾽ ἴσως ἐνίους ταράττει θάτερον τῆς ἀστρολογικῆς μέρος, ὃ περὶ τὸ προγνωστικὸν τῶν ἐσομένων καὶ ἀποτελεσματικὸν καταγίνεται καὶ τὰς κινήσεις καὶ τὰς δραστικὰς ποιότητας τῶν ἀστέρων καὶ τὰς ἐποχὰς ἐπὶ τοῦ ζωδιακοῦ καὶ τοὺς πρὸς ἀλλήλους σχηματισμοὺς αἴτιά πως εἶναι τῶν γιγνομένων ἁπάντων σπουδάζει δεικνύναι, καὶ συνδιατιθέναι κατ᾽ αὐτὰς πάντα τὰ ξυμβαίνοντα περὶ τὸν κόσμον καὶ τἀνθρώπινα, καλά τε καὶ φλαῦρα, καὶ κατὰ φύσιν καὶ κατὰ γνώμην καὶ κατὰ τύχην.

Τοῦτό γε μὴν προδήλως πάνυ τοι καὶ ἀναντιρρήτως λυμαίνεται τῇ πίστει καὶ τῇ καθ᾽ ἡμᾶς χριστιανικῇ θεοσεβείᾳ· καὶ αὐτὸς ἐγὼ καὶ φρονεῖν οὕτως ἐμαυτὸν πείθω πάνυ τοι καρτερῶς καὶ ἀτρέπτως, καὶ λέγειν παρρησίᾳ καὶ κηρύττειν πρὸς πάντας ἀξιῶ, τὴν ἀσφάλειαν ἐμαυτῷ τῶν δογμάτων τῆς ἀληθείας καὶ τῆς πίστεως παριστάνων, καὶ τοῖς νῦν καὶ τοῖς ἔπειτα, καὶ καταμεμφόμενος εὖ μάλα καὶ καταγελῶν ὅστις ἄλλως καὶ φρονεῖ καὶ λέγει.

Εὖ γὰρ δῆλον ὡς εἰ πάντα τῆς ἀνάγκης ταύτης τῶν ἀστέρων ἀπαραιτήτως αἰτιατὰ τὰ γιγνόμενα, πάντως ἀνεμέσητα ὅπως ἄρ᾽ ἔχοι καὶ οὐ μή ποτέ τι κατ᾽ ἀνθρώπους ἐπαινετόν, οὐδ᾽ ὥστε θαυμάζειν οὐδ᾽ ὥστε τιμᾶν ἄξιον οὐδὲ τοὐναντίον φαῦλον οὐδ᾽ ὥστε καταδικάζειν καὶ ὑπεύθυνον ἀκούειν δίκαιον (ἀναγκαστῶς οὕτω γιγνόμενον καὶ τῇ τῶν ἀστέρων περιφορᾷ συστρεφόμενον, ἀρετή τε πᾶσα καὶ κακία ἐκποδὼν ἐσεῖται) καὶ σπουδαστέον καὶ φευκτέον οὐδέν (ἀναγκαστῶς γὰρ καὶ ἀπαραλογίστως ἐπὶ πᾶν ὁτιοῦν ἱκνούμεθα, κἂν θέλωμεν, κἂν μὴ θέλωμεν), μᾶλλον δὲ καὶ τὸ θέλειν καὶ προαιρεῖσθαι οὐχ ἡμῶν αὐτῶν, ἀλλὰ καὶ τοῦτο πᾶσα ἀνάγκη καὶ ἄκοντες ὡς εἰπεῖν θέλομεν καὶ δυναστεία τίς ἐστιν ἡ προαίρεσις καὶ τυραννὶς ἀλόγιστος καὶ οὐκ ἔστιν ἡτισοῦν κρίσις οὐδ᾽ αἵρεσις ἅ τε πράττειν χρὴ καὶ ἃ μή· πράττειν γὰρ χρὴ καὶ μὴ ποιεῖν χρή, καὶ ἄλλως οὐκ ἔστιν.

155 ἄλλως : ἄλλος V **158** τι C^{sl} : om. V C^{ac} **163** ἀπαραλλογίστως C

It is obvious that things are so. However there may be some who are bothered by the other part of astrology, that part which has to do with predicting future events, i.e. *apotelesmatics*. This part aims to show that the motions and unusual qualities of the stars, their longitudes in the zodiac and their relative positions are responsible for everything that happens. They try to explain everything in the world and among human beings — the good and the wicked — the natural, the logical and everything that happens by chance.

It is indeed evident that this is without doubt harmful to the faith and our Christian (faith) worship; and I also believe and am convinced myself strongly and unequivocally and deem worthy to state honestly and declare firmly to everybody the truth of our doctrines and faith for now and ever. Finally, I blame strongly and laugh with scorn at anybody who thinks and speaks otherwise.

It is also plainly clear that if everything that happens is a consequence of stars, then there is no cause for blame or praise among human beings. Then all virtue and admirations and respect, or [on the contrary] vice, will be abolished and there will be nothing to strive for or shun away. [On this premise] there is no reason to convict someone and bring him to justice since we arrive upon everything inevitably, whether we wish it or not. In other words, it is not in our power to wish and choose, for we wish without wishing, by necessity so to speak. Now, choice is some kind of irrational compulsion, i.e. it is impossible to choose what is necessary to do or not to do because there is no other choice.

Ἀλλὰ ταῦτα μαινομένων εἰσὶ λόγοι καὶ παιζόντων ἐν οὐ παικτοῖς καὶ πολὺ τὸ σαθρὸν ἔχει καὶ πολὺ τὸ ἄπιστον. καὶ μὴν οὐδ᾽ αὐτοί γε οἱ περὶ ταῦτα σπουδάζοντες καὶ προσέχειν σφίσι τὸν νοῦν τοὺς ἄλλους ἀξιοῦντες, ὅσοις ἄρα τοι καὶ νοῦς σωφρονικὸς ἔνεστι, καὶ τῆς ἀληθείας καθάπαξ οὐκ ἀλογοῦσιν, οὐδ᾽ αὐτοῖς, οἶμαι, τὰ τοιαῦτα παντάπασιν ἄσειστα δοκεῖ καὶ βέβαια καὶ ἀεὶ καὶ διὰ πάντων τἀληθὲς ἔχοντα, οὐδ᾽ ὥσπερ θεοπρόπι᾽ ἡμῖν τὰ τοιαῦτα καὶ σεμνὰ καὶ ἀναντίρρητα φθέγγονται.

Αὐτίκα γὰρ Πτολεμαῖος αὐτὸς ὁ Κλαύδιος, ὃς τὰ μάλιστα καὶ ὑπὲρ πάντας πρὸ αὐτοῦ τε καὶ μετ᾽ αὐτὸν τὰ τῆς ἀστρονομικῆς ἐπιστήμης ἐσπούδασε καὶ ἠκριβώσατο καὶ οὗ πολὺ καὶ μέγιστον τῆς περὶ ταῦτα σοφίας τὸ κλέος, οὗτος δὴ ἐν τῷ πρώτῳ τῆς ἀστρονομικῆς αὐτοῦ Τετραβίβλου περὶ τῶν φυσικῶν εὐθὺς ἐν προοιμίοις τὸ προγνωστικὸν τῆς ἐπιστήμης ἐκ δυοῖν τῶν τρόπων κατασκευάζων, ‘ἑνὸς μὲν ὅπερ καὶ πρῶτον τῇ τάξει καὶ τῇ δυνάμει,’ φησί, ‘δι᾽ οὗ καταλαμβανόμεθα τοὺς σχηματισμοὺς ἡλίου καὶ σελήνης καὶ τῶν ἀστέρων, καθὼς πρός τε ἀλλήλους καὶ τὴν γῆν συσχηματίζονται, δευτέρου δὲ καθ᾽ ὃ τὰς μεταβολὰς ἐπισκεπτόμεθα τῶν περιεχομένων ἃς ποιοῦσιν οἱ συσχηματισμοὶ διὰ τῆς φυσικῆς αὐτῶν ἰδιοτροπίας.’

‘Περὶ μὲν τοῦ πρώτου’, φησίν, ‘ὡς ἦν μάλιστα δυνατὸν ἐν τῇ Συντάξει ἀποδεικτικῶς εἴρηται. ἔστι γὰρ τοῦτο καὶ χωρὶς τῆς ἐπιζεύξεως τοῦ δευτέρου ἀναγκαῖον καὶ καθ᾽ ἑαυτὸ ἰδίως. νῦν δὲ περὶ τοῦ δευτέρου καὶ μὴ ὁμοίως τελείου καθ᾽ ἑαυτὸ ποιησόμεθα τὸν λόγον φιλαλήθως καὶ ὡς ἁρμόζει τῇ φιλοσοφίᾳ μὴ πρὸς τὸ βέβαιον καὶ τὴν τοῦ πρώτου κατάληψιν καὶ ἔχοντος ἀεὶ ὡσαύτως τοῦτο παραβάλλοντες’. καὶ πάλιν μετ᾽ ὀλίγα φησίν· ‘ἐπεὶ δὲ πᾶν τὸ δυσκατάληπτον οἱ πολλοὶ διαβάλλουσι, τούτων δὴ τῶν δύο καταλήψεων οἱ μὲν τὴν προτέραν διαβάλλοντες τυφλοὶ παντελῶς εἰσιν, οἱ δὲ τὴν δευτέραν ἔχουσί πως ἀφορμήν’.

182-93 [Procl.], *Par.Ptol.* 1.7-2.15 **194-96** [Procl.], *Par.Ptol.* 2.24-3.1

182 post ὅπερ habet ἐστὶ [Procl.] **183** καὶ[1] post πρῶτον transp. [Procl.] **184** καθὼς Vat. gr. 1453 (ap. Bydén) : καθ᾽ ὡς Allatius **185** post δὲ habet ὄντος [Procl.] **188** τοῦ Vat. gr. 1453 (ap. Bydén) : om. Allatius **192** μὴ : οὔτε [Procl.]

These are arguments of mad men and those who play with absurd, unsound and unbelievable things. Not even those who study them and demand the attention of others, in particular those who have sound minds and are never irrational, not even those think that the predictions are unshakable, or certain and always true. Nor do they proclaim them to be prophecies, or solemn or without objections.

For example, Claudius Ptolemy, who studied and determined accurately the field of Astronomy better than anybody before or after him and is very famous for his wisdom on this topic, describes in the preface of the first book of his *Physical Tetrabiblos* two methods for making predictions: "one method", he says, "is the first in importance and order (among the two). In the first method we explain the positions of the Sun, the Moon and stars relative to each other and relative to the Earth. In the second method we examine the changes of certain objects enclosed inside their natural orbits.

"The first method", he says, "is presented in the *Syntaxis*, whenever possible with proofs, and stands by itself and without any reference to the second one. Now, concerning the second method, we cannot honestly say that it is rigorous to the same extent, and according to the standards of philosophy it is not as certain as the first one". A little later he states "since the common people criticize everything that is difficult to comprehend, those who criticize the first of the two mentioned methods are absolutely blind, but those who criticize the second are partly justified".

Οὕτω μὲν αὐτὸς ὁ Πτολεμαῖός φησι, μηδόλως παραβάλλειν ἀξιῶν τῇ πρώτῃ κατ᾽ αὐτὸν σπουδῇ, τῇ τῆς Μαθηματικῆς δηλαδὴ Συντάξεως ὑπὲρ τῶν κινήσεων καὶ σχηματισμῶν τῶν ἀστέρων, τὴν δευτέραν αἰτιολογικὴν τοῦ προγνωστικοῦ τῆς ἀστρονομίας, περὶ ἧς ἐν τῇ εἰρημένῃ Τετραβίβλῳ τῶν φυσικῶν δηλονότι σπουδάζει, ὡς μὴ τὸ βέβαιον ἔχουσαν κατ᾽ ἐκείνην καὶ ὡσαύτως ἔχον ἀεί, κἀντεῦθεν πάντως μηδὲ καταληπτὴν βεβαίως καὶ θεωρουμένην ἀσάλευτον καὶ ἀφορμὴν δικαίως διδοῦσαν τοῖς αὐτὴν διαβάλλουσιν.

Ἐμοὶ δ᾽ ὅτι μὲν ἐκ τῆς κινήσεως τῶν ἀστέρων καὶ τῆς τοιᾶσδε πρὸς ἀλλήλους καταστάσεως 'διαβαίνει' τις, κατὰ τὸν Πτολεμαῖον αὐτὸν ἐρεῖν, εἰς τὰ ὑπὸ τὸ οὐράνιον σῶμα φυσικὴ συμπάθεια καὶ συνδιατίθησί πως αὐτά καὶ μυρίαν ἐν αὐτοῖς ἐμποιεῖ μεταβολήν, πάνυ τοι δῆλον δοκεῖ καὶ ἀληθέστατον, κἂν εἰ μή τις τυφλώττοι πρὸς αὐτὸ τὸ τῆς ἀληθείας φῶς, οὐκ ἄλλως γε ἐρεῖ. καὶ τὰ μὲν ἄλλα παρήσειν μοι δοκῶ· τίς δὲ οὐκ οἶδε καὶ τῶν πάντα ἀβελτερωτάτων καὶ μόναις αἰσθήσεσι χρωμένων, ὡς σελήνη ἐγγυτάτω τῶν ἄλλων τῆς γῆς, ὡς ἔοικεν, οὖσα τοῦτο παρίστησιν ἐπιδηλότερον; 'πολλὴν γὰρ διαδίδωσι τὴν ἀπόρροιαν' τοῖς ὑπ᾽ αὐτὴν στοιχείοις καὶ περιγείοις ἅπασι 'καὶ συμπάσχουσιν αὐτῇ', καί, καθά φησι πάλιν αὐτὸς Πτολεμαῖος, 'καὶ συντρέπονται καὶ ἔμψυχα καὶ ἄψυχα, καὶ ποταμοὶ μὲν συναύξονται καὶ συμμειοῦνται πρὸς τὸ ἐκείνης φῶς, θαλάσσης δὲ ὁρμήματα συντρέπονται ταῖς ἀνατολαῖς αὐτῆς καὶ ταῖς δύσεσι, καὶ φυτὰ δὲ καὶ ζῷα συμμειοῦται αὐτῇ καὶ συμπληροῦται ἢ ὅλα ἢ μερικῶς.' καὶ οὐκ ἐκεῖνος μὲν οὕτω φησίν, ἀγνοεῖ δέ τις τῶν ἁπάντων σχεδὸν τοῦτο καὶ τῶν πάντα ἀμαθῶν τε καὶ ἀλογίστων, ἀλλ', ὡς ἔφην, πρόδηλος καὶ ἀναντίρρητος ἡ τῆς σελήνης ἰσχὺς ἐν ταῖς διαφόροις αὐτῆς μεταβολαῖς καὶ σχηματισμοῖς, ὅσα γε πρὸς τὴν γενητὴν φύσιν καὶ μεταβλητὴν καὶ τρεπτήν· οὐκ αὐτῆς δὲ μόνης, ἀλλὰ καὶ τῶν ἄλλων πάντως ἀστέρων παραπλησίως, κἂν μὴ τὰ κατ᾽ αὐτοὺς ἐπισημαίνηται σαφῶς οὕτω δὴ καὶ προδήλως.

206 [Procl.], *Par.Ptol.* 3.20 **213-19** [Procl.], *Par.Ptol.* 4.20-29

213 γὰρ om. [Procl.] **215** post φησι add. καὶ C : rasura in V **217** θαλάττης C **217** αὐτῆς : αὐταῖς V C : ῆς add. V[sl] C[sl] **218** συμπληροῦνται C **222** αὐτῆς : αὐτοῖς C **223** γεννητὴν V[ac] C

This is what Ptolemy says, refusing to compare the value of the studies in the first method — that is, the one in the *Mathematical Syntaxis* describing the movements and positions of the stars — to the second arguments on the predictability of Astronomy included in *Tetrabiblos*. This is so because the predictions are not certain, do not always happen and therefore are not considered certain or unshakable, thus being open to criticism.

In my opinion it is evidently very true that the motions of the stars and their formations relative to each other cause something, as is also proclaimed by Ptolemy, which naturally influences and rearranges objects below bringing innumerable changes in them. All these in my opinion are obvious and very true so that nobody should be blind to the truth and disagree. I will omit the rest, but who does not recognize the most evident using simply his senses, that the Moon being closer to Earth than the other heavenly bodies influences considerably all sub-lunar elements and everything close to the Earth? Again, according to Ptolemy, "both animate and inanimate things vary with the Moon: rivers rise and sink with the moonlight, currents of the sea surge to the East and the West and plants and animals wane and wax with it, some completely and other partially." He does not say that one may ignore all these and remain ignorant and illiterate, but as I said the strength of the Moon is evident and without doubt causes various changes and formations in natural things that are generated and grow and change. In all these the Moon is not alone but all other stars participate even though their effects are not as profound or clear.

Ἡλίου δὲ πέρι καὶ περιττὸν ἄν, ἐμοὶ δοκεῖν, εἴη διεξιέναι καὶ δεικνύειν ὡς ἡ ἐτησία περίοδος αὐτοῦ πάντα συγκινεῖ τὰ ὄντα καὶ τὴν γενητὴν φύσιν τρέπει παντοίαν κατάστασιν καὶ ποιεῖ πάνθ' ἕκαστα καὶ μεταποιεῖ καὶ 'δεύτερος' ὡσπερεὶ 'δημιουργὸς' τῶν φυσικῶν ἁπάντων ὁρᾶται, κατὰ τὸν παλαιὸν ἐκεῖνον λόγον—μᾶλλον δ' ὡς ἀληθῶς τε καὶ εὐσεβῶς ἐρεῖν, τοῦ πρώτου καὶ μόνου δημιουργοῦ καὶ ποιητοῦ τῶν ἁπάντων κάλλιστον καὶ τίμιον δημιούργημα καὶ κράτιστον καὶ δραστικὸν καθυπουργεῖν τῷ λόγῳ καὶ τῇ τάξει τῆς προνοίας πρὸς τὴν κίνησιν τῶν ὄντων καὶ διοίκησιν τῆς γενητῆς φύσεως—τῇ κατ' αὐτὸν κινήσει καὶ περιόδῳ καὶ τοῖς πρὸς τὴν οἰκουμένην ἀναλόγως ἐντεῦθεν προσεγγισμοῖς καὶ ἐπιδημίαις αὐτοῦ καὶ ἀποδημίαις αὖθις ἀναλόγως καὶ ἐκτοπισμοῖς, τὴν καλλίστην καὶ ἐναρμόνιον καὶ παντάπασι σύμφωνον χορείαν τῶν ὡρῶν αὐτὸς ἀποκαθιστάνων καὶ τακτοῖς ὅροις διευθετῶν κἀντεῦθεν βασιλικῶς καὶ νομοθετικῶς εὐτακτῶν τὴν γενητήν, ὡς εἴρηται, φύσιν καὶ τὰς προόδους τῶν γιγνομένων καὶ μεταβολὰς καὶ φθίσεις, ἕκαστα κατὰ καιρὸν συγκυκλῶν ὅροις ἀσαλεύτοις καὶ τὸ ἑστὸς καὶ ἄτρεπτον ἔχουσιν ἐν τῇ συνεχεῖ τροπῇ καὶ κινήσει.

Καὶ τί ἂν ἐνταῦθα τρίβοιμεν ἔτι τὰ πᾶσι δῆλα διεξιόντες; ἀλλ' ὅπερ ἦν τοῦ λόγου σκοπός, τὸ μὲν τὰς περιφορὰς ἡλίου τὲ καὶ σελήνης καὶ τῶν ἄλλων ἀστέρων καὶ τὰς κατὰ τόπους μεταβάσεις καὶ τοὺς πρὸς ἀλλήλους λόγους καὶ σχηματισμούς, πολλὴν τὴν δύναμιν ἔχειν καὶ ἡγεμονικὴν αἰτίαν ἐν τοῖς οὖσι καὶ τῇ γενητῇ φύσει, ἀληθέστατόν τέ ἐστι καὶ δῆλον παντὶ ξυνορᾶν βουλομένῳ. καὶ οὐδὲν ἐπιζήμιον ἐντεῦθεν τῷ καθ' ἡμᾶς τῆς θεοσεβείας δόγματι, τὸ φρονεῖν οὕτω περὶ τούτων καὶ λέγειν καὶ προλέγειν ἐνίοτε τὸν ἐπιστήμονα τὰς περὶ τὰ στοιχεῖα καὶ τὴν γενητὴν φύσιν καὶ σύνθετον κατὰ μέρος καταστάσεις ἑκάστοτε καὶ συμπτώματα ἐκ τῆς τῶν ἀστέρων ταύτης κινήσεως, οὐκ ἄλλως ἢ λειτουργικῶς ὑπηρετούντων τῇ θείᾳ προνοίᾳ, καὶ κατὰ τὴν βασίλειον ἐντολὴν τοῦ πρώτου καὶ μόνου δεσπότου καὶ παντοκράτορος ἐπιτροπευόντων καὶ διοικούντων τὴν φύσιν, τῆς ὅλης κατ' αὐτοὺς

229-32 cf. Herm., in *Phaedr.* p. 17.2-4 Lucarini-Moreschini

228 γεννητὴν V[ac] C **234** γεννητῆς V[ac] C **239** γεννητὴν V[ac] C **247** γεννητῇ V[ac] C[ac] **251** γεννητὴν V[ac] C

It would be superfluous, I think, to argue and to demonstrate that the annual cycle of the Sun influences all beings and modifies the entire state of the created world, making and remaking everything. As the old saying goes, the Sun is viewed as "second creator of all natural things". Or, to say it correctly and with reverence, [the Sun] is the perfect and most valued creation of the first and maker of everything, which He provides with reason and order for the motion of things and the organization of the created world. With its own movement and period and the corresponding local or global changes it restores the perfect, harmonious and always consistent sequence of the hours. As a result, like a king, and in agreement with precise laws, it brings order to the developments dealing with the growth and decay [of created things]. Each being completes its cycle at its own time and according to unchanging rules, remaining unperturbed during the continuous revolution and motion.

Why waste space on what is obvious to everyone? My point is that the revolutions of the Sun and Moon and the other stars with their passage, their mutual relations and formations exert a great power and create major influences on themselves and on the created world — something which is true and evident to everyone who wishes to observe them. This [thinking] is not at all harmful to our devout dogma, i.e. for a scholar to have such an opinion for the consequences resulting from movements of the stars and to describe or predict the state of the elements, or of the composed substances or of the created nature. They only do service to Divine Providence and the royal commands of the first and only Lord who controls and governs nature, which

δυνάμεως εἰς τὴν ἐκείνου κυριότητα καὶ νομοθεσίαν καθάπαξ ἐξηρτημένης, ὥσπερ ἂν εἴ τινες ὑποστράτηγοι μεγίστου βασιλέως καὶ ἁρμοσταὶ καὶ ἡγεμόνες πόλεων καὶ χωρῶν, δρᾶν μὲν δύναιντ᾽ ἂν ἐφ᾽ ἑκάστοις, δρᾶν δὲ ὡς ἄρ᾽ αὐτὸς νομίζοι. καὶ τὸ δραστικὸν γὰρ αὐτῶν ἐκείνου τῆς αἰτίας ἐξῆπται καὶ ὅλως ἐστὶν ἐκείνου.

Καὶ περὶ μὲν τῶν τῆς φύσεως οὕτως ἐγὼ κρίνω, καὶ οἶμαί γε ἀσφαλῶς, καὶ τῶν προσηκόντων οὐκ ἀποτυγχάνω. ὅσοι δὲ τὰς ἐκ τῆς τῶν ἀστέρων κινήσεως αἰτίας καὶ τὸ ἐντεῦθεν προγνωστικὸν φιλονεικοῦσι καὶ βιάζονται κἀν τοῖς κατὰ τύχην γιγνομένοις κἀν τοῖς κατὰ γνώμην τὲ καὶ προαίρεσιν ληρεῖν μοι δοκοῦσι, καὶ καταγέλαστά τινα καὶ μάταια πονεῖν καὶ ἀπατᾶσθαι ἢ ἀπατᾶν κακοὶ κακῶς καὶ δυσνοϊκῶς τοὺς ἄλλους πειρᾶσθαι. οὗτοί γε μὴν καὶ λυμαίνονται τοῖς εὐσεβέσι δόγμασι καὶ πονηρῶν νομίμων καὶ σπερμάτων εἰσὶ γεωργοὶ καὶ μυσταγωγοί· καὶ σφόδρ᾽ ἔγωγε τοὺς τοιούτους ἀποστρέφομαι καὶ ἀποτροπιάζομαι καὶ τοὺς ἄλλους οὕτω φρονεῖν ἀξιῶ. οὕτω μέντ᾽ ἂν παρεσκευασμένοι σωφρόνως τὲ καὶ φιλευσεβῶς οἷοί τ᾽ ἂν εἶεν βιοῦν καὶ τῇ σοφίᾳ κατὰ καιρὸν χρῆσθαι καὶ τούτων τῆς μερίδος εἰμί τε καὶ εἴην κρίνων ἐπ᾽ ἀληθείας ὀρθῶς.

Τὸ μὲν γὰρ ὑπὸ τῆς τῶν ἀστέρων κινήσεως καὶ τῆς αὐτῶν κατὰ φύσιν ἐνεργείας, δεδομένης κατ᾽ ἐντολὴν θείαν, καὶ τῆς διαφόρου κράσεως ἔκ τε τῶν τοπικῶν ἑκάστοτε ἐπιδημιῶν καὶ τῶν τοῦ ζωδιακοῦ τμημάτων καὶ τοῦ συνεγγισμοῦ καὶ τῆς ἀποστάσεως καὶ ὅλως τῶν λόγων καὶ τῶν σχηματισμῶν τῶν πρὸς ἀλλήλους, γίγνεσθαί τινα ἀπορροὴν εἰς τὰ ὑποκείμενα καὶ συμπαθεῖν ὁπῃοῦν καὶ συνδιατίθεσθαι κατὰ φύσιν, προηγουμένως μὲν τὰ ὑπ᾽ αὐτῶν περιεχόμενα στοιχειώδη καὶ ἁπλᾶ σώματα καὶ λοιπὸν ἔπειθ᾽ ἑξῆς τὰ ὑπ᾽ αὐτῶν περιεχόμενα σύνθετα καὶ γιγνόμεν᾽ ἑκάστοτε καὶ φθίνοντα, μάλ᾽ ἔγωγε τίθεμαι, καὶ πολὺ τὸ ἀληθεῦον ἔχει καὶ πολὺ τὸ εὔλογον, καὶ πᾶς ἄν, οἶμαι, σωφρονῶν καὶ τὸν νοῦν προσέχειν εὖ μάλ᾽ ἀξιῶν καὶ μὴ τυφλώττειν, ἑκών γε εἶναι, πρὸς οὕτω δὴ πρᾶγμα κατάδηλον συνερεῖ καὶ οὐκ ἄλλως ἔχειν φιλονεικήσειε.

263 κινήσεως C^{ac} (ut vid.) : κινήσεων V C^{pc}

depends on His complete power, authority and laws. They do this as if they are generals of a Great King, or vassals or governors of cities and countries who may act by themselves, but acting in accord as He decrees. Their actions are at His service and are completely His.

This is, then, how I judge matters regarding natural events and I think with confidence that I do not fail on these matters. But people who predict and constrain events happening by chance, or by choice as the consequence of the stars, are silly and deceitful. They practice laughable and trifling things. They fool themselves and attempt maliciously with ill-reasoning to fool others. These people indeed do harm to our reverend doctrines, giving with their behaviour bad examples by sowing evil seeds. These people I strongly avoid and condemn and demand others to do the same. [On the contrary], those who are equipped with wisdom and devotion, when occasionally encountering these others, use at that time wise reasoning. I am part of this group who chooses truth correctly.

I definitely endorse the view that the movement of the stars and their physical energy, endowed to them by the Almighty, have influences on things below, depending on various local conditions, signs of the zodiac, the relative distances [of the stars], as well as their mutual positions. I am of the opinion that it is true and reasonable that [the stars] affect and arrange physically together first the elementary and simple objects below, then the composite which have the ability to grow and decay. Thus anybody who is prudent and whose mind appreciates values and is not blind to events considers these things as evident, agrees and does not argue about them.

Τὸ δ᾽ ὑπὸ τῆς αὐτῆς ταύτης αἰτίας τῆς ἀπὸ τῶν ἀστέρων ἀνάγκης εἶναι κἀντεῦθεν ἄρα καὶ προλέγειν ἐξεῖναι, τὸν μέν τινα δεσπότην γίγνεσθαι, τὸν δὲ δοῦλον, καὶ δεσπότην ὄντα τινὰ πρὸς μὲν τόνδε τιν᾽ ἀμέλει τῶν ὑπηκόων εὐεργετικῶς διατίθεσθαι, πρὸς δὲ τὸν τοὐναντίον, καὶ τὸν μὲν εὖ ποιεῖν, τὸν δὲ κακῶς, καὶ τόνδε τιν᾽ αὖθις πονηρᾷ συνοικῆσαι γυναικὶ καὶ δυστρόπῳ καὶ φιλέριδι, τόνδ᾽ ἀγαθῇ τινι καὶ χρηστῇ τὸ ἦθος, καὶ τοὺς μὲν δή τινας ἀστυγείτονας ἢ ὅλως ἐν γειτόνων ὄντας ἐκπολεμωθῆναι πρὸς ἀλλήλους τέως εἰρηνεύοντας, τοὺς δ᾽ ἐκ πολέμων διαλῦσαι τὴν μάχην καὶ ὑποσπόνδους γενέσθαι καὶ εἰρηνικὰ καὶ φίλια φρονεῖν, καὶ τὸν μὲν ἁλῶναι λωποδυτοῦντα, τὸν δ᾽ ἱεροσυλοῦντα, καὶ τὸν μὲν χρηστὰ βουλεύειν, τὸν δὲ δυσνοϊκά τε καὶ φαῦλα, καὶ ἄλλα τοιαῦτα παραπλήσια πλεῖστα, ταῦτα δ᾽ οὖν οὕτω φρονεῖν καὶ προλέγειν ἀξιοῦν ὡς ἐκ τῆς τῶν ἀστέρων ἀνάγκης, ἔν γε τοῖς κατὰ βούλευσιν δηλονότι καὶ προαίρεσιν καὶ ἐν τοῖς κατὰ τύχην εὐπραγήμασί τε καὶ δυσπραγήμασιν, ἃ θεϊκὴ πάντως πρόνοια ξὺν λόγῳ πάνθ᾽ ἑκάστοτε τελεσιουργεῖ, οὐ μόνον ταῖς εὐσεβέσιν ἐναντιοῦται δόξαις καὶ τὴν παντοκρατορικὴν θείαν δεσποτείαν ἀφαιρεῖ κακῶς καὶ ἀρετὴν καὶ κακίαν ἀναιρεῖ καὶ τοὺς ἐπ᾽ αὐταῖς εὐλόγους ἐπαίνους καὶ τὰς καταδίκας, ἀλλὰ καὶ παντάπασι τόδ᾽ ἐστὶν ἀλογιστότατον καὶ ἠλιθιώτατον.

Τὸ μὲν γὰρ τόδε τι θερμοποιὸν ἴσως ἢ ψῦχον ἀναλόγως πρὸς τόδε τι παθητὸν παρακείμενον ἔγγιστ᾽ ἢ μακρόθεν πάθος ἐμποιεῖν καὶ θερμαίνειν ἢ ψύχειν αὐτό, καὶ νῦν μὲν ὁλοσχερῶς καὶ ἀκράτως μονομερὲς ὂν καὶ ἁπλοῦν, νῦν δὲ καὶ μετριώτερον καὶ κεκραμένως διά τιν᾽ ἄλλην ἐναντίου κοινωνίαν ἔγγιστ᾽ οὖσαν καί, ἁπλῶς εἰπεῖν, τὰς ποιητικὰς παθῶν ποιότητας, αἳ κατεθεωρήθησαν εἰς τοὺς ἀστέρας ἢ καὶ τοὺς οὐρανίους τόπους, ἐμποιεῖν πάθη καὶ διαφόρους κινήσεις καὶ ἀλλοιώσεις εἰς τὴν γενητὴν καὶ μεταβλητὴν φύσιν, πρᾶγμα πολὺ τὸ εὔλογον ἔχον καὶ πάνυ τοι πιστόν, ἐμοὶ δοκεῖν, καὶ μάταιον ἢ φιλόνεικον ἢ βάσκανον εἰς τοῦτ᾽ ἀντιλέγειν.

314 γεννητὴν V^{ac} C^{ac}

On the other hand, it is argued that many [other] events are the necessary consequence of the stars. For example, that one will become master and the other slave, with one master being benevolent to his subjects and the other bad. One will be living together with a wily, difficult and quarrelsome wife and another has an honest wife with honourable character; or to have neighboring cities or simply neighbors fighting each other even though they were in peace before. Some cancel the fight and come together, declaring peace and friendship. One is caught stealing and another robbing the temples, or one considers honest things, while another considers malicious and base things, and there are many other such examples. Believing and attributing all these events — the intentional and those occurring by chance, or fortune and misfortune — as the necessary consequence of the stars and not the result of holy providence and reason, is not only contrary to our pious doctrines but also diminishes Divine Providence. In addition, it refutes virtue and vice, together with praise and blame that go together with them. All this is illogical and very foolish.

[As an example], an object that warms up or cools down influences another object which is nearby or farther away, by heating or cooling it, either directly and completely when it is simple, like a single unit, or partially and temperately whenever there is an opposing agent. To say it simply, [in a similar manner] qualitative effects are also attributed to stars or celestial regions generating changes and various motions on the created and changeable world. This fact appears to me to be reasonable and accurate and it is useless to contest, slander, or argue against it.

Τὸ δ᾽ ἐκ τῆς αὐτῆς παραθέσεως ἀστέρων ὡντινωνοῦν ἢ πόρρωθεν ἢ ἔγγιστα ἢ μονομερῶς ἢ καὶ κεκραμένως μετ᾽ ἄλλων ἐναντίως ἐχόντων καὶ μετὰ τῆς ἑκάστοτε διαφορᾶς καὶ ἑτερότητος τῶν οὐρανίων τόπων ἢ καὶ τῶν τῆς κινήσεως χρόνων, ὡς οἱ ταῦτα τεχνιτεύοντες διαιροῦσι τὲ καὶ λογολεσχοῦσι, τόνδε τινὰ εὐπραγεῖν, τόνδε τοὐναντίον, καὶ τόνδε τινὰ πονηρὸν εἶναι τοὺς τρόπους, τόνδε χρηστόν, καὶ τόνδε τινὰ πατράσιν ἐπαναστῆναι καὶ φονικὴν κατ᾽ αὐτῶν ἐντείνασθαι χεῖρα, τόνδε φιλοδωρίαν καὶ ἐλευθεριότητα καὶ ὅλως φιλανθρωπίας ἐνδείξασθαι ἔργα, καὶ τόνδε τινὰ ὑπὸ λῃστῶν ἐπ᾽ ἐρημίας ἐλλοχηθέντα κτανθῆναι, τόνδ᾽ ἐν μέσῃ πόλει δικασταῖς καὶ δεσπόταις παρρησίᾳ ταὐτὰ παθεῖν—τούτων δὴ τῶν ξυμπτωμάτων τίς ἀνάγκη φυσικὴ ἢ τίς παρὰ τῶν ἀστέρων αἰτία ξυνέλκει καὶ βιάζεται, ἢ τί τὸ κοινὸν ἢ τίς δεσμὸς συμπαθείας ἐν τούτοις καὶ λόγος ἀκόλουθος; πρόδηλον τοῦτο παντὶ νοῦν ἔχοντι καὶ σωφρονοῦντι, ὡς φαῦλος ὁ πόνος οὗτος καὶ μάταιος καὶ οὐδὲν ἔχων ὑγιές.

Ἀτὰρ δὴ περὶ τούτων ἔστι μὲν πλείω λέγειν ἔτι καὶ πάνυ τοι πλεῖστα, ἀλλὰ τὸ μῆκος εἴργει, καὶ παρήσω λοιπὸν ἀποχρώντως ἔχειν ἡγησάμενος καὶ μόνα τὰ εἰρημένα, ὥστε παριστάνειν τὴν ἡμετέραν γνώμην καὶ ὅπως ἄρα περὶ τῶν τοιούτων δοκιμάζω καὶ φρονῶ. ἐπεὶ δ᾽ ἱκανῶς, ἐμοὶ δοκεῖν, ἅττα προειπεῖν ἐχρῆν καὶ προκαταστήσασθαι τῆς ὅλης σπουδῆς καὶ προθέσεως, εἴρηται, φέρε δὴ λοιπὸν ἐναρξόμεθα τοῦ προκειμένου καὶ τῶν ἀστρονομικῶν ὑποθέσεων ἁψόμεθα, ὡς ἄρα τὴν ἀρχὴν ἐπιβάλλειν ἀξιοῦμεν.

Now, from the formations of the stars being nearby or far away, or standing by themselves or being mixed with others, and by referring to the occasional differences and peculiarities of the celestial formations, or the time intervals of their motions, some make up, predict and babble away that this one [person] will be acting properly and the other not, and this will have devious behaviour while the other kind, or that this one will revolt against his parents and raise a murderous hand or one is gratuitous, liberal and engaging in charitable deeds. [Sometimes they say] that someone will be attacked by bandits at a deserted location and will be killed, or will experience the same fate in the hands of judges or lords and in the middle of a town — tell me what is the reason for such events, or what cause from the stars attracts and compels them? It is evident to everyone who has a brain that this type of activity is trifling and there is nothing healthy about it.

Well, much more can be said about them and repeated many times but its length is prohibitive and thus I stop, considering well that what has been said represents my opinion and how I stand on these matters. Furthermore, I considered useful to say in advance my attitude and intension toward the entire study. Come let us start with the subject and state the astronomical hypotheses and deem it worthy to begin.

6 περὶ τῆς πρώτης φορᾶς τῶν οὐρανίων σωμάτων τῆς ἀπὸ ἀνατολῶν εἰς δυσμάς

Διτταὶ τοίνυν περιφοραὶ περὶ τὸ οὐράνιον θεωροῦνται σῶμα ἐναντίως πρὸς ἀλλήλας ἔχουσαι, ἡ μὲν ἀπ᾽ ἀνατολῶν ὡς πρὸς δυσμάς, ἥτις καὶ ὡς πρὸς τὰ ἡγούμενα καλεῖται παρὰ τῶν ἀστρονόμων, ἡ δὲ ἀπὸ δυσμῶν εἰς ἀνατολάς, ἥτις καὶ ὡς πρὸς τὰ ἑπόμενα καλεῖται. ἡ τοίνυν προτέρα καὶ ὑπερτέρα καὶ κρείττων καὶ ταχυτέρα καὶ κυριωτέρα τοῦ οὐρανίου σώματος κίνησίς ἐστιν, ἡ ἀπ᾽ ἀνατολῶν εἰς δυσμάς, ἥτις δὴ περιφορὰ ὑπερβάλλοντι καὶ σχεδὸν ἀρρήτῳ κράτει τάχους καὶ ἀσυγκρίτῳ πρὸς τὴν ἄλλην ἀπαρτίζεται, ἀπὸ τοῦ αὐτοῦ σημείου εἰς τὸ αὐτὸ σημεῖον περιφέρουσα τὸ οὐράνιον ἅπαν σῶμα, ἄνωθεν ἀρξαμένη ἀπ᾽ αὐτῶν τῶν πρώτων αὐτοῦ καὶ ὑπερτάτων μερῶν καὶ περιεχόντων τἄλλα πάντα ἐντός. ἅτινα δὴ καθ᾽ ἑαυτὰ μὲν φέρονται τὴν ἐναντίαν καὶ εἰς τὰ ἑπόμενα φοράν, βραδύτερον μὲν τὰ πάντα τῆς εἰρημένης πρώτης φορᾶς, ἄλλα δὲ ἄλλων βραδύτερον καὶ ταχύτερον, ὡς προϊόντες ἐροῦμεν περὶ ἑκάστων κατὰ μέρος· ἀναγκαστῶς δὲ καὶ βιαίως συμπεριφέρονται τῇ πρώτῃ κινήσει, καὶ νικᾷ τὴν αὐτῶν οἰκείαν κίνησιν τὸ ἀπαράμιλλον τάχος τῆς πρώτης κινήσεως, ὡς λανθάνειν σχεδὸν τὴν εἰς τἀναντία προχώρησιν αὐτῶν καὶ μάλιστα τῶν βραδύτερον κινουμένων.

Τὸ γὰρ οὐράνιον σῶμα κοῦφον μέν ἐστιν οἷον ἄρρητον καὶ διαφανὲς καὶ λεπτότατον, σῶμα δὲ ὅμως ὂν τριχῇ διαστατὸν καὶ βάθος ἔχει. καὶ μία μὲν ἐστιν ὁλοσχερῶς θεωρουμένη σφαῖρα, βάθος δὲ ἔχουσα ὑπερβάλλον καὶ σχεδὸν ἀλόγιστον καὶ ἀμέτρητον δοκεῖ πως τέμνεσθαι διόλου σφαιρικῶς ἐντὸς καὶ κυκλικῶς ἐκ τῆς ἄνω καὶ πρώτης καὶ κυρτῆς ἐπιφανείας εἰς αὐτὴν ἐντὸς τὴν ἐσχάτην καὶ κοίλην ἐπιφάνειαν πλείσταις κατατομαῖς, περὶ ὧν κατὰ μέρος μετὰ ταῦτ᾽ ἐροῦμεν.

Ἑκάστη δὲ τῶν τοιούτων κατατομῶν σφαιρικὴ πάντως ἐστίν, ὡς ἔφην, καὶ κυκλοτερής· καὶ ἀλλήλαις εἰσὶ συνεχεῖς καὶ ὑπ᾽ ἀλλήλων περιεχόμεναι καὶ παρ᾽ ἑκάστην πεπήγασιν, ὥσπερ ἀστέρων σώματα. κἂν μὲν τῇ πρώτῃ σφαιρικῇ ὡς εἰπεῖν ἀποτομῇ, τῇ μετὰ τὴν πρώτην

4 προτέρα V[sl] : πρώτη V[ac] C **5** ὑπερτέρα V[sl] : ὑπερτάτη V[ac] C

6 On the first revolution of the celestial bodies from east to west

Two rotations are contemplated for the celestial body in opposite directions: one from east to west, which the astronomers call in reverse sequence [to the signs of the zodiac] and the other from west to east, which follows the sequence [of the signs of the zodiac]. Moreover, the greater, faster and more important rotation is the former, from east to west; its speed defies any description, being incomparable relative to the other [rotations]. It is a revolution from a point back to the same point, carrying along the entire celestial body. The first rotation originates at the highest parts and contains inside everything else. The lower parts rotate in the opposite direction, following the sequence [of the signs in the zodiac]. Everything else is slower than the above-mentioned first revolution; among the other [celestial objects] some are slower and others faster and, as we progress, we will describe each of them separately. By necessity, they are forced to participate in the first rotation, which is stronger than their own. The extreme speed of the first motion nearly hides their own motion in the opposite direction, especially for those moving slowly.

The celestial body is light, indescribable, transparent and very delicate. It is still a three-dimensional object, possessing depth. Altogether, it is considered to be one sphere, having a surpassing depth of almost incomprehensible and immeasurable dimensions; it is asserted that it is subdivided in the inner part into many spherical and circular shells, from the first and uppermost, down to the innermost and last concave surface, which we will describe in detail later on.

As I mentioned, each section is spherical and circular being continuous relative to each other — one enclosing the other and supporting each other with the bodies of the stars attached to them. On the first spherical shell, below

καὶ πάσας περιέχουσαν καὶ πάσας συγκινοῦσαν ἑαυτῇ, ἐνθεωροῦνται ἅπαντες οἱ ἀπλανεῖς λεγόμενοι ἀστέρες, ταῖς ἐντὸς δὲ καὶ ἑξῆς κατωτέρω ἥλιος καὶ σελήνη καὶ οἱ πέντε καλούμενοι πλανῆται ἀστέρες ἄλλος ὑπ᾽ ἄλλων, ὡς μετὰ ταῦτ᾽ ἐροῦμεν τοὺς ἑκάστων βαθμοὺς καὶ τὴν τάξιν, πλήν γε ὅτι διεστήκασιν ἀλλήλων οὐκ ὀλίγον τοῦ οὐρανίου βάθους ἕκαστος ἀποτεμνόμενοι. ὅπερ δὲ ἐλέγομεν, ὅτι αἱ πᾶσαι τοιαῦται σφαῖραι, αἱ θεωρούμεναι ἑξῆς ἀλλήλων συνεχεῖς καὶ περιεχόμεναι ὑπ᾽ ἀλλήλων μέχρι καὶ τῆς ἐσχάτης, ἥτις ἐστὶν ἡ σεληνιακή, αἱ πᾶσαι δ᾽ οὖν αὗται σφαῖραι συμπεριφέρονται καθάπερ βίᾳ τινὶ καὶ ἀηττήτῳ καὶ ἀναπαντήτῳ κράτει τῇ πρώτῃ καὶ κυριωτάτῃ τοῦ οὐρανίου ἄνωθεν σώματος περιστροφῇ, τῇ ἀπ᾽ ἀνατολῶν εἰς δυσμάς.

Σὺν αὐταῖς δὲ πάντως ταῖς σφαίραις ὡσαύτως συμπεριφέρονται καὶ οἱ ἐν αὐταῖς ἀστέρες, οἵ τε ἀπλανεῖς καὶ πεπηγέναι δοκοῦντες καὶ οἱ ἄλλοι, οἳ μὴ μόνον κατὰ μῆκος ἔχουσι διαστάσεις, οἱ μὲν ἔμπροσθεν φαινόμενοι, οἱ δ᾽ ὄπισθεν, ἔτι δὲ καὶ κατὰ βάθος, οἱ μὲν ὑψηλοτέραν λαχόντες θέσιν ἐν τῷ οὐρανίῳ βάθει καὶ ἀπογειότεροι ὄντες, οἱ δὲ προσγειότεροι· ἀλλὰ καὶ κατὰ πλάτος ὡσαύτως ἔχουσι διαστάσεις, οἱ μὲν βορειότεροι φαινόμενοι τῶν ἄλλων, οἱ δὲ νοτιώτεροι. οὗτοι δὴ πάντες, ὡς ἔφην, οἱ ἀστέρες συμπεριφέρονται τῇ πρώτῃ καὶ ταχυτάτῃ τοῦ οὐρανίου σώματος περιστροφῇ κατὰ παραλλήλων ὡς εἰπεῖν κύκλων, οἱ μὲν μειζόνων, οἱ δὲ ἡττόνων, ἕκαστος ἰσοταχῶς συμπεριαγόμενοι τῇ κοινῇ καὶ καθόλου ταύτῃ περιφορᾷ. πᾶσαν γὰρ δὴ σφαῖραν δυνατόν ἐστιν εἰς μυρίους ὅσους διαιρουμένην νοεῖν παραλλήλους κύκλους, περὶ τοὺς αὐτοὺς πόλους ἅπαντας στρεφομένους τοὺς τῆς ὅλης σφαίρας, ὧν ἀκινήτων μενόντων πάντως ἡ σφαῖρα εἰς ἑαυτὴν περιστρέφεται· καὶ ἔστι τῶν τοιούτων παραλλήλων μέγιστος πάντως τις τῶν ἄλλων πάντων κύκλος, ὁ εἰς δύο ἶσα τέμνων τὴν ὅλην σφαῖραν καὶ διὰ τοῦ κέντρου αὐτῆς περιαγόμενος. τὸ γὰρ κέντρον τῆς σφαίρας κέντρον ἐστὶ καὶ τοῦ μεγίστου ἐν αὐτῇ κύκλου καὶ τοῦτό ἐστι τὸ ἴδιον καὶ χαρακτηριστικὸν τοῦ μεγίστου κύκλου. οἱ δὲ ἐφ᾽ ἑκάτερα τούτου παράλληλοι συνεχεῖς ἀεὶ ἥττους ἀλλήλων ἑξῆς θεωροῦνται, μέχρι καὶ ἐς αὐτοὺς καταντῆσαι τοὺς ἐκ διαμέτρου πόλους τῆς σφαίρας, περὶ οὓς

46 ante οἱ[1] habet νῦν C^{sl} **46** ante οἱ[2] habet νῦν C^{sl} **53** πάντων C^{pc}

the first revolution, which encloses and moves all others, are located all the so-called fixed stars. On the lower shells inside are located the Sun, the Moon and the five planets, one below the other as we shall describe later the position and order of each one in turn, with the proviso that they are distinct from each other, as each one occupies a significant part of the celestial depth. Indeed we have mentioned that all of them are considered to be adjacent to each other and containing one another, down to the last one — the sphere of the Moon. All these spheres participate, by virtue of some unbeatable force which meets no resistance, in the first very fast revolution of the upper part of the sky from east to west.

The stars rotate together with these spheres — both the stars impended in them (the fixed stars attached to the spheres) and the others, which do not have only a longitude (some appearing to be ahead and others behind), but also in depth (some possessing a higher position in celestial depth, being further away from the Earth, and others closer); they also have a dimension of latitude — some appearing more to the North, others further to the South. As mentioned, all these stars participate in the first and very fast revolution of the celestial body along parallel circles, some of them along great circles, others along smaller circles, each revolving with equal angular velocity carried along by the common rotation. It is possible to divide each sphere into an infinite number of parallel circles, all rotating around the same poles, with the poles remaining completely at rest as the sphere rotates. Among this type of parallels, one is the largest among all circles, which divides the sphere into two equal [parts] and rotates around the centre of the sphere. The centre of the sphere is the centre of the great circle, a property characterizing the great circle. The parallels on either side of the great circle are continuously smaller

ἀκινήτους αὕτη στρέφεται. οἱ δὴ παράλληλοι πάντες οὔτε εἰς δύο ἶσα τέμνοντες τὴν σφαῖραν οὔτε τὸ ταύτης κέντρον ἴδιον ἔχοντες, ὥσπερ ὁ μέγιστος ἐν αὐτῇ κύκλος, ὅμως οἱ πάντες ἐν τῇ ταύτης περιφορᾷ τοὺς αὐτοὺς ἔχοντες πόλους ἰσοταχῶς περιστρέφονται, κἂν ὁ μὲν μέγιστος, οἱ δὲ μείζους ἄλλοι ἄλλων, οἱ δὲ ἥττους ὦσι. τοιγαροῦν καὶ ἡ οὐράνιος σφαῖρα καὶ μέγιστον ἔχουσα κύκλον—ἤτοι τὸν ἰσημερινὸν ἢ ὁντιναοῦν ἄλλον εἰς ἶσα ταύτην τέμνοντα καὶ τὸ κέντρον αὐτῆς κέντρον ἴδιον ἔχοντα καὶ ἐφ᾽ ἑκάτερα τούτου παραλλήλους μέχρι καὶ τῶν πόλων αὐτῆς, πάντας μὲν αὐτοῦ ἥττους, ἀλλήλων δὲ μείζους καὶ ἥττους ἅπαντας—ἰσοταχῶς καὶ τὸν μέγιστον καὶ τοὺς ἄλλους μείζονάς τε καὶ ἥττονας ἐν τῇ οἰκείᾳ περιφορᾷ ἀποκαθίστησιν ἐν τοῖς οἰκείοις ἕκαστον σημείοις, ἀφ᾽ ὧν ἤρξαντο.

Ἐπὶ τῶν τοιούτων γοῦν παραλλήλων τῆς οὐρανίου σφαίρας, ἀπείρων ὡς εἰπεῖν νοουμένων, τοὺς ἀστέρας πεπηγότας ὥσπερ ἕκαστον θεωροῦντες κατὰ τὰς πλατικὰς αὐτῶν ἀποστάσεις ἀπ᾽ ἀλλήλων ὁρῶμεν τῇ πρώτῃ περιφορᾷ τοῦ παντὸς οὐρανίου σώματος, τῇ ἀπ᾽ ἀνατολῶν εἰς δυσμάς, ἰσοταχῶς συμπεριαγομένους καὶ ἀποκαθισταμένους τοῖς σημείοις ἀφ᾽ ὧν ἤρξαντο τῆς περιστροφῆς· οἷον οἵτινες ἀστέρες ὡς ἐπὶ τοῦ ἰσημερινοῦ μεγίστου τῶν παραλλήλων κύκλου θεωροῦνται ὡσπερεὶ πεπηγότες. καὶ αὖ ἄλλοι ἐπὶ τῶν ἐλαχίστων παραλλήλων τῶν σύνεγγυς τῶν πόλων θεωροῦνται, κατὰ ταὐτὸν ἐπὶ τοῦ ἀνατολικοῦ ὁρίζοντος γιγνόμενοι, ἰσοταχῶς ἐπὶ τοῦ δυτικοῦ ὁρίζοντος κατὰ ταὐτὸν αὖθις ἐπὶ τῆς ὀρθῆς γίγνονται σφαίρας. κἀντεῦθεν ἰσοταχῶς περιίασιν αὖθις τὰ κάτωθεν ἡμικύκλια καὶ κατὰ ταὐτὸν ἐπὶ τοῦ ἀνατολικοῦ ὁρίζοντος, ὥσπερ πρότερον φαίνονται. οἱ γὰρ κύκλοι καὶ τὰ αὐτῶν ἡμικύκλια, ἐφ᾽ ὧν δοκοῦσι περιφέρεσθαι οἱ ἀστέρες τῇ πρώτῃ τοῦ οὐρανίου σώματος κινήσει καὶ περιφορᾷ, ἄνισα μὲν (ἄλλα γὰρ ἄλλων καὶ μείζω καὶ ἥττω), ἀλλά γε πάνθ᾽ ὅμοια (παράλληλα γάρ)—περὶ γὰρ τοὺς αὐτούς, ὡς ἔφημεν, περιφέρεται πόλους. καὶ τὰ μὲν τῆς πρώτης καὶ καθόλου τοῦ οὐρανίου σώματος περιφορᾶς οὕτω πως ἔχει καὶ τάχει ἀπαραμίλλῳ καὶ σχεδὸν ἀρρήτῳ καταλαμβάνεται. τοσοῦτος γὰρ ὁ χρόνος τῆς τοῦ μεγίστου ὅλου οὐρανίου σώματος πρώτης περιφορᾶς,

71 καθίστησιν C[ac] **90** post καθόλου add. τῆς πρώτης C[ac]

until they reach the poles of the sphere, which remain stationary, while the sphere revolves around them. All the parallels do not divide [the sphere] into two [equal] parts, nor do they have the same centre as the great circle, but all rotate at the same speed because they have the same poles, even though one is a great circle while some are larger and others smaller. Therefore, the celestial globe has a great circle — the equator — one that divides it into two equal parts, and its centre is the same centre [with that of the stellar sphere]. On each side there are parallels up to the poles, all of them smaller [than the equator], but relative to each other, some are larger and others smaller. They all return simultaneously to their own starting positions.

We consider the stars to be embedded in the parallel circles of the celestial sphere and describe each of them with its latitude. We observe the parallels, which are infinite in number, to participate in the first rotation of the entire celestial body from east to west, rotating with the same speed and returning to their starting positions. This way, some stars are thought to be attached to the equator — the greatest of the parallel circles — and others are considered to be located on the parallel circles closer to the poles. On their revolutions they appear on the eastern horizon, and with equal speed proceed to the western horizon of *sphaera recta*; from then on they travel anew with equal velocity to the lower semicircles and this way they appear as before on the eastern horizon. Now the circles and their semicircles, on which we consider the stars to rotate on the first motion and revolution of the celestial globe, are unequal, some being larger and others smaller, but each rotating, as we said, with the same [angular] speed around the same poles. The first revolution of the entire celestial sphere is incomprehensibly fast and almost beyond description. The time elapsed for a complete revolution of the entire

ὅσον ἐστι τὸ ἡμερονύκτιον, μᾶλλον δὲ καὶ βραχύτερος τούτου. ἡμέραν γὰρ ἀπεργάζεται ἥλιος ἐν τῷ ὑπὲρ γῆν ὢν ἡμισφαιρίῳ, νύκτα δὲ ἐν τῷ ὑπὸ γῆν. ἐπεὶ γοῦν ὁ ἥλιος συμπεριφέρεται τῇ καθόλου περιφορᾷ τοῦ παντὸς οὐρανίου σώματος, ἔχει δὲ καὶ αὐτὸς μεταβατικὴν ἰδίαν ἐναντίαν κίνησιν εἰς τὰ ἑπόμενα, ἤτοι ἀπὸ δυσμῶν εἰς ἀνατολὰς ἐν τῇ περιφορᾷ τοῦ παντός, γενομένου τοῦ σημείου ἐν ᾧ ηὑρίσκετο τῇ προλαβούσῃ περιόδῳ κατὰ τὸν ἀνατολικὸν ὁρίζοντα, τοσοῦτον ὑπολείπεται αὐτὸς ὁ ἥλιος τῆς ἐνάρξεως τῆς δευτέρας αὐτῆς περιστροφῆς, ὅσον κεκίνηται εἰς τὰ ἑπόμενα κατὰ τὴν ἰδίαν αὐτοῦ κίνησιν κατὰ τὸ παρελθὸν ἤδη ἡμερονύκτιον· ὥστε ἐπεὶ τὴν ἀρχὴν πάλιν τῆς ἡμέρας μέλλει ἐνεργῆσαι, ὅταν γένηται ἐπὶ τοῦ ἀνατολικοῦ ὁρίζοντος, ἡ τοῦ παντὸς περιστροφὴ ἐν βραχυτέρῳ ἡμερονυκτίου χρόνῳ ἀπαρτίζεται καὶ τοσοῦτον βραχυτέρῳ ὅσον ἀναφέρεται τὸν ἀνατολικὸν ὁρίζοντα ἡ περιφέρεια τοῦ κύκλου, ἣν προὐχώρησεν εἰς τὰ ἑπόμενα ὁ ἥλιος, ἥτις καὶ πλείων καὶ ἐλάττων ἐστὶ καὶ διὰ τοῦτο καὶ μείζονα καὶ ἐλάττονα ποιεῖ τὰ ἡμερονύκτια. ὅπως δὲ τοῦτο γίνεται καὶ κατὰ τίνα τὸν λόγον ἐν τοῖς ἑξῆς ἐροῦμεν κατὰ καιρόν.

94-95 πῶς ἄν τις ἀπὸ τῶν φαινομένων πιστεύσειε κινεῖσθαι τὸν ἥλιον εἰς τὰ ἑπόμενα τοῦ παντός sch. in mg. C (Chort.) **96** ἐναντίαν C[pc] (in mg.) **100** τῆς δευτέρας αὐτῆς : αὐτῆς ante τῆς transp. C

celestial body is a day and a night, and very likely somewhat shorter. The Sun produces the day when it is on the half of the sphere above the Earth and the night when it is below the Earth. As the Sun takes part in the revolution of the entire celestial globe it has, in addition, its own transitory motion in the opposite direction toward the subsequent stars, that is from west to east. In the revolution of the universe, when the Sun returns to the eastern horizon, where it was located during the previous rotation, it lags behind in the beginning of its second revolution by the amount it moved toward the subsequent stars during the last day and night. Thus when the day is about to begin, i.e. the Sun is in the eastern horizon, the revolution of the universe is completed in less time than a day and a night, shorter by the amount that the Sun advanced to the subsequent stars. This amount is sometimes slightly more and other times a bit less. For this reason, the *nychthemera* are larger or smaller, and this happens for reasons that we shall explain at a later time.

7 περὶ τῆς δευτέρας φορᾶς τῶν οὐρανίων σωμάτων τῆς ἀπὸ δυσμῶν εἰς ἀνατολάς

Ἀλλ᾽ οὕτω μὲν ἔχει τὰ τῆς ἀνωτάτω καὶ πρώτης φορᾶς, τῆς ἀπὸ ἀνατολῶν εἰς δυσμάς. ἡ δ᾽ ἑτέρα περιφορὰ τῶν οὐρανίων εἰς τἀναντία τῆς πρώτης ἐστίν, ἀπὸ δυσμῶν δηλονότι πρὸς ἀνατολὰς εἰς τὰ ἑπόμενα, καθὼς ἀστρονόμοι καλοῦσιν ὡς ἔφημεν. ἄνευ γὰρ τῆς πρώτης καὶ ἀνωτάτω σφαίρας καὶ πάσας περιεχούσης, ἥτις καὶ ἄναστρος νοεῖται καὶ βίᾳ τὰς ἄλλας ὅλας τὰς ὑπ᾽ αὐτὴν τῷ ἀνυπερβλήτῳ κράτει καὶ τάχει συμπεριφέρει ἑαυτῇ εἰς τἀναντία τῶν οἰκείων κινήσεων, αἱ ἄλλαι πᾶσαι αἱ μετ᾽ αὐτὴν σφαῖραι, αἱ καὶ τοὺς ἀστέρας ἐν αὐταῖς ὥσπερ πεπηγότας ἔχουσαι, καθ᾽ ἑαυτὰς αἱ πᾶσαι εἰς τὰ ἐναντία τῇ πρώτῃ φέρονται εἰς τὰ ἑπόμενα, ὡς ἀπὸ δυσμῶν εἰς ἀνατολάς, αἱ μὲν θᾶττον, αἱ δὲ βραδύτερον, συμπεριφέρουσαι καὶ τοὺς ἐν αὐταῖς ὥσπερ πεπηγότας ἀστέρας ἅπαντας· ἥ τε δηλαδὴ πρώτη μετὰ τὴν πρώτην τὴν ἄναστρον καὶ πάντας τοὺς ἀστέρας τοὺς ἀπλανεῖς λεγομένους ἔχουσα ἐν ἑαυτῇ καὶ αἱ ταύτης ἑξῆς ἑπτά, ἑτέρα ὑφ᾽ ἑτέρας περιεχομένη, αἵ τε τοὺς πέντε πλανωμένους ἀστέρας ἐν ἑαυταῖς ἔχουσαι, μία τις ἑκάστη ἕνα, καὶ ἡ τὸν ἥλιον ὡσαύτως ἔχουσα καὶ ἡ τὴν σελήνην ἄλλη ἐσχάτη γε πασῶν οὖσα.

Αὗται γὰρ αἱ ὀκτὼ σφαῖραι τὰς οἰκείας περιόδους ἐναντίας, ὡς ἔφην, τῇ πρώτῃ ποιοῦνται. καὶ αὗταί γέ εἰσιν αἱ ὀκτὼ σεπταὶ περίοδοι, ὧν, ὡς ὀλίγῳ πρότερον ἐλέγομεν, ὁ θαυμάσιος Πλάτων ἐν τῇ Ἐπινομίδι μνημονεύει. ἥ τε γὰρ πρώτη τῶν αὐτῶν ὀκτὼ περιόδων, ἡ καὶ τοὺς ἀπλανεῖς καὶ πεπηγότας ἅπαντας ἑαυτῇ συμπεριάγουσα, εἰς τὰ ἑπόμενα φέρεται, καὶ αἱ ἄλλαι ὡσαύτως αἱ ἑξῆς. βραδύτατα δὲ ὑπερβαλλόντως κινουμένη ἡ τοὺς εἰρημένους ἀπλανεῖς καὶ πεπηγότας ἔχουσα σφαῖρα δοκεῖ πως ἀκίνητος εἶναι καὶ οἱ ἐν αὐτῇ ἀστέρες ἑστῶτες καὶ πεπηγότες—καὶ διὰ τὴν βραδυτῆτα τῆς αὐτῶν κινήσεως καὶ τὸ τάχος

20 [Pl.], *Epin.* 986a-987c

9 τὰ ἐναντία : τἀναντία C

7 On the 2nd revolution of the celestial bodies from the West to the East

Such is the highest and first rotation from east to west. The other rotation of the celestial [spheres] is in the opposite direction — from west to east, which the astronomers call motion toward the subsequent [signs of the zodiac]. [It is] independent from the first and highest sphere, which is thought to be without stars and encloses all lower spheres carrying them with great force and speed; all lower spheres have stars attached to them and all rotate in the opposite sense to the first sphere, toward the subsequent stars. Some of them move faster and others slower carrying with them all the stars fixed upon them. After the first sphere, which is without stars [anastros], comes the next sphere carrying the so-called fixed stars, and after that there are seven additional spheres, one inside the other, carrying the five planets, one sphere having the Sun and the last one among them carrying the Moon.

As we mentioned, the eight spheres have their own periods in a direction opposite to that of the first sphere. These are the eight venerable periods which the admirable Plato mentions in *Epinomis*. The first of these eight periods contains the fixed stars, all being carried toward the subsequent [signs of the zodiac], and the remaining seven spheres rotate in the same way. Some think that the sphere which rotates very slowly together with the stars attached to it remains motionless. Due to the slowness of the motion and the speed

τῶν ἑξῆς μετ᾽ αὐτοὺς κινουμένων σφαιρῶν καὶ συμπεριφερουσῶν τοὺς ἐν αὐταῖς ἀστέρας, τούς τε πέντε πλανωμένους καὶ τὸν ἥλιον καὶ τὴν σελήνην, ἔδοξαν οὗτοι δὴ οἱ ἀστέρες καὶ τοῖς πολλοῖς ἐκλήθησαν ἀκίνητοι καὶ ἑστῶτες καὶ πεπηγότες ἐπὶ τῶν αὐτῶν· τῇ δ᾽ ἀληθείᾳ καὶ αὐτοὶ κινοῦνται εἰς τὰ ἑπόμενα πάνυ τοι κατ᾽ ὀλίγον καὶ τοῖς πολλοῖς οὐκ ἐπίδηλον καὶ συμπεριφέρονται τῇ αὐτοὺς περιεχούσῃ σφαίρᾳ.

Πλανῆται δὲ κυρίως παρὰ τοῖς ἐπιστήμοσιν οἱ πέντε λέγονται μόνοι, ὅ τε Κρόνος, ὁ Ζεύς, ὁ Ἄρης, ἡ Ἀφροδίτη καὶ ὁ Ἑρμῆς. ὅτι δὴ οὐ τὴν αὐτὴν ἀεὶ ποιοῦνται κίνησιν εἰς τὰ πρόσω ἄτρεπτον καὶ ὁμοίαν καὶ ἀμετάβλητον, ἀλλ᾽ ὁτὲ μὲν εἰς τὰ πρόσω φέρονται καὶ λέγονται προποδίζειν, ὁτὲ δὲ δοκοῦσιν ἑστάναι καὶ ὥσπερ ἀκίνητοι μένειν, ὁτὲ δὲ καὶ εἰς τὰ ἐναντία καὶ ὄπισθεν φέρονται καὶ ὑποποδίζειν λέγονται καὶ διὰ ταύτην ἄρα τὴν αἰτίαν πλανῆται τὲ καλοῦνται καὶ πλάνον ποιεῖσθαι κίνησιν. ἥλιος δὲ καὶ σελήνη καὶ πάντες οἱ δοκοῦντες ἀστέρες πεπηγέναι οὐ καλοῦνται παρὰ τοῖς ἐπιστήμοσι πλανῆται, ὅτι δὴ μὴ μεταβάλλουσι τὰς οἰκείας κινήσεις, ὡς ἄρ᾽ ἐκεῖνοι δοκοῦσι, οὔτ᾽ ἄλλοτ᾽ ἄλλας ταύτας ποιοῦνται. οὔτε γὰρ ἑστάναι δοκοῦσι καὶ στηρίζειν οὔτε προποδίζουσιν οὔτε ὑποποδίζουσι. καθ᾽ οὓς δὲ λόγους ταῦτ᾽ ἐστὶ καὶ αἰτίας καὶ ὅπως ἐκεῖνοι μὲν οἱ πέντε πλανητικὴν ποιοῦνται τὴν κίνησιν, ἥλιος δὲ καὶ σελήνη καὶ οἱ ἄλλοι πάντες ἀστέρες οὐχ᾽ ὡσαύτως ποιοῦνται τὰς οἰκείας κινήσεις, ἀλλὰ προχωρητικὰς ἀεὶ καὶ παραπλησίας, ἐν τοῖς ἑξῆς ἐροῦμεν πλατυκώτερον καὶ κατὰ μέρος σαφέστερον καὶ αἰτιολογικώτερον, νυνὶ δέ, ὅπερ ἐλέγομεν, ὅτι αἱ πᾶσαι ὀκτὼ σφαῖραι αὗται καὶ οἱ ἐν αὐταῖς ἀστέρες ἄλλη ἄλλης βραδύτερον καὶ ταχύτερον, ὅμως δ᾽ οὖν εἰς τὰ ἑπόμενα αἱ πᾶσαι καὶ ὡς πρὸς ἀνατολὰς ποιοῦνται τὰς καθ᾽ ἑαυτὰς κινήσεις καὶ ἐναντίας τῇ πρώτῃ καὶ ἀνωτάτῳ φορᾷ τῇ ἀπ᾽ ἀνατολῶν εἰς δυσμάς. κατελήφθησαν δ᾽ οὕτω κινούμεναι ἀπὸ τῶν κατ᾽ αἴσθησιν φαινομένων καὶ τῶν ὁρωμένων αὐτῶν· τὰ γὰρ ὁρώμενα χορηγεῖ τῇ ἐπιστήμῃ καὶ τοῖς συλλογισμοῖς τὰς ἀρχάς, καὶ τοῦτ᾽ ἐν πᾶσι μὲν τοῖς μαθηματικοῖς, μάλιστα δὲ ἐν τοῖς ἀστρονομικοῖς.

53 κατελείφθησαν V[ac]

of the subsequent spheres which carry the five planets, the Sun and the Moon, many people suppose that they are stationary and they say so. But the truth is that they move very slowly with their sphere toward the subsequent stars, something that is not evident to most people.

According to scientists, "planets" are called only the following five: Saturn, Jupiter, Mars, Venus and Mercury. In fact they do not always move forward and without many changes, but sometimes they do move forward and we say they advance. Sometimes they stand still and remain motionless and sometimes they move in the opposite direction — backwards — and we say they retrograde. For this reason they are called planets, because they perform a wandering motion. The scientists do not refer to the Sun, the Moon and the fixed stars as planets because they do not change their revolutions whimsically, moving differently at different times, nor do they remain stationary or advance or retrograde. We shall discuss clearly and in greater detail later on the origins and reasons for these events, that is how the five planets perform their wandering motion and how the Sun, the Moon and all the other stars, which have different revolutions, always move forward. For now, we have said there are eight spheres and the stars are attached to them and move slower or faster. They all move toward the east in the opposite direction to the first and highest revolution from east to west. We perceived that they move this way from empirical phenomena and observations. The observations provide the principles and syllogisms for science, in particular to the mathematicians and astronomers.

Ὕλη γὰρ τῆς ἀστρονομικῆς ἐπιστήμης, ὡς ὁ παλαιὸς λόγος, 'τὰ φαινόμενα' καὶ τῶν ὁρωμένων αἱ ὑποθέσεις. καὶ γὰρ δὴ καὶ Πτολεμαῖος αὐτὸς φησὶν ἐν τῷ πρώτῳ τῶν Ἁρμονικῶν: 'ἁρμονικοῦ δ' ἂν εἴη πρόθεσις τὸ διασῶσαι πανταχῇ τὰς λογικὰς ὑποθέσεις μηδαμῇ μηδαμῶς ταῖς αἰσθήσεσι μαχομένας, ὡς ἀστρολόγου τὸ διασῶσαι τὰς τῶν οὐρανίων κινήσεων ὑποθέσεις συμφώνους ταῖς τηρουμέναις παρόδοις, εἰλημμένας μὲν καὶ αὐτὰς ἀπὸ τῶν ἐναργῶν καὶ ὁλοσχερέστερον φαινομένων, εὑρούσας δὲ τῷ λόγῳ τὰ κατὰ μέρος ἐφ' ὅσον δυνατὸν ἀκριβῶς. ἐν ἅπασι γὰρ ἴδιον τοῦ θεωρητικοῦ καὶ ἐπιστήμονος τὸ δεικνῦναι τὰ τῆς φύσεως ἔργα μετὰ λόγου τινὸς καὶ τεταγμένης αἰτίας δημιουργούμενα καὶ μηδὲν εἰκῇ μὴδ' ὡς ἔτυχεν ἀποτελούμενον ὑπ' αὐτῆς.' καὶ ὁ μὲν Πτολεμαῖος οὕτω καὶ ἁρμονικὸν καὶ ἀστρολόγον τὰς ἀρχὰς τῶν ὑποθέσεων λαμβάνειν ἀπὸ τῶν ἐναργῶν καὶ ὁλοσχερέστερον φαινομένων φησί, τουτέστι μὴ ἠκριβωμένως, ἀλλὰ παχυμερῶς διὰ τῆς αἰσθήσεως καταλαμβανομένων, εὑρίσκοντος ἐντεῦθεν τοῦ λόγου τὸ ἀκριβές, ὅτι δὴ καὶ ἀδύνατόν ἐστιν ἐντελῶς τὰς αἰσθήσεις τὰ κατὰ πρόθεσιν λαμβάνειν, ἀλλὰ τὰς ἀφορμὰς διδόναι τῇ διανοίᾳ ἐξ ὧν αὕτη τοὺς συλλογισμοὺς τῶν ὄντων εὑρίσκει καὶ συμπεραίνει. καὶ Ἀριστοτέλης γάρ φησιν ἐν τοῖς Λογικοῖς καὶ ἐν τοῖς Μετὰ τὰ φυσικά, ὡς αἴσθησις μὲν ἐμπειρίαν ἐμποιεῖ, ἐμπειρία δὲ τὰς ἀρχὰς δίδωσι τῇ ἐπιστήμῃ. καὶ αὐτὸς ὁ Πτολεμαῖος πάλιν εὐθὺς ἐν προοιμίοις τοῦ πρώτου τῶν Ἁρμονικῶν 'τὰς αἰσθητικὰς διαλήψεις ὁρίζεσθαι καὶ περαίνεσθαι ταῖς λογικαῖς, ὑποβαλλούσας μὲν πρώτως ἐκείναις τὰς ὁλοσχερέστερον λαμβανομένας διαφορὰς ἐπί γε τῶν δι' αἰσθήσεως νοητῶν, προαγομένας δ' ὑπ' ἐκείνων ἐπὶ τὰς ἀκριβεῖς καὶ ὁμολογουμένας.' καὶ γὰρ ὁ αὐτὸς μικρὸν πρὸ τούτου φησίν ὅτι 'καθόλου τῶν μὲν αἰσθήσεων ἴδιόν ἐστι, τὸ τοῦ μὲν σύνεγγυς εὑρετικόν' (ταὐτὸ

58 Cf. Sext. *Adv. Math.* 7.140 ὄψις γὰρ τῶν ἀδήλων τὰ φαινόμενα, ὥς φησιν Ἀναξαγόρας. Cf. Christidis (1979) 560-63. **59-68** Ptol. *Harm.* 1.2.3-13 **75-76** Cf. Arist. *APo.* 100a.3-8; *Metaph.* 981a.3 **77-82** Ptol. *Harm.* 1.1.11-14

60 post ὑποθέσεις habet τοῦ κανόνος Ptol. **61** post μαχομένας habet κατὰ τὴν τῶν πλείστων ὑπόληψιν Ptol. **65** post ἴδιον habet ἐστι Ptol. **67** μήδ' : μηδέ Ptol. **77-78** εὐθὺς ἐν προοιμίοις : εὐθὺς post προοιμίοις transp. C **79** λογικαῖς in ras. C **79** πρώτας Ptol. **81** προσαγομένας Ptol. **81** δ' : δε Ptol.

According to old tradition, the topics of astronomy are the phenomena and the formulation of hypotheses for observations. Indeed, this is professed by Ptolemy in the first [book] of *Harmonics*: "the purpose of *Harmonics* is to completely save the logical hypotheses without contradicting in any way the observations; in the same way the *Astrologer* aims to save the hypotheses for the celestial motions in agreement with the observed orbits collected from many reliable observations, finding out the details as precisely as possible. The aim of the theoretician and the scientist is to describe the works of nature logically and with sound reasoning leaving nothing to chance". Furthermore, Ptolemy says in the *Harmonics* and in the *Astrologer* that he draws his hypotheses from reliable and sound phenomena, which are not the most accurate, but those which broadly fit with other observations, thus deducing from them their precise cause. It is indeed impossible to understand the ultimate design relying simply on observations. They stimulate the mind to find syllogisms and arrive at conclusions. As Aristotle says in the *Syllogism* and in the *Metaphysics* observations create empirical knowledge that lead to the principles of science. Again the same Ptolemy in the introduction to the first book of Harmonics professes and concludes that "empirical knowledge determines and stimulates logical thinking, subjecting to scrutiny any important empirical differences, thus advancing some among them to precise and acceptable ones". A little earlier he said that "it is a general property that those observations which are

λέγων τὸ σύνεγγυς τῷ ὁλοσχερεστέρῳ καὶ μὴ ἠκριβωμένῳ), 'τοῦ δι' ἀκριβοῦς παραδεκτικόν' (παραδέχεται γὰρ πάντως τὸ ἀληθὲς ἡ αἴσθησις, ἀπὸ τοῦ λόγου τὴν ἀρχὴν καὶ τὸ ὁλοσχερέστερον εὑροῦσα), 'τοῦ δὲ λόγου τὸ τοῦ μὲν σύνεγγυς παραδεκτικόν, τοῦ δ' ἀκριβοῦς εὑρετικόν' (ἀντιστρόφως γὰρ ὁ λόγος παραδεχόμενος τῆς αἰσθήσεως τὸ ὁλοσχερέστερον καὶ σύνεγγυς καὶ μὴ ἠκριβωμένον ἐφευρίσκει τὸ ἀληθὲς αὐτὸς καὶ τὸ ἀκριβὲς καὶ διὰ τῆς ἀλλήλων κοινωνίας συμπεραίνεται ἡ κατάληψις, ἄνευ δὲ ἀλλήλων οὐχ οἷον τέ ἐστιν ἐντελῆ ταύτην εἶναι). 'οὔτε γὰρ αἰσθητόν', ἐν τοῖς τοιούτοις ἐπεξηγούμενος ὁ Πορφύριος, 'δύναται', φησί, 'συστῆναι καθ' αὑτό', προσθείην δ' ἂν ἔγωγε τελείως, 'δίχα λόγου οὔτε λόγος ἰσχυρότερός ἐστι παραστῆσαί τι μὴ τὰς ἀρχὰς λαβὼν παρὰ τῆς αἰσθήσεως καὶ τὸ τέλος τοῦ θεωρήματος ὁμολογούμενον πάλιν τῇ αἰσθήσει ἀποδιδούς.'

Τὰ μὲν τῶν σοφῶν ἐκείνων ἀνδρῶν τοιαῦτα, παρίστησι δὲ τοὺς λόγους αὐτῶν ἀληθεῖς ἄλλα τὲ μυρία. καὶ πάντες ἴσασιν, ὅσοις νοῦς ἐστιν ἐπόπτης μετ' αἰσθήσεως ὑπουργοῦ, καὶ ὃ νῦν προέκειθ' ἡμῖν πρὸ βραχέος διεξιοῦσι περὶ τῆς δευτέρας φορᾶς τῶν οὐρανίων σωμάτων καὶ ἀντιστρόφου τῇ πρώτῃ, τῆς ἀπὸ δυσμῶν δηλονότι πρὸς ἀνατολάς. τὴν γὰρ πρώτην καὶ καθολικωτέραν καὶ κοινὴν φορὰν καὶ παντὸς οὐρανίου σώματος, τὴν ἀπ' ἀνατολῶν εἰς δυσμάς, πρόδηλον οὕτω καὶ εὐκατάληπτον καὶ τοῖς πάμπαν ἀμαθεστάτοις ἔχοντες οἱ ἄνθρωποι, εἶτα ἐφορῶντες ὡς ἔνιοι τῶν ἀστέρων καὶ ἥλιος αὐτὸς καὶ μάλιστα προδηλότατα ἡ σελήνη ἐν ταῖς περιφοραῖς τοῦ παντὸς οὐρανίου σώματος ταῖς ἀπὸ ἀνατολῶν εἰς δυσμάς, οὐ τοὺς αὐτοὺς ἀεὶ τόπους ἔχουσιν ἔν γε τῷ ἀνατολικῷ ὁρίζοντι γινόμενοι ἢ ἐν τῷ δυτικῷ, ἀλλ' ὑπολειπόμενοι φαίνονται, ἵν' ἐρῶ ἀστρονομικῶς—μᾶλλον δέ, ἵν' ἐρῶ σαφέστερον, μεταβαίνοντες φαίνονται ἀφ' ὧν ἦσαν πρότερον τόπων καὶ προχωροῦντες εἰς τὰ ἑπόμενα—, ἐντεῦθεν ἐπενόησαν καὶ

87-88 Ptol. *Harm*. 1.1.6-8 **92-96** Porph. in *Harm*. 2 (p. 25.19-22 Düring)

85 δι' : δε Ptol. **107-112** sch. in mg. C (Chort.) ἑτέρα πάλιν ἀπὸ τῶν φαινομένων ἀπόδοσις τοῦ κινεῖσθαι τόν τε ἥλιον καὶ τὴν σελήνην καὶ τοὺς ἄλλους ἀστέρας εἰς τὰ ἑπόμενα τοῦ κόσμου.

close to each other lead to discoveries" (meaning important observations not yet established but supporting each other) and "those that are precise are acceptable" (among observations acceptable are those that are close to reason and are very reliable). On the other hand, "in logical reasoning statements close to each other are acceptable but precise reasoning leads to discovery" (to say it in another way, logical thinking together with important observations that are not extremely precise but are close to each other lead to the discovery of true facts. Combining the two, one arrives at conclusions, but without their interplay one can never reach perfection). On this topic Porphyry remarks that "empirical knowledge by itself can not establish anything" and I may add "[observation cannot establish something] without logical thinking, nor is logic able to establish something that does not originate from experience. Finally, the results of a theorem must be verified by experiment".

These are the opinions of those wise men and the correctness of their remarks is supported in many other ways. For all those who set as a priority in their mind to explain the observations must deal with a topic that we formulated earlier concerning the second rotation of the celestial bodies from west to east, which is in opposite sense to the first. The first revolution is a universal and common rotation for the entire celestial body from East to West, as is evident and understandable to the most uneducated person. Then observing some of the stars or the Sun, and more clearly the Moon, in the revolution of the entire celestial body, they do not always maintain the same [relative] positions when they appear on the eastern horizon or at other times on the western. They appear, astronomically speaking, to be left behind. To say it more precisely, as they rotate from one location to the next location, [the

συνελογίσαντο καὶ δευτέραν ἄλλην εἶναι φορὰν ἐν τοῖς οὐρανίοις σώμασι, βραδυτέραν μὲν καὶ οὐκ ἰσοταχῆ τῇ πρώτῃ, εἶναι δ' ὅμως καθ' ἑαυτὴν ἐναντίαν τῇ πρώτῃ. καὶ τὸ μὲν ἀπαράμιλλον τάχος καὶ κράτος τῆς ἀνωτάτω καὶ πρώτης περιφορᾶς συμπεριφέρει βίᾳ καὶ τὰ ὑπ' αὐτὸ πάντα. οὐ μὴν ἀλλ' ἔχει καὶ αὐτὰ ἰδίαν κίνησιν, βραδυτέραν μὲν αὐτοῦ, βραδυτέραν δὲ καὶ ἄλλην ἄλλης.

Καὶ οὕτω δὴ ἡ ἀστρονομικὴ ἐπιστήμη ἀπὸ τῆς δι' αἰσθήσεως ἀρχῆς καὶ τῶν φαινομένων συμπεραίνει καὶ συλλογίζεται καὶ τὰς ἑπτὰ σφαιρικὰς ἀποτομὰς τοῦ παντὸς οὐρανίου βάθους, ἐν αἷς ἥλιος καὶ σελήνη καὶ οἱ πέντε καλούμενοι πλανῆται ἀστέρες ὡσπερεὶ πεπήγασι, ταύτας δὴ τὴν ἐναντίαν τῇ πρώτῃ φορᾷ περιστροφὴν ποιεῖσθαι καὶ συμπεριφέρειν οὕτω ὃν ἑκάστη ἔχει ἀστέρα ἐν ἑαυτῇ. ἔπειθ' οὕτω κατασκοπησαμένη προσεξευρίσκει καὶ τὴν ὑπὲρ ταύτας τὰς ἑπτὰ τῶν ἀπλανῶν ἁπάντων σφαῖραν βραδύτατα μὲν καὶ σφόδρα βραδύτατα ὡς πρὸς τὰς ἄλλας, ὅμως δ' οὖν παραπλησίως καὶ αὐτὴν περιφερομένην καὶ συμπεριφέρουσαν τοὺς ἐν αὐτῇ πάντας ἀστέρας εἰς τἀναντία τῇ ὑπερτάτῃ καὶ πρώτῃ φορᾷ.

astronomers] inferred and invented a second rotation of the celestial body, being slower and with a different speed than the first one. It is also in opposite sense to the first, whose incomparable speed and power carries everything that is below. [The second sphere] has its own rotation being slower than any other rotation.

In this manner the field of Astronomy, starting from observations and phenomena, arrives at the conclusion and the framework [that there are] seven spherical regions throughout the [entire] celestial depth, onto which are attached the Sun, the Moon and five planets. These regions rotate in opposite sense to the first revolution and carry with them their own stars. Thereafter pondering carefully, one realises that above the seven spheres there is another sphere rotating slowly, in fact very very slowly by comparison. However, it is very close and rotates in opposite sense to the first and highest sphere, carrying along all stars attached to it.

8 ὅτι λοξὸς νοεῖται κύκλος ἐπὶ τῆς οὐρανίας σφαίρας, μέγιστος καὶ αὐτὸς ὢν τῶν ἐν αὐτῇ κύκλων καὶ εἰς ἶσα δύο ἡμισφαίρια τέμνων αὐτὴν καὶ τὸν ἰσημερινὸν κύκλον

Καὶ οὕτω δὴ κατελήφθησαν καὶ αἱ ὀκτὼ σφαῖραι, αἱ τοὺς πάντας ἀστέρας ἐν ἑαυταῖς ἔχουσαι, τοῖς ἐπιστημονικῶς τὰ τῆς αἰσθήσεως συλλογιζομένοις περιφερόμεναι τὴν ἐναντίαν τῇ πρώτῃ φορᾷ, ἀπὸ δυσμῶν εἰς ἀνατολάς. κατελήφθησαν δὲ οὐ περὶ τοὺς πόλους, οὓς ἔφημεν, τῆς πρώτης φορᾶς κινούμεναι, τουτέστι τοῦ ἰσημερινοῦ μεγίστου κύκλου καὶ τῶν αὐτῷ παραλλήλων, ὧν ἐλέγομεν, ἄλλων κύκλων, ἀλλὰ περὶ ἄλλους πόλους, ὧν ὁ μέγιστος κύκλος καταλαμβάνεται ἐγκεκλιμένος καὶ λοξὸς πρὸς τὸν ἰσημερινόν, ὥστε μὴ γίνεσθαι τὴν τῶν εἰρημένων σφαιρῶν ἐν τῇ περιαγωγῇ κίνησιν πρὸς τὴν πρώτην καὶ ἀνωτάτω φορὰν κατ᾽ εὐθυωρίαν, ἀλλ᾽ ἐναντίαν μέν, λοξὴν δὲ καὶ περὶ πόλους ἄλλους, ὡς ἔφημεν.

Ὁ γοῦν τοιοῦτος λοξὸς κύκλος, ἐφ᾽ οὗ φέρεσθαι νοοῦνται καὶ ἀντιπεριάγεσθαι τῇ πρώτῃ φορᾷ ὁ ἥλιος καὶ ἡ σελήνη καὶ οἱ πέντε πλανῆται, μέγιστός ἐστι κύκλος καὶ αὐτὸς τῶν ἐν τῇ οὐρανίᾳ καθόλου σφαίρᾳ. τέμνει γὰρ καὶ αὐτὸς εἰς δύο ἶσα ἡμισφαίρια ταύτην, καὶ τὸ κέντρον τῆς σφαίρας κέντρον ἐστὶ καὶ αὐτοῦ. τέμνει δὲ καὶ τὸν ἰσημερινὸν εἰς ἶσα δύο, ἐπειδὴ καὶ τοῦτο ἴδιον τῶν μεγίστων ἐν σφαίρᾳ κύκλων, τὸ τέμνειν ἀλλήλους εἰς ἡμικύκλια. κατὰ γὰρ δύο σημεῖα ἅπτεται ὁ λοξὸς τοῦ ἰσημερινοῦ, κατά τε τὴν ἀρχὴν τοῦ Ζυγοῦ καὶ κατὰ τὴν ἀρχὴν τοῦ Κριοῦ, καὶ εἰσὶ ταῦτα κατὰ διάμετρον, ὥστε ἶσα ἡμικύκλια ποιοῦσιν ἑκατέρωθεν τοῦ λοξοῦ. ἐπεὶ δὲ δύνοντος τοῦ ἑνὸς τούτων σημείου, κατὰ ταυτὸν ἀνατέλλει τὸ ἕτερον ἐν τῇ πρώτῃ περιφορᾷ τοῦ οὐρανίου σώματος, εὔδηλον ὅτι τὰ τοιαῦτα τμήματα καὶ ἐπὶ τοῦ ἰσημερινοῦ κύκλου κατὰ διάμετρόν εἰσι καὶ τέμνουσιν αὐτὸν εἰς ἡμικύκλια. καὶ οὕτω γε δὴ λοιπὸν δῆλον κατὰ τὸ πεντεκαιδέκατον

25-26 Theodos. *Sph.* 1.15

4 φορᾷ : περιφορᾷ C[ac] **17** ἶσα δύο : ἶσα post δύο transp. C

8 The inclined circle of the celestial sphere is a major circle dividing the celestial sphere into two hemispheres and the equator into two semicircles

The eight spheres, which carry all the stars, were discovered by scientists by analysing empirical data; they were found to rotate in the opposite direction to the first sphere, i.e. from west to east. It has also been established that they do not rotate around the poles of the first revolution, that is, parallel to the great circle of the equator and its parallels, but around other poles whose great circle is inclined and oblique to the equator, so that its rotation around the Earth is not directly along that of the first and highest [sphere], but in the opposite sense, oblique and around other poles, as we mentioned.

In any event, this inclined circle, on which revolve and are carried along the Sun, the Moon and the five planets is again a great circle of the celestial sphere. It divides the globe into two hemispheres and its centre is the centre of the globe. It also divides the equator into two equal [sections], because it is a property of major circles on a sphere to divide each other into semicircles. The inclined circle intersects the equator at two points, at the beginning of Libra and at the beginning of Aries, which are diametrically opposite to each other, thus defining equal semicircles on each side of the inclined. In addition, because during the first revolution of the celestial globe, when one [point] is at sunset the other is at sunrise, it is evident that these points lie also diametrically opposite on the equator. They divide it into semicircles as

τοῦ πρώτου τῶν Θεοδοσίου Σφαιρικῶν, ὡς ἀλλήλους κατὰ τὰ κοινὰ σημεῖα τῆς ἁφῆς αὐτῶν τέμνοντες εἰς ἡμικύκλια ἐν τῇ αὐτῇ σφαίρᾳ ὅ τε ἰσημερινὸς καὶ ὁ λοξὸς μέγιστοί εἰσι καὶ ἀμφότεροι τῶν ἐν τῇ αὐτῇ οὐρανίᾳ σφαίρᾳ κύκλων.

Οὗτος δὴ ὁ λοξὸς καὶ ζωδιακὸς καλεῖται παρὰ τοῖς ἀστρονόμοις. τέμνουσι γὰρ αὐτὸν εἰς τμήματα δώδεκα, ἃ καὶ ζώδια καλοῦσι. καὶ εἰσὶ τῶν τοιούτων δωδεκατημορίων, εἴτουν ζωδίων, ἀφ᾽ ὧν καὶ ζωδιακὸς κύκλος καλεῖται, τὰ ὀνόματα ταῦτα· Κριός, Ταῦρος, Δίδυμοι, Καρκῖνος, Λέων, Παρθένος, Ζυγὸς ἢ Χηλαὶ (κατ᾽ ἀμφότερα γὰρ καλεῖται τὸ ζώδιον καὶ μάλιστα Χηλαὶ παρὰ τῷ Πτολεμαίῳ καὶ τοῖς παλαιοῖς), Σκορπίος, Τοξότης, Αἰγόκερως, Ὑδροχόος καὶ Ἰχθύες. ταῦτα μὲν τὰ ὀνόματα, αὕτη δὲ καὶ ἡ τάξις καὶ ἡ διαδοχὴ αὐτῶν κατὰ τὸ εἰκὸς ὡς πρὸς τὰ ἑπόμενα.

26 τῶν : τοῦ C[ac]

follows from the fifteenth [theorem] of the first book of Theodosios' *Sphaerica*. Since the equator and the inclined lie on the same sphere and divide each other into semicircles at the two points of intersection, they are major circles of the celestial sphere.

The astronomers call the inclined [circle] "zodiac" and divide it into twelve parts which they call "signs of the zodiac" and it is from these twelve sectors that it was christened the "zodiac circle". The names of the twelve signs are Aries, Taurus, Gemini, Cancer, Leo, Virgo, Libra or Claws (this sign has 2 names - in particular called Claws by Ptolemy and the ancients), Scorpio, Sagittarius, Capricorn, Aquarius and Pisces, from which the name zodiac derives. These are their names, and this is the order and the true sequence toward the subsequent stars.

9 ὅτι περὶ τοὺς τοῦ τοιούτου λοξοῦ κύκλου πόλους ἀπὸ δυσμῶν εἰς ἀνατολὰς πάντες φέρονται οἱ ἀστέρες

Περὶ γοῦν τοὺς τοῦ τοιούτου λοξοῦ κύκλου πόλους αἱ πᾶσαι τῶν ἀστέρων σφαῖραι νοοῦνται καὶ ἀντιπεριάγονται, ὡς ἔφην, τῇ ἀνωτάτω καὶ πρώτῃ φορᾷ, συμπεριφέρουσαι καὶ συμπεριάγουσαι τοὺς ἐν αὐταῖς ἀστέρας. πρώτη μὲν δὴ ἡ τοὺς ἀπλανεῖς ἅπαντας ἔχουσα ἐν ἑαυτῇ, πεπηγέναι μὲν δοκοῦντας διὰ τὴν βραδυτῆτα τῆς περιφορᾶς τῆς κατ' αὐτοὺς σφαίρας, κινουμένους δὲ καὶ αὐτοὺς ἀναμφιβόλως, μᾶλλον δὲ συγκινουμένους τῇ κατ' αὐτοὺς σφαίρᾳ κατ' ὀλίγον πάνυ τοι καὶ σχεδὸν οὐκ ἐπίδηλον εἰς τὰ ἑπόμενα. μετ' αὐτὴν δὲ αἱ ἕτεραι ἑπτά, αἱ νοούμεναι κατατέμνειν σχεδὸν τὸ τοῦ οὐρανίου ὅλου σώματος βάθος ἐξ αὐτῆς τῆς περιφερούσης τοὺς ἀπλανεῖς ὅλους ἀστέρας μέχρι καὶ τῆς ἐσχάτης τῆς περιφερούσης τὴν σελήνην.

Ἑκάστη δ' οὖν τῶν αὐτῶν ἑπτὰ κατὰ τάξιν καὶ βαθμὸν ὃν ἐροῦμεν συμπεριφέρει, ἡ μὲν ἥλιον, ἡ δὲ σελήνην, ἡ δ' ἕνα τινὰ τῶν πέντε πλανωμένων. καὶ ὁ μὲν ἥλιος κατείληπται μὴ μόνον ἐν τῇ μεσαιτάτῃ κατὰ βάθος τῶν ἑπτά, ἀλλὰ καὶ τὸ μεσαίτατον αὐτοῦ περιοδεύων τοῦ ζῳδιακοῦ, ἴσην ἀεὶ καὶ βασιλικὴν ὄντως ὁδὸν καὶ ἁπλουστέραν τῶν ἄλλων οὐδὲν ὁτιοῦν παρεκκλίνων ὅλως τοῦ μέσου. ἡ σελήνη δὲ καὶ οἱ πέντε πλανώμενοι ἐφ' ἑκάτερα τοῦ μέσου, τούτου δὴ τοῦ λοξοῦ καὶ ζῳδιακοῦ, παρεκκλίνουσι καὶ φέρονται καὶ βόρεια δηλαδὴ καὶ νότια, κατὰ κύκλων παροδεύοντες, συνδέσμους πάντως ἐχόντων πρὸς αὐτὸν τὸν μέσον τοῦ ζῳδιακοῦ καὶ κατὰ δύο σημεῖα τεμνόντων αὐτόν.

Ταῦτα δὲ δὴ καθάπαξ ἠκριβωμένως κατελήφθησαν τοῖς μαθηματικοῖς δι' ἀστρολαβικῶν ὀργάνων. καὶ ὅσον δὲ ἕκαστος αὐτοῦ τοῦ μέσου τοῦ ζῳδιακοῦ τὸ πλεῖστον ἀφίσταται, καὶ αὐτὸ ἀσφαλέστατα κατείληπται, ὡς ἐν τοῖς ἑξῆς κατὰ καιρὸν ἐροῦμεν. νῦν δὲ τοσοῦτο λέγομεν, ὡς ἡ σελήνη τὲ καὶ οἱ πέντε πλανώμενοι οὐκ ἐπ' αὐτοῦ φέρονται τοῦ μέσου τοῦ ζῳδιακοῦ, ἀλλ' αὐτοῦ καὶ νοτιώτεροι καὶ βορειότεροι παροδεύοντες τοὺς οἰκείους κύκλους γίνονται, ὥστ'

3 φορᾷ : περιφορᾷ C

9 The motion of all the stars on the inclined [circle] is from west to east

Thus we envision all spheres of the stars to revolve around the poles of the inclined, participating at the same time in the first and highest revolution and carrying with them their own stars. The first sphere has all fixed stars and they do not appear to move, due to the slowness of the revolution of their sphere; however, there is no doubt that they move very slowly towards the subsequent stars. After this sphere there are seven additional spheres, which are thought to occupy the entire depth of the celestial sphere, from the one that carries the fixed stars all the way down to the lowest, the one that carries the Moon.

We mention each of the seven spheres according to their order and rank. Each sphere carries along its own star: the Sun, the Moon and the other five planets. The Sun, in particular, not only occupies the middle distance among the seven, but has the median period in the zodiac — always having the same royal trajectory, being much simpler than that of the others, without any deviation whatsoever. The Moon and the five planets wander around the median path of the ecliptic and deviate sometimes to the north and other times to the south – their circular path intersecting the zodiac at two points: the nodes.

These properties have been established by mathematicians using astronomical instruments. How much each of them deviates from the median of the ecliptic is known with certainty and we shall mention them at the appropriate time. Here it suffices to say that the Moon and the five planets do not move through the middle of the ecliptic, but their trajectories are sometimes to the south and sometimes to the north. As a consequence, in each revolution they cross the median of the zodiac at two nodes.

ἐξανάγκης ἐν ἑκάστῃ αὐτῶν περιστροφῇ δὶς αὐτοὺς ἅπτεσθαι τοῦ μέσου τοῦ ζωδιακοῦ κατὰ τοὺς εἰρημένους πρὸς αὐτὸν δύο συνδέσμους.

Αὐτός γε μὴν ὁ λοξὸς καὶ ζωδιακὸς κύκλος, περὶ ὅν, ὡς ἔφην, οἱ πέντε πλανώμενοι καὶ ἥλιος καὶ σελήνη τὰς οἰκείας καταλαμβάνονται ποιούμενοι περιόδους ἐν τῇ αὐτοῦ σφαίρᾳ ἕκαστος ἐγκεκλιμένος ὢν πρὸς τὸν ἰσημερινόν, τὸν μέγιστον, ὡς προείπομεν, τῶν ἐν τῇ ὀρθῇ σφαίρᾳ παραλλήλων, ἐξανάγκης καὶ τοὺς οἰκείους πόλους διϊσταμένους ἔχει τῶν πόλων τοῦ ἰσημερινοῦ, ἤτοι τῆς ὀρθῆς σφαίρας τοσοῦτον, ὅσον καὶ ἐγκέκλιται αὐτὸς ὁ λοξὸς πρὸς τὸν ἰσημερινόν. ἐὰν γὰρ νοήσωμεν ἄλλον μέγιστον κύκλον διὰ τῶν πόλων αὐτῶν ἀμφοτέρων τῶν τοιούτων μεγίστων κύκλων, τοῦ τε λοξοῦ δηλονότι καὶ τοῦ ἰσημερινοῦ, διϊόντα καὶ ἀμφοτέρους πάντως εἰς δύο ἡμικύκλια τέμνοντα κατὰ τὸ τῶν μεγίστων, ὡς ἔφημεν, ἴδιον κύκλων—καταλαμβάνεται δὲ οὗτος ἐνίοτε ὁ δι᾽ ἀμφοτέρων τῶν πόλων αὐτῶν ὁ αὐτὸς τῷ μεσημβρινῷ, ἐφ᾽ οὗ δηλαδὴ γινόμενος ὁ ἥλιος ἐν τῇ περιαγωγῇ τῆς πρώτης φορᾶς κατὰ κορυφὴν γίνεται τῶν οἰκήσεων ἑκάστων καὶ μεσημβρίαν, ἤτοι μέσον, ἤγουν ἥμισυ ἡμέρας δηλονότι ποιεῖ—ἐὰν γοῦν νοήσωμεν μέγιστον κύκλον, ὡς λέγομεν, δι᾽ ἀμφοτέρων τῶν πόλων διϊόντα, ὁπόσον δὴ τμῆμα περιφερείας ἀπὸ τοῦ τοιούτου κύκλου μεταξύ ἐστι τοῦ ζωδιακοῦ καὶ τοῦ ἰσημερινοῦ, τοσοῦτον δὴ καὶ ἶσον πάντως ἔσται καὶ μεταξὺ τῶν πόλων αὐτῶν.

Ἐπεὶ γοῦν πάντα κύκλον οἱ ἀστρονόμοι εἰς τριακόσια ἑξήκοντα τμήματα διαιροῦσιν, ἤτοι μοίρας, καὶ κατὰ τοῦτο καὶ αὐτὸν τὸν διὰ τῶν πόλων ἀμφοτέρων διϊόντα νοοῦσιν ὡσαύτως διῃρημένον εἰς τριακόσια ἑξήκοντα τμήματα, εὕρηται καὶ κατείληπται τοῖς ἀστρονόμοις διὰ τῶν εἰρημένων ἀστρολαβικῶν ὀργάνων τὸ τμῆμα τῆς περιφερείας αὐτοῦ, τῆς μεταξὺ τῶν δύο κύκλων τοῦ λοξοῦ καὶ τοῦ ἰσημερινοῦ καὶ μεταξὺ τῶν πόλων αὐτῶν, τοιούτων μοιρῶν εἴτουν τμημάτων εἰκοσιτεσσάρων ἔγγιστα, οἵων ἐστὶν ὁ ὅλος κύκλος τριακοσίων ἑξήκοντα, ὡς εἶναι τὰς ἐφ᾽ ἑκάτερα βόρειά τε καὶ νότια ἐγκλίσεις, ἤτοι ἀποστάσεις τοῦ ζωδιακοῦ πρὸς τὸν ἰσημερινόν, μοιρῶν τοιούτων εἰκοσιτεσσάρων ἔγγιστα, ὡς εἴρηται. ἐπίσης γὰρ κατ᾽ ἀμφότερα τὰ μέρη τά τε νότια

36 post ἤτοι add. ἐπὶ V[sl] **58** νότιά τε καὶ βόρεια C[ac]

The inclined circular annulus of the ecliptic (the zodiac) on which the five planets, the Sun and Moon are considered to perform their periodic motions is inclined relative to the equator, which is the great circle among the parallels of *sphaera recta*. Consequently, their poles deviate from the poles of the equator, that is those of *sphaera recta*, by an amount equal to the angle between the inclined and the equator. If we imagine next another great circle drawn through poles of the ecliptic and the equator, crossing and dividing both of them in two semicircles, then this is the meridian. When the Sun during its first revolution is located directly above on the meridian of a specific location then it is noon, i.e. the middle of the day. Consider next a major circle passing through both poles; the arc on this circle between the ecliptic and the equator is equal [to the angle] between the poles.

Since the astronomers divide each circle into 360 degrees, they also consider the meridian to be divided into 360 degrees. The astronomers, measuring with standard astronomical instruments, found the arc between the two circles — ecliptic and equator — and also the angle between their poles. Its value is approximately 24° for a circle of 360 degrees, as are also the deviations of the ecliptic to the north and south of the equator, i.e., approximately 24 degrees. The ecliptic deviates from the equator on both sides, to the south and

καὶ τὰ βόρεια διίσταται τοῦ ἰσημερινοῦ ὁ λοξός, ὅτι δὴ καὶ ἀμφότεροι μέγιστοί εἰσιν, ὡς ἔφην, κύκλοι, ὡς εἶναι τὸ συναμφότερον τμῆμα τῆς περιφερείας τοῦ εἰρημένου δι᾽ ἀμφοτέρων τῶν πόλων κύκλου, τὸ μεταξὺ τῶν τε βορειοτάτων καὶ τῶν νοτιωτάτων τοῦ μέσου τοῦ ζωδιακοῦ, τοιούτων τεσσαρακονταοκτὼ μοιρῶν εἴτουν τμημάτων ἔγγιστα, ἤτοι παρὰ πολλοστόν τι μόριον μοίρας, οἵων ἐστὶν ὁ τοιοῦτος κύκλος ὁ διὰ τῶν πόλων ἀμφοτέρων τριακοσίων ἑξήκοντα, ὡς εἴρηται. ὥστε, ἐπεὶ τὸ μεσαίτατον τοῦ τοιούτου ζωδιακοῦ κύκλου ὁ ἥλιος παροδεύει, τοσοῦτον πλάτος τοῦ οὐρανίου σώματος ἀπολαμβάνει ἡ ἡλιακὴ λοξὴ κίνησις, ὅσον ἐστὶ τῶν τεσσαρακονταοκτὼ ἔγγιστα μοιρῶν πρὸς τὰς τριακοσίας ἑξήκοντα. εὕρηταί γε μὴν τὸ τοιοῦτον διάστημα τῶν τεσσαρακονταοκτὼ τοιούτων τμημάτων Πτολεμαίῳ καὶ τοῖς ἀστρονόμοις ἀσφαλέστατα καὶ εὐμεθόδως, ὡς αὐτὸς φησὶν ἐν τῷ πρώτῳ τῆς Συντάξεως, δι᾽ ἀστρολάβου κύκλου, ὡς ἔφημεν, ὃν ὅμοιον τῷ παντὶ καὶ ὁμόκεντρον καταστησάμενος ἐν ἐπιπέδῳ ἴσῳ καὶ κατατεμὼν εἰς τριακόσια ἑξήκοντα τμήματα καὶ δι᾽ ἐποπτείας ἠκριβωμένης ἀνευρὼν μέχρι τίνος εἰς τὰ βορειότατα καὶ εἰς τὰ νοτιώτατα γίνεται ὁ ἥλιος, ἐπελογίσατο τὸ ἐν μέσῳ διάστημα καὶ ἠρίθμησε τεσσαρακονταοκτὼ τμημάτων ἔγγιστα, ὡς ἔφημεν, τοιούτων οἵων ἐποιήσατο τὴν κατατομὴν τοῦ ὅλου κύκλου τριακοσίων ἑξήκοντα. ἀλλὰ καὶ ἐν πλινθίδι λιθίνῃ ἢ ξυλίνῃ ἰσοπέδῳ διὰ σκιῶν τὰς ἀκροτάτας πλατικὰς παρόδους τοῦ ἡλίου ἐπ᾽ ἀμφότερα καὶ βόρεια καὶ νότια ἀκριβέστατα κατοπτεύσας, πάλιν τὸν αὐτὸν ἀριθμὸν συνελογίσατο τοῦ μεταξὺ ἀμφοτέρων διαστήματος τὸν τῶν τεσσαρακονταοκτὼ μοιρῶν ἔγγιστα. ἀποδεικνύει δὲ ὅτι καὶ οἱ πρὸ αὐτοῦ δι᾽ ἄλλων μεθόδων συμφώνως αὐτῷ τὴν περὶ τούτου εὕρεσιν ἐποιήσαντο.

73-74 Cf. Ptol. *Alm.* 1.12

to the north, because both of them are great circles with the deviations being measured on the meridian. Between the northern-most and the southern-most points of the ecliptic there are 48 degrees, this being an approximate value to within a fraction of a degree, again counting the circle between the poles to be 360 degrees. Thus as the Sun travels through the middle of the ecliptic, its deviations in latitude are approximately 48 degrees. The value of 48° was found with certainty by Ptolemy and other astronomers using reliable measurements and methods, as he says in the first chapter of the *Syntaxis*. As we mentioned, he used the astrolabe and located it on a plane imitating the universe and divided it in three hundred and sixty parts; then, with precise observations he determined positions of the Sun furthest to the north and to the south, computed the interval between them and measured approximately 48 parts of the complete circle. As a second method he used a plane surface made of stone or wood, where he very carefully observed the shadows of the extreme latitudes of the Sun in the north and in the south, and again he arrived at the same number for the difference between the two, that of approximately 48°. He also showed that earlier [scientists] using other methods for the same discovery, had found the same values.

10 ὅτι τὸ ὑπὸ τὸν ἰσημερινὸν τμῆμα τῆς γῆς ἀοίκητόν ἐστι, καὶ δι᾽ ἥντινα τὴν αἰτίαν

Καὶ ἡ μὲν ἔγκλισις τοῦ ζωδιακοῦ τοιαύτη τις καὶ τοσαύτη πρὸς τὸν ἰσημερινὸν εὕρηται καὶ ἡ τῶν πόλων αὐτοῦ ἀπόστασις πρὸς τοὺς πόλους τοῦ ἰσημερινοῦ. οἱ μέν γε πόλοι τοῦ ἰσημερινοῦ νοοῦνται ἐπὶ τῆς ὀρθῆς σφαίρας ἐφαπτόμενοι καὶ ὡσπερεὶ προσηρμοσμένοι τῷ κατ᾽ αὐτὴν ὁρίζοντι. ἔστι δὲ καὶ καταλαμβάνεται καὶ αὐτὸς ὁ ὁρίζων μέγιστος ἐν τῇ σφαίρᾳ κύκλος εἰς δύο ἶσα τέμνων τὴν σφαῖραν καὶ τοὺς ἐν αὐτῇ μεγίστους ἄλλους κύκλους εἰς δύο ἡμικύκλια, ἤτοι τὸν ἰσημερινόν, τὸν ζωδιακόν, τὸν μεσημβρινόν, καὶ ὑπ᾽ αὐτῶν πάντων εἰς δύο ἶσα τμήματα καὶ αὐτὸς τεμνόμενος.

Ἔστι μὲν οὖν καὶ τῆς ὀρθῆς σφαίρας ὁρίζων, ἔστι δὲ καὶ καθ᾽ ἑκάστην αὐτῆς ἔγκλισιν καὶ καθ᾽ ἑκάστην δηλαδὴ οἴκησιν ὡσαύτως ὁρίζων μέγιστος ἀεὶ κύκλος ἐν τῇ σφαίρᾳ καταλαμβανόμενος, περικυκλῶν οἱονεὶ τὸ ἐπίπεδον ἑκάστης οἰκήσεως καὶ νοούμενος ὀρθὸς καὶ ἀπαρέγκλιτος, ὡσπερεὶ ἡνωμένος καὶ περιηρμοσμένος αὐτῷ δὴ τῷ ἐπιπέδῳ ἑκάστης οἰκήσεως, καὶ εἰ ἄρα ἐγκεκλιμένος καταλαμβάνεται πρὸς τὴν ὀρθὴν σφαῖραν καὶ τὸν τῆς ὀρθῆς σφαίρας ὁρίζοντα. διαιρεῖ μέν γε καὶ αὐτὸς ἕκαστος ἡστινοσοῦν οἰκήσεως ὁρίζων τό τε ὑπὲρ γῆν ἡμισφαίριον τὸ οὐράνιον καὶ τὸ ὑπὸ γῆν, ὥσπερ ὁ μεσημβρινὸς τὸ ἑῷον τὲ καὶ τὸ δυτικὸν ἡμισφαίριον συνημμένως ὑπὲρ γῆν τε καὶ ὑπὸ γῆν. καὶ γὰρ δὴ καὶ μεσημβρινὸς ἑκάστης οἰκήσεως καταλαμβάνεται, ὥσπερ καὶ ὁρίζων ἑκάστης οἰκήσεως, διαιρῶν, ὡς ἔφην, ἐφ᾽ ἑκάστης αὐτῆς οἰκήσεως δίχα, τά τε ἑῷα αὐτῆς ὑπὲρ γῆν καὶ τὰ κατὰ γῆν καὶ τὰ πρὸς δύσιν ὡσαύτως τῆς σφαίρας. ἔτι γὲ μὴν αὐτὸς ὁ ἰσημερινὸς διαιρεῖ τὴν σφαῖραν κατὰ τὰ βόρεια αὐτῆς μέρη καὶ νότια ὡσαύτως ὑπὲρ γῆν τε καὶ ὑπὸ γῆν.

Ἀλλ᾽, ὅπερ ἔλεγον, οἱ τοῦ ἰσημερινοῦ πόλοι προσηρμοσμένοι εἰσὶ καὶ νοοῦνται ἐπὶ τῆς ὀρθῆς σφαίρας τῷ κατ᾽ αὐτὴν ὁρίζοντι καὶ κατ᾽ εὐθυωρίαν τῷ ἐπιπέδῳ αὐτοῦ προσκεκολλημένοι. ἔστι δὲ καὶ

5 καταλαμβάνεται : λαμβάνεται C

10 The region of the Earth below the [celestial] equator is uninhabited and its justification

It has been established that the inclination of the ecliptic relative to the equator is equal to the angle between the pole of the inclined and the pole of the equator. The poles of the equator are understood to be tangent to *spheara recta* and just as if they are fitted to the local horizon. The horizon is understood to be a major circle of *sphaera recta*, intersecting the sphere and its great circles in two equal parts, i.e. dividing the equator, the zodiac and the meridian, while at the same time the horizon is bisected by them.

There is a horizon for *sphaera recta* and for every latitude and every location being always a great circle of the sphere, always surrounding the plane of each inhabited region, being understood to be a flat plane attached and tangent to the surface of each location. Thus it is understood to be inclined to the *sphaera recta* and its horizon. The horizon of each location divides [the celestial sphere] in two hemispheres the one above the Earth toward the sky and the one below the Earth; in the same way that the meridian divides the major sphere into the morning (east) and the western hemispheres. Furthermore, the equator divides the sphere into the northern and southern parts in the same manner as the division above and below the Earth.

As I mentioned already, the poles of the equator are understood to be located on *sphaera recta*, tangent to its horizon and directly attached to the plane. The *sphaera recta* is conceived and understood this way,

καταλαμβάνεται ἡ ὀρθὴ σφαῖρα ἡ ἀπαρεγκλίτως καὶ ὁμαλῶς καὶ ἀεὶ ὡσαύτως κινουμένη κατὰ τὴν πρώτην καὶ ὑπερτάτην φοράν, τὴν ἀπ᾽ ἀνατολῶν εἰς δυσμάς, καὶ συγκινοῦσα, ὡς ἔφημεν, τὸ οὐράνιον ὅλον ὡσαύτως βάθος, αὐτὸν ἔχουσα μέσον μέγιστον κύκλον τὸν ἰσημερινὸν καὶ ἐφ᾽ ἑκάτερα αὐτοῦ παραλλήλους ἥττονας αὐτοῦ κύκλους ἔχουσα καὶ συνεχῶς ἥττονας ἀλλήλων ἀεὶ παραλλήλους, μέχρις ἂν καταντήσωσιν εἰς αὐτοὺς τοὺς πόλους τῆς τοιαύτης ὀρθῆς σφαίρας· οἵτινες, ὡς ἔφημεν, ἐπ᾽ αὐτοῦ νοοῦνται τοῦ ὁρίζοντος τῆς οἰκήσεως, μᾶλλον δὲ τοῦ τῆς ὅλης γῆς τμήματος, ὅπερ δὴ νοεῖται κατὰ κορυφὴν ἔχον τὸν ἰσημερινὸν αὐτόν. ἔστι δὲ τὸ τοιοῦτο τμῆμα τῆς γῆς ἀοίκητον. δὶς γὰρ αὐτοῦ κατὰ κορυφὴν ὁ ἥλιος γίνεται, ὅταν τὲ παροδεύῃ τὴν ἀρχὴν τοῦ Κριοῦ καὶ ὅταν παροδεύῃ τὴν ἀρχὴν τοῦ Ζυγοῦ. κατ᾽ αὐτὰ γὰρ ἀμφότερα τὰ σημεῖα τοῦ λοξοῦ καὶ ζωδιακοῦ γινόμενος κύκλου ὁ ἥλιος, ἐπ᾽ αὐτοῦ ἐστι τοῦ ἰσημερινοῦ, ἐπειδὴ καὶ κατ᾽ αὐτὰ τὰ σημεῖα ἀλλήλους, ὡς προέφημεν, τέμνουσιν ὁ ζωδιακὸς καὶ ὁ ἰσημερινός. ὅπερ δὲ ἐλέγομεν, τὸ τμῆμα τῆς γῆς, οὗ κατὰ κορυφήν ἐστιν ὁ ἰσημερινός, ἀοίκητόν ἐστι, δι᾽ ὑπερβολὴν θέρμης τὲ καὶ ξηρότητος· δίς, ὡς ἔφημεν, τοῦ ἡλίου κατὰ κορυφὴν αὐτοῦ γινομένου, ἐν μέσῳ ὄντος καὶ ὑποκειμένου τῷ ὅλῳ πλάτει, ὅπερ τεσσαρακονταοκτὼ μοιρῶν ἔγγιστα ἔφημεν, τῆς ἡλιακῆς παρόδου. ἐφ᾽ ἑκάτερα γὰρ ἀεὶ γινόμενος αὐτοῦ τοῦ τμήματος τῆς γῆς καὶ περιλαμβάνων αὐτὸ ἐκκαίει τὲ καὶ ὑπερβαλλόντως ὑπὸ ξηρότητος ἄκρατόν τε καὶ ἀοίκητον ἀπεργάζεται.

Γίνεται μὲν γὰρ καὶ ἐπ᾽ ἄλλων τινῶν οἰκήσεων, ὡς ἑξῆς ἔσται δῆλον, δὶς κατὰ κορυφὴν ἐν τῇ μιᾷ περιόδῳ αὐτοῦ ὁ ἥλιος, ἀλλ᾽ οὐχ οὕτως ὡς ἔχει ἐπὶ τοῦ εἰρημένου τμήματος τῆς γῆς, ὃ ὑπόκειται τῷ ἰσημερινῷ, καὶ οὐ ἴσην ἀπόστασιν ἀφίσταται ὁ ἥλιος ἐν ταῖς ἐφ᾽ ἑκάτερα αὐτοῦ μεγίσταις πλατικαῖς παρόδοις, εἰς τὰ βορειότατα δηλαδὴ τῆς αὐτοῦ κινήσεως καὶ τὰ νοτιώτατα. ἐν γὰρ τοῖς ἄλλοις τῆς γῆς τμήμασιν, ἐν οἷς δὶς ὁ ἥλιος κατὰ κορυφὴν γίνεται, ὧν δή τινα καὶ μετρίως πως κέκραται, ἄνισός ἐστιν ἡ ἐφ᾽ ἑκάτερα αὐτῶν πλατικὴ ἀπόστασις τοῦ ἡλίου, κατὰ τὰ βόρεια δηλονότι καὶ τὰ νότια, οἷον καθ᾽ ὑπόθεσιν τῷ ὑποκειμένῳ τμήματι τῆς γῆς, τῷ ἐν τῇ οὐρανίᾳ σφαίρᾳ παραλλήλῳ,

42-43 ὡς προέφημεν : ὥσπερ ἔφημεν C **55** post πλατικαῖς add. αὐτοῦ V

always rotating uniformly on the first and highest revolution, from East to West, carrying with it the entire celestial body. Its major circle is the equator and on each side of it there are smaller and smaller parallels, until they end up at the poles which are visualised to be on the horizon of the habitation or rather on that section of the Earth. The region of the Earth where the celestial equator is directly above is uninhabited. The Sun stands twice on the top of this area — once when it passes the beginning of Aries and again at the beginning of Libra. When the Sun arrives at these two points of the ecliptic, it is also over the equator, because at these points the ecliptic and the equator intersect each other. For this region we were saying that the part of the Earth lying below the celestial equator is uninhabited, because of the high temperature and the extreme drought. The Sun finds itself twice above these places, being in the middle of its range of latitudes which is approximately 48°. When the Sun is at each of these two places of the Earth, it burns and makes them very dry, excessively hot and uninhabitable.

As is evident, there are other regions of the Earth where the Sun, during its revolution, stands directly above them twice, but not in the same manner as it occurs in regions of the Earth below the equator. At these other locations the Sun is not at equal distances from its extreme latitudes to the north and to the south. At these other locations of the Earth, where the Sun stands twice directly overhead, it heats them with some moderation, because the distances from the positions of maximum latitudes to the north and to the south are

τῷ διὰ τῆς ἀρχῆς τοῦ δωδεκατημορίου τῶν Διδύμων καὶ τῆς ἀρχῆς τοῦ Λέοντος. ὁ γὰρ αὐτὸς παράλληλός ἐστιν, ὡς ἔστι ῥάδιον κατανοῆσαι, διὰ τῆς ἀρχῆς τῶν δύο εἰρημένων τμημάτων τοῦ ζωδιακοῦ, τῶν Διδύμων καὶ τοῦ Λέοντος. δὶς ὁ ἥλιος κατὰ κορυφὴν γίνεται, ὡς ἔστι πρόδηλον, ἀλλ᾽ οὐκ ἔχει τὴν ἴσην ἀπόστασιν ὁ ἥλιος ἀπ᾽ αὐτῆς τῆς οἰκήσεως, ὅταν τὲ εἰς τὰ βορειότατα αὐτοῦ γένηται, ἤτοι περὶ τὰ τέλη τῶν Διδύμων καὶ τὴν ἀρχὴν τοῦ Καρκίνου, καὶ ὅταν εἰς τὰ νοτιώτατα αὐτοῦ πλάτη γένηται, ἤτοι περὶ τὰ τέλη τοῦ δωδεκατημορίου τοῦ Σκορπίου καὶ τὴν ἀρχὴν τοῦ Αἰγοκέρωτος. ἐν μὲν γὰρ τῷ βορειοτάτῳ πλάτει τῆς αὐτοῦ παρόδου γινόμενος ὁ ἥλιος ἔγγιστά ἐστι τῆς ὑπὸ τὸν εἰρημένον παράλληλον οἰκήσεως, ἐν δὲ τῷ νοτιωτάτῳ γινόμενος πορρωτάτω ἀφίσταται ταύτης. καὶ διὰ ταύτην δὴ τὴν αἰτίαν ἔστι καὶ ὅλως κεκρᾶσθαι πρὸς οἴκησιν τὸ ὑπὸ τὸν εἰρημένον παράλληλον τμῆμα τῆς γῆς, κἂν καὶ δὶς ὁ ἥλιος κατὰ κορυφὴν αὐτοῦ γίνηται, τὸ δὲ ὑπὸ τὸν ἰσημερινὸν ἀοίκητον παντάπασι καταλαμβάνεται.

61 τῶν Διδύμων : τοῦ Διδύμου C[ac] **75** παντάπασι : καθόλου C

unequal; for example, this happens for the regions below the parallels at the beginning of the constellation of Gemini and the beginning of Leo. It is evident the Sun stands twice overhead, but it is not at equal distance from the northernmost position at the end of Gemini and the beginning of Cancer or the southernmost position at the end of Scorpio and the beginning of Capricorn. In the northernmost latitude of its revolution the Sun is closest to the parallel under discussion, but in the southernmost [part of its trajectory] it is farthest away from it. For this reason, the region of Earth below the aforementioned parallel is more moderate for habitation, in spite of the fact that the Sun stands there twice at zenith. The region below the equator is completely uninhabited.

11 ὅτι τὰ παρ’ ἑκάτερα τούτου τῆς γῆς τέτταρα τμήματα, τὰ βόρεια καὶ νότια, δύο ἄνω καὶ δύο κάτω, πέφυκεν εἶναι οἰκούμενα ὡς εὔκρατα

Καθόλου μὲν γὰρ δὴ πᾶσα ἡ γῆ σφαιρική ἐστι κατὰ τὸ οὐράνιον καὶ αὕτη σῶμα. καὶ ὡς μὲν πρὸς τὸ μέγεθος τοῦ οὐρανοῦ καὶ τὴν ἀπ’ αὐτῆς ἀπόστασιν, τῶν ὑψηλοτέρων αὐτοῦ λέγω μερῶν, κέντρου καὶ σημείου λόγον ἔχει πρὸς αὐτό, καθὼς ἀποδείκνυται σαφῶς τε καὶ ἀναντιρρήτως τοῦτο Πτολεμαίῳ τὲ καὶ τοῖς ἄλλοις μαθηματικοῖς. σφαιροειδὴς δὲ οὖσα, ὡς εἴρηται, καὶ ὁμόκεντρος τῷ παντὶ καὶ οὐρανίῳ σώματι, καὶ εἰς ὅμοια λοιπὸν κατ’ αὐτὸ τὸ οὐράνιον σῶμα καὶ αὐτὴ τριακόσια ἑξήκοντα τμήματα διαιρουμένη καὶ ὑπ’ αὐτοῦ περιεχομένη, τὰ κατ’ αὐτὴν τριακόσια ἑξήκοντα περιφερῆ τμήματα τὰ μὲν ἔχει ὑποκείμενα τοῖσδε τοῖς οὐρανίοις βορείοις τυχὸν καὶ νοτίοις τμήμασι, τὰ δὲ τοῖσδε· ὡς εἶναι τὰ τοιαῦτα οὐράνια τμήματα ταῦτα μὲν κατὰ κορυφὴν τοῖς ὁμοίοις τοῖσδε τμήμασι τῆς γῆς, ταῦτα δὲ ἐκείνοις καὶ ἄλλα ἄλλοις. τὰ γοῦν ὑποτείνοντα τὸν ἰσημερινὸν καὶ ὑποκείμενα αὐτῷ πάντα ἀοίκητά εἰσι διὰ θέρμην, ὡς εἴρηται, καὶ ξηρότητα, τὰ δὲ ἐφ’ ἑκάτερα αὐτοῦ μέρη τμήματα τῆς γῆς ὑποχαλώσης ἀεὶ πρὸς τὸ μετριώτερον καὶ κεκραμένον πως τῆς θέρμης τῆς ἀπὸ τῶν τοῦ ἡλίου παρόδων, ἄλλα ἄλλων, ὡς τὸ εἰκὸς ἀπαιτεῖ τῆς φύσεως, εὐκρατότερά τέ ἐστι καὶ οἰκεῖσθαι δυνατά.

Τοιγαροῦν εἰς δύο πρῶτον διαιρουμένου τοῦ ὅλου σώματος τῆς γῆς κατὰ τὸ τοῦ ὁρίζοντος ἐπίπεδον, εἴς τε τὸ ἀνωτέρω μέρος τοῦ ὁρίζοντος καὶ ὑποκάτω πάλιν αὐτοῦ τοῦ ὁρίζοντος, καὶ κατὰ δευτέραν τομὴν ἑκατέρου τῶν τοιούτων δύο τμημάτων εἰς ἕτερα δύο αὖθις διαιρουμένου, εἴς τε τὰ βόρεια τοῦ ἰσημερινοῦ καὶ πρὸς τὸ ἓν μέρος αὐτοῦ τὸ δεξιὸν καὶ εἰς τὰ νότια καὶ πρὸς τὸ ἕτερον, ἕπεται κατὰ πᾶν τὸ εἰκός, ἐπειδὴ πᾶν τὸ ὑπὸ τὸν ἰσημερινὸν τό τε ἄνω τοῦ ἐπιπέδου τοῦ ὁρίζοντος καὶ τὸ κάτω ἀοίκητόν ἐστι διὰ τὴν αὐτὴν αἰτίαν τῆς ἐκπυρώσεως, τῆς ἀπὸ τῆς ἡλιακῆς ἐνταῦθα χρονίου κινήσεως, τὰ

5 Cf. Ptol. *Alm.* 1.5-6 **5** Cf. Cleom. 1.5, 1.8; Theon Sm. p. 120.1-124.6 Hiller (= p. 157-160 Petrucci), p. 128.1-129.2 Hiller (= p. 161 Petrucci)

11 The four regions of the Earth on each side of the equator, two in the north and two in the south, are habitable because they are temperate

The entire Earth viewed as a celestial body is spherical. Comparing the size of the sky to that of the Earth and its distance from the highest parts (of the sky), the Earth is a point at the centre of the [celestial] sphere as proven clearly and beyond any doubt by Ptolemy and other mathematicians. The Earth itself is spherical and homocentric to the universe. As is the practice with the celestial body, the Earth is divided in 360 degrees and is contained within [the stellar globe]. These 360 parts lie below the corresponding celestial regions in the north or south, with the coordinates on the surface of the Earth being extensions and projections of the coordinates on the celestial sphere. Therefore, the regions around and below the equator are all uninhabited, as mentioned, because of the high temperature and dryness. In the regions of the Earth beyond the zone of the equator the passing of the Sun produces more temperate and moderate temperatures, and these regions are capable of supporting life.

Furthermore, the body of the Earth is divided first by the plane of the horizon into two regions: one region above and another below the horizon. In each of these two parts a second incision subdivides them into two parts: one occurring to the north of the equator (to the right) and another to the south. The entire region close to the equator, the one above the horizon as well as the one below, is uninhabited, because of the heat produced by the slow passage

παρ' ἑκάτερα τοῦ ὑποκειμένου τούτου μέρους τῆς γῆς, τοῦ ἀοικήτου τέτταρα τμήματα, τό τε ἄνω βόρειον καὶ τὸ κάτω βόρειον καὶ τὸ ἄνω νότιον καὶ τὸ κάτω νότιον, κατὰ φύσιν εὔκρατα εἶναι καὶ οἰκούμενα.

of the Sun. On each side of the uninhabited region there are the four zones, the upper- and the lower-northern, as well as the upper- and lower-southern, which are habitable due to their temperate nature.

12 ὅτι ἡ καθ' ἡμᾶς οἰκουμένη ἐν τῷ ἑνὶ τῶν τεττάρων τμήματι καταλαμβάνεται, τῷ βορείῳ τῷ ἄνω, καὶ πόθεν ἡ ταύτης καταλαμβάνεται ἀρχή, καὶ περὶ τῶν ἐπ' αὐτῆς νοουμένων κύκλων

Τούτων δὴ τῶν τεττάρων τμημάτων τῶν εὐκράτων ἡ καθ' ἡμᾶς οἰκουμένη ἐν τῷ ἑνὶ καταλαμβάνεται, τῷ ὑπεράνω τοῦ ὁρίζοντος τῷ βορείῳ, ὅπερ δὴ τεταρτημόριον ἀπὸ τοῦ βορειοτέρου νοεῖται μέρους τοῦ ἀοικήτου τοῦ ὑπὸ τὸν ἰσημερινόν, διῆκον μέχρι τοῦ πόλου τοῦ βορείου καὶ ἀρκτικοῦ, ὑποκείμενον τοῖς ἑξῆς παραλλήλοις κύκλοις τῆς οὐρανίας σφαίρας ἀπὸ τοῦ ἰσημερινοῦ πρὸς βορρᾶν. ἀφ' ὧν δὴ πρῶτον ὡς εἰπεῖν παχυμερέστερον καὶ ὁλοσχερέστερον τὸ διὰ Μερόης ἐν τῷ Προχείρῳ Κανόνι ὁ Πτολεμαῖος καὶ οἱ μαθηματικοὶ ὑποτίθενται.

Τὸ γὰρ κατὰ Μερόην κλίμα πρῶτον ὑπόκειται ἐγνωσμένον οἰκεῖσθαι ἀπὸ τῆς καθ' ἡμᾶς οἰκουμένης τοῖς τὰ τῆς οἰκουμένης αὐτῆς γεωγραφοῦσι καὶ ἱστοροῦσι. καὶ τοῦτό ἐστι τῶν ἄλλων κλιμάτων τὸ νοτιώτερον καὶ ἐγγύτερον τῷ διακεκαυμένῳ καὶ ἀοικήτῳ τῆς γῆς τμήματι, καὶ ὁ διὰ τούτου παράλληλος ἐγγύτερός ἐστι τοῦ ἰσημερινοῦ κατὰ τὸ βόρειον αὐτοῦ μέρος τῶν ἐν τῇ οὐρανίᾳ σφαίρᾳ παραλλήλων ἄλλων, τῶν δι' ἄλλων ὡντινωνοῦν κλιμάτων καὶ οἰκήσεων νοουμένων. οὗτος δὲ ὁ διὰ Μερόης παράλληλος καὶ ἐν τῷ μεταξὺ διαστήματι καταλαμβάνεται τοῦ τε βορειοτάτου μέρους τοῦ ζῳδιακοῦ, ἤτοι τοῦ περὶ τὰ τέλη τῶν Διδύμων καὶ τὴν ἀρχὴν τοῦ Καρκίνου, καὶ τοῦ κατὰ τὸν ἰσημερινόν, ἤτοι τοῦ περὶ τὴν ἀρχὴν τοῦ Κριοῦ καὶ τῶν Χηλῶν.

Καὶ τοῦτό γε δῆλον ἐκ τῶν ἐν ταῖς μεσημβρίαις γινομένων σκιῶν ἐν ἐκείνῃ τῇ οἰκήσει. ὅταν γὰρ ὁ ἥλιος ἐν αὐτοῖς τοῖς βορειοτάτοις γένηται, ἤτοι τῷ τέλει τοῦ δωδεκατημορίου τῶν Διδύμων καὶ τῇ ἀρχῇ τοῦ Καρκίνου, ὡς ἔφημεν, αἱ κατὰ τὴν μεσημβρίαν σκιαὶ τῶν σωμάτων ἐν τῷ τοιούτῳ κλίματι τῷ κατὰ Μερόην καὶ τῇ οἰκήσει πρὸς νότον

8 Cf. Ptol. *Proch. Kan.* p. 101-104 Tihon

19 τοῦ[1] V[sl] : om. C **24** ante Μερόην add. τὴν C

12 Our ecumene is one of the four upper regions to the north with a description of its beginning and of its parallel circles

Our ecumene is one of the four temperate regions and is located above the horizon toward the north. It is defined by the quadrant that starts at the very end of the uninhabited region north of the equator and extends all the way to the North Pole and the Arctic. It is between the parallel circles of the stellar globe starting from the equator toward the north. Ptolemy, in the *Handy Tables*, and the mathematicians describe it as the very broad and diverse zone of Meroë.

The first zone of Meroë is known to be our inhabited region where geography and history developed. Relative to the other zones is further to the south and very close to the very hot and uninhabited region of the Earth with its parallel being closer to the northern boundary of the equator than any other parallel of the stellar globe (i.e. of the inhabited zones). The northern parallel of Meroë is the parallel at the end of Gemini and the beginning of Cancer. The southern is the equator, i.e. the parallel at the beginning of Aries and of Libra.

This is also evident by the shadows produced at noon in that habitation. When the Sun is in its northern-most position (that is at the end of the constellation of Gemini and the beginning of Cancer), the shadows of objects at noon on the habitation of Meroë are toward the south.

ἐκπίπτουσιν· ὡς ἐντεῦθεν δῆλον εἶναι, νοτιώτερον εἶναι τὸν κατὰ κορυφῆς παράλληλον τοῦ τοιούτου κλίματος τοῦ κατὰ Μερόην τοῦ παραλλήλου τοῦ διὰ τοῦ βορειοτάτου εἰρημένου τμήματος αὐτοῦ τοῦ ζωδιακοῦ νοουμένου. αἱ γὰρ σκιαί, ὡς ἔστιν εὔδηλον, ὅταν μὲν τὰ φωτίζοντα σώματα βορειότερα ὦσι τῶν φωτιζομένων, πρὸς νότον ἐκπίπτουσιν, ὅταν δὲ νοτιώτερα ὦσι τὰ φωτίζοντα σώματα τῶν φωτιζομένων, πρὸς βορρᾶν ἐκπίπτουσιν. ὅταν δὲ κατὰ κορυφὴν ὦσι τῶν φωτιζομένων, κατ᾽ οὐδέτερα μέρη ἐκπίπτουσιν αἱ σκιαί, ὥσπερ δὴ τοῦτο συμβαίνει καὶ κατὰ τοὺς τρεῖς τρόπους τοῖς ἐν τῷ κατὰ Μερόην τμήματι τῆς γῆς οἰκοῦσιν.

Ὅταν γὰρ ἐν τοῖς νοτίοις τμήμασι καὶ ζωδίοις τοῦ λοξοῦ, ἔτι γὲ μὴν καὶ ἐν αὐτοῖς τοῖς ἰσημερινοῖς καὶ μικρόν τι πρός, ὁ ἥλιος παροδεύῃ, αἱ σκιαὶ τοῖς ἐν τῷ τοιούτῳ κλίματι οἰκοῦσι πρὸς βορρᾶν ἐν ταῖς μεσημβρίαις ἐκπίπτουσιν. ὅταν δὲ ἐν αὐτῷ τῷ τοῦ ζωδιακοῦ τμήματι ὁ ἥλιος παροδεύῃ, καθὸ ὁ διὰ Μερόης παράλληλος θεωρεῖται τηνικαῦτα ἐν ταῖς μεσημβρίαις, ἄσκιοί εἰσιν οἱ ἐν τῷ τοιούτῳ κλίματι τῷ κατὰ Μερόην οἰκοῦντες. ὅταν δὲ παροδεύῃ ὁ ἥλιος ἐν αὐτῷ τῷ προειρημένῳ βορειοτάτῳ τμήματι τοῦ ζωδιακοῦ, τηνικαῦτα αἱ ἐν ταῖς μεσημβρίαις σκιαὶ ἐν τῷ τοιούτῳ κλίματι τῷ κατὰ Μερόην πρὸς νότον φέρονται. διὰ τοῦτο καὶ ἀμφίσκιοι οἱ τοιοῦτοι καλοῦνται· κατ᾽ ἀμφότερα γὰρ τὰ μέρη, τά τε βόρεια καὶ τὰ νότια, ἐν ταῖς μεσημβρίαις τὰς σκιὰς ἐκπέμπουσι, διὰ τὸ καὶ παρ᾽ ἀμφοτέρων τῶν μερῶν φωτίζεσθαι παρὰ τοῦ ἡλίου. οὐκ αὐτοὶ δὲ μόνοι οἱ κατὰ Μερόην οἰκοῦντες οὕτω καὶ εἰσὶ καὶ καλοῦνται ἀμφίσκιοι, ἀλλὰ καὶ ἄλλοι ὅσοι οἰκοῦσιν ὑπὸ τὸ προειρημένον μεταξὺ διάστημα, τοῦ τε διὰ τοῦ βορειοτάτου τμήματος τοῦ ζωδιακοῦ περὶ τὰ τέλη τῶν Διδύμων καὶ τὴν ἀρχὴν τοῦ Καρκίνου νοουμένου παραλλήλου καὶ τοῦ ἰσημερινοῦ. ὅσοι δὲ ὑπερεκπίπτουσι κατὰ τὴν οἴκησιν τοῦ τοιούτου μεταξὺ διαστήματος καὶ παντελῶς ἔξω τοῦ ζωδιακοῦ εὑρίσκονται, οὗτοι ἀπὸ τῶν νοτίων μερῶν διαπαντὸς ὑπὸ τοῦ ἡλίου φωτιζόμενοι πρὸς βορρᾶν ἐκπέμπουσι τὰς σκιὰς καὶ ἑτερόσκιοι καλοῦνται.

32 post σκιαὶ add. ἐν ταῖς μεσημβρίαις V^{mg} **42** προειρημένῳ om. C **48** ἄλλοι : ὅλοι C

It is very obvious from this that the zenith of Meroë is to the south of the northern-most region of the ecliptic. Because when the source of light is to the north of the objects they illuminate, the shadows fall to the south; but when the source of light is to the south of the objects they illuminate, the shadows fall to the north, and whenever the source of light is directly above the bodies they illuminate there are no shadows. These three cases occur in the inhabited zone of Meroë.

When the Sun is at the southern part of the inclined zodiac, close to the equator and a little further, the shadows at noon for the people living in this zone are toward the north. When it travels on that part of the inclined which is on the parallel of Meroë, there are no shadows at noon for those living there. Whenever the Sun travels on the most northern part of the inclined, the shadows at noon for the zone of Meroë are to the south. For this reason we call these places with "double shadows", because at noon the shadows can fall either to the north or to the south, depending from which direction the Sun shines on the objects. This happens not only for the inhabitants of Meroë, but for all people living in the region we already mentioned, whose northern boundary is at the end of Gemini and the beginning of Cancer all the way to the equator. The places outside this inhabited region, being completely outside the ecliptic, are always illuminated by the Sun from the south; thus their shadows are always to the north and are called "single-shadowed".

Καὶ πρῶτον μὲν τὸ κατὰ Μερόην κλίμα οὕτως ἔγνωσται ὑπὸ παράλληλον ἀπέχοντα τοῦ ἰσημερινοῦ μοίρας ις ζʹ κζʹʹ ἐπὶ τοῦ πρότερον ἤδη εἰρημένου μεγίστου καὶ αὐτοῦ κύκλου ἐν τῇ σφαίρᾳ τοῦ διὰ τῶν πόλων νοουμένου ἀμφοτέρων, τοῦ τε ζωδιακοῦ καὶ τοῦ ἰσημερινοῦ. δεύτερον δὲ κλίμα τό τε κατὰ Συήνην ἔγνωσται ὑπὸ παράλληλον ἀπέχοντα ὡσαύτως τοῦ ἰσημερινοῦ μοίρας κγ ναʹ. ὃς δὴ παράλληλος κατὰ τὸ βορειότατόν ἐστι δηλαδὴ κλίμα τοῦ ζωδιακοῦ, τὸ περὶ τὴν ἀρχὴν τοῦ Καρκίνου, ἐπειδὴ καὶ οὕτω προέφημεν εἰκοσιτεσσάρων ἔγγιστα μοιρῶν εἶναι τὴν ἀπὸ τοῦ ἰσημερινοῦ ἔγκλισιν τοῦ ζωδιακοῦ ἐπὶ τὰ βόρεια καὶ τὰ νότια· ὥστε τοῖς ὑπὸ τὸν τοιοῦτον παράλληλον οἰκοῦσιν αὐτοῖς, δηλαδὴ τοῖς ἐν τῷ κατὰ Συήνην κλίματι, ἅπαξ τοῦ ἔτους τὸν ἥλιον κατὰ κορυφῆς γίνεσθαι καὶ τηνικαῦτα ἐν ταῖς μεσημβρίαις ἀσκίους αὐτοὺς εἶναι. πρῶτοι δὲ οὗτοί εἰσι καὶ ἑτερόσκιοι ἀεὶ τῶν σκιῶν κατ᾽ αὐτοὺς πρὸς βορρᾶν ἐκπιπτουσῶν διὰ τὸ τὸν ὅλον ζωδιακὸν κύκλον, ὃν ὁ ἥλιος παροδεύει, νοτιώτερον αὐτῶν εἶναι. καὶ οἱ ἐν τοῖς ἑξῆς δὲ κλίμασιν οἰκοῦντες ἔτι ἐπιπλέον ἀφιστάμενοι τοῦ ζωδιακοῦ οἱ ὅλοι ἑτερόσκιοί εἰσιν. οἷον τρίτον κλίμα ἐστὶ τὸ τῆς κάτω χώρας Αἰγύπτου λεγόμενον, ὑπὸ παράλληλον ἀφιστάμενον τοῦ ἰσημερινοῦ μοίρας λ κβʹ καὶ τελείως ἐκπίπτοντα τοῦ ζωδιακοῦ. τέταρτον τὸ κατὰ Ῥόδον ὑπὸ παράλληλον ἀφιστάμενον τοῦ ἰσημερινοῦ μοίρας λϛ. πέμπτον τὸ καθ᾽ Ἑλλήσποντον ὑπὸ παράλληλον ἀφιστάμενον τοῦ ἰσημερινοῦ μοίρας μ νστʹ. ἕκτον τὸ κατὰ μέσον τὸν Πόντον ὑπὸ παράλληλον ἀφιστάμενον τοῦ ἰσημερινοῦ μοίρας με. καὶ ἕβδομον τὸ κατὰ τὰς ἐκβολὰς τοῦ Βορυσθένους ποταμοῦ ὑπὸ παράλληλον ἀφιστάμενον τοῦ ἰσημερινοῦ μοίρας μη.

Καὶ ὁλοσχερέστερον μὲν ἐν τοῖς Προχείροις Κανόσι, ὡς εἴρηται, οὕτως αὐτὰ μόνα τὰ ἑπτὰ διαλαμβάνονται κλίματα ἐν τῇ καθ᾽ ἡμᾶς οἰκουμένῃ, πρῶτον τὸ κατὰ Μερόην καὶ ἕβδομον τὸ κατὰ τὰς ἐκβολὰς τοῦ Βορυσθένους ποταμοῦ. καὶ οἱ δι᾽ αὐτῶν παράλληλοι τοσαύτας ἔχοντες τὰς ἀποστάσεις ἀπὸ τοῦ ἰσημερινοῦ κανονικῶς ἐκτέθεινται καταληφθέντες γραμμικαῖς δείξεσι τῷ Πτολεμαίῳ, ἐν τῷ δευτέρῳ τῆς Συντάξεως, ἀπὸ τοῦ λόγου τῶν σκιῶν τῶν γνωμόνων τῶν ἐν τοῖς

81 Cf. Ptol. *Proch. Kan.* p. 101-128 Tihon **86-87** Cf. Ptol. *Alm.* 2.5-6

First we define the zone of Meroë which is beneath the parallel circle whose latitude distance from the equator is 16° 7′ 27″ measured on the great circle that passes through the poles of both the equator and the ecliptic. The second zone is that of Syene, being below the parallel 23° 51′ from the equator. This parallel is beneath the northern-most boundary of the ecliptic, at the beginning of Cancer. Since we said that the inclination of the ecliptic is approximately 24° to the north and to the south, for those who inhabit the zone of Syene the Sun comes to the zenith once each year, and on that day at noon there are no shadows. The zones directly to the north are the first ones to experience the shadows falling always in one direction, because as the Sun travels the entire ecliptic it is located to the south [of those places]. Those who live in the subsequent zones are outside the zodiac and have shadows in only one direction. The third zone is the one called Lower Egypt at the parallel of 30° 22′, being completely outside the ecliptic. The fourth zone is that of Rhodes with a parallel 36° from the equator. Fifth is that of the Hellespont with a parallel 40° 56′; sixth is the region passing through the middle of the Black Sea with 45°, and the seventh at the mouth of the river Borysthenes at a distance of 48° from the equator.

These seven zones are described in great detail in the book of *Handy Tables*, as belonging to our ecumene, with the first one of Meroë and the seventh at the mouth of river Borysthenes. The distances of their parallels are recorded in sequel being calculated by Ptolemy in the second book of the

ὑποκειμένοις ἑκάστοις κλίμασι. καὶ παρηύξηνται αἱ ἡμεριναὶ ὧραι ὡς ἰσημεριναὶ θεωρούμεναι ἑκάστῳ τῶν τοιούτων ἑπτὰ κλιμάτων καθ' ἡμιώριον.

Ἐν δέ γε τῇ Γεωγραφίᾳ καὶ τῷ δευτέρῳ τῆς Συντάξεως πλείονάς τε καταγράφει ὁ Πτολεμαῖος τοὺς παραλλήλους κατὰ τεταρτημόριον ὥρας διαιρῶν τὰς ὑπ' αὐτοὺς οἰκήσεις καὶ καθ' ἥττονας ἀποστάσεις ἀπὸ τοῦ ἰσημερινοῦ. καὶ πρόεισιν ἐκτιθεὶς τοὺς τοιούτους παραλλήλους καὶ ἐπέκεινα τοῦ ἑβδόμου κλίματος τοῦ κατὰ τὰς ἐκβολὰς Βορυσθένους μέχρι καὶ εἰς αὐτὴν τὴν οἴκησιν τὴν ὑπὸ τὸν βόρειον πόλον. ὑποτίθεται μὲν γάρ, καθὼς καὶ ἡ ἀλήθεια καὶ ἡ ἀπόδειξις ἀπαιτεῖ, τὴν ὑπὸ τὸν ἰσημερινὸν οἴκησιν, ἤτοι κατὰ τὴν περιφορὰν τῆς ὀρθῆς σφαίρας, ιβ ὡρῶν εἶναι διὰ παντὸς τοῦ ἔτους ἐν παντὶ τμήματι τοῦ ζῳδιακοῦ γινομένου τοῦ ἡλίου, καὶ ἴσας εἶναι τὰς ἡμέρας ταῖς νυξὶν ἀείποτε. καὶ γὰρ ὁ ὁρίζων τῆς ἐν τῇ τοιαύτῃ οἰκήσει, μᾶλλον δὲ θέσει τῆς γῆς, θεωρουμένης σφαίρας, ἤτοι τῆς ὀρθῆς καὶ τῆς κοσμικῆς, ὡς ἂν εἴποι τις, κατ' ἰσότητα ἀπαρεγκλίτου θέσεως, αὐτός ἐστιν ὁ κυρίως ὁρίζων τῆς σφαίρας καὶ οὐ μόνον τὸν ἰσημερινόν, ἀλλὰ καὶ πάντας τοὺς ἐφ' ἑκάτερα αὐτοῦ καὶ βόρεια καὶ νότια παραλλήλους εἰς ἶσα τέμνει, ἤτοι εἰς δύο ἡμικύκλια, ὡς εἶναι τὰ μὲν ὑπὲρ γῆν, τὰ δὲ ὑπὸ γῆν.

Οἱ δὲ ἄλλοι πάντες ὁρίζοντες οἱ ἑξῆς τῆς καθ' ἡμᾶς οἰκουμένης τῆς ἐν τῷ βορείῳ τεταρτημορίῳ τῆς ὅλης γῆς τῷ ἄνω, ὡς εἴρηται, πρὸς τὰς διαφόρους θέσεις καὶ ἐγκλίσεις τοῦ τῆς γῆς σφαιρικοῦ σώματος καταλαμβάνονται ὁρίζοντες. καὶ τὸν μὲν ἰσημερινὸν ἀείποτε καὶ οἱ πάντες εἰς ἡμικύκλια τέμνουσι δύο—μέγιστοι γὰρ οἱ πάντες ὄντες κύκλοι τῆς σφαίρας, μέγιστον καὶ αὐτὸν ὄντα κύκλον τῆς σφαίρας ἐξανάγκης ὡς εἴρηται τέμνουσιν εἰς ἶσα—τοὺς δὲ ἄλλους παραλλήλους εἰς ἄνισα δύο μείζονα καὶ ἐλάττονα. καθ' ὅσον γὰρ ἐγκλίνεται ἡ οἴκησις ἡτισοῦν ἑκάστη πρὸς τὰ βορειότερα μέρη, σὺν αὐτῇ δὲ πάντως καὶ ὁ πρὸς τῷ ἐπιπέδῳ αὐτῆς ἕκαστος ὁρίζων ἐγκλίνεται τοῦ τῆς ὀρθῆς

91 Cf. Ptol. *Geog.* 1.23; *Alm.* 2.6

91 post συντάξεως add. αὖθις V[sl] **102** εἴποι : εἴπῃ C **105** βόρεια καὶ νότια : νότια καὶ βόρεια C[ac] **114-116** σημείωσαι ἀναγκαῖον sch. in mg. C

Syntaxis by using shadows of the gnomons, and the day-time for each of the seven zones increases by a half equinoctial hour.

In the *Geography* and in the second book of *Syntaxis* Ptolemy records many more parallels, dividing the inhabited region into quarter-hours and into smaller distances from the equator. He then extends the parallels beyond the seventh zone at the mouth of the Borysthenis all the way up to the habitation of the North Pole. He assumes, as also the truth demands, that for the zone of the equator a [daily] rotation of the *sphaera rectra* happens in 12 hours for any time of the year and independent of the position of the Sun on the ecliptic; thus the days and nights are always equal. In fact, the horizon of this zone or rather this location of the Earth has no inclination. This is the main horizon of the sphere, which bisects not only the equator but all parallels to the north and to the south. It divides them into two semi-circles, one above and the other below the Earth.

All other horizons of our ecumene are located in the northern quadrant. They are at different positions and inclinations on the body of the spherical Earth. They always intersect the equator in two semicircles, because [all horizons] are major circles of the sphere and the equator is a major circle. They divide all other parallels into two unequal parts, a larger and a smaller, because the more each habitation is inclined to the north, and at the same time its horizon inclines relative to the horizon of the *sphaera recta*,

σφαίρας ὁρίζοντος, κατὰ τοσοῦτον ἀναλόγως καὶ τὸ κατὰ κορυφὴν ταύτης δὴ τῆς οἰκήσεως ἑκάστης τμῆμα καὶ σημεῖον τοῦ οὐρανίου σώματος ἀφίσταται τοῦ κατὰ κορυφὴν τῆς ὑπὸ τὴν ὀρθὴν σφαῖραν γῆς, ἤτοι τοῦ ἰσημερινοῦ. καὶ ὁ βόρειος πόλος τοῦ ἰσημερινοῦ καὶ τῆς ὀρθῆς σφαίρας ἐξαίρεται πρὸς ὕψος ἀπὸ τοῦ ἐπιπέδου τῆς τοιαύτης ἑκάστης θέσεως, ὁ δὲ νότιος πόλος τὸ ἐναντίον ὑπὸ γῆν κρύπτεται. κἀντεῦθεν τέμνονται οἱ παράλληλοι τῷ ἰσημερινῷ οἱ ἐφ᾽ ἑκάτερα τά τε βόρεια καὶ τὰ νότια αὐτοῦ παρὰ τῶν τοιούτων ὁριζόντων εἰς ἄνισα τμήματα, οἱ μὲν ἐπὶ τὰ νότια αὐτοῦ ἔχοντες τὰ ὑπὲρ γῆν τμήματα ἥττονα, τὰ δὲ ὑπὸ γῆν μείζονα, οἱ δὲ ἐπὶ τὰ βόρεια ἔχοντες τὰ ὑπὲρ γῆν τμήματα μείζονα, τὰ δὲ ὑπὸ γῆν ἥττονα.

Ὡς ἐντεῦθεν πάντως ἀκολουθεῖν, τῶν ὑπὲρ γῆν τμημάτων τῶν τοιούτων παραλλήλων τοῖς βορειότερον οἰκοῦσι μειζόνων ὄντων, τῶν δὲ ὑπὸ γῆν ἡττόνων, τὰς μεγίστας τοῦ ἔτους ἡμέρας μείζονας ἀποτελεῖσθαι ἢ τοῖς νοτιώτερον οἰκοῦσι καὶ αὐτοῦ ἐγγύτερον τοῦ ἰσημερινοῦ καὶ τὰ ὑπὲρ γῆν τμήματα τῶν παραλλήλων ἔχουσιν ἥττονα. τὰ γὰρ μείζονα ὑπὲρ γῆν τμήματα τῶν κύκλων ἐν τῇ πρώτῃ καὶ καθολικῇ περιφορᾷ τοῦ παντὸς καὶ βραδύτερον καὶ χρόνῳ πάντως πλείονι διεξέρχονται τὸν δυτικὸν ὁρίζοντα. διὰ τοῦτο ἄλλα παρ᾽ ἄλλαις οἰκήσεσι τὰ μεγέθη τῶν ἡμερῶν γίνεται καὶ τῶν νυκτῶν ἀνάπαλιν· καὶ μείζονα μὲν ἀεὶ ἀναλόγως ταῖς βορειοτέραις οἰκήσεσι τὰ τῶν ἡμερῶν τῶν μεγίστων ἐν τῷ παντὶ δηλονότι ἔτει, καὶ ἐλάττονα τὰ τῶν νυκτῶν τῶν ἐλαχίστων ὡσαύτως ἐν τῷ παντὶ δηλονότι ἔτει. τῇ μὲν οὖν ὑπὸ τὸν ἰσημερινὸν ὡς εἴρηται νοουμένῃ οἰκήσει, ὡς ἔστιν ἐκ τῶν εἰρημένων λοιπὸν δῆλον διαπαντός, ιβ ὡρῶν εἰσιν αἱ ἡμέραι καὶ ιβ αἱ νύκτες. ταῖς δὲ ὑπὸ τοὺς ἐφεξῆς αὐτῷ παραλλήλους πρὸς τὸ βορειότερον μέρος, ἤγουν ἐν τῇ καθ᾽ ἡμᾶς οἰκουμένῃ, διὰ τὴν εἰρημένην αἰτίαν τοῦ ἐξάρματος τοῦ βορείου πόλου καὶ τῶν μειζόνων ὑπὲρ γῆν τμημάτων τῶν παραλλήλων, μείζονες καὶ ἡμέραι τινὲς καταλαμβάνονται καὶ πλειόνων ὡρῶν ἢ ιβʹ. καὶ τοῦτο συνεχῶς ἀεὶ γίνεται πλεῖον, ἐφ᾽ ὅσον αἱ οἰκήσεις καὶ τὰ ἐγκλίματα τῆς οἰκουμένης βορειότερα γίνονται.

125-26 τὰ ὑπὲρ γῆν ... ἔχοντες om. C^{ac} (homoeotel.) **138** δηλονότι post ἔτει transp. C

the zenith of each habitation deviates correspondingly from the zenith of the *sphaera recta*, i.e. the zenith of the equator. At the same time, the North Pole of the equator and of the *sphaera recta* is elevated above the plane of the location; on the contrary the South Pole is concealed. Consequently, the parallels to the equator are divided by the horizons into two unequal parts: the places to the south have smaller parts above the horizon and larger below. [The opposite happens for habitations] in the north, having larger parts [of the parallels] above their horizons, and smaller below.

From all this it follows that for habitations further to the north, the sections of the parallels above the horizons become larger and the ones below smaller, so that the longest days for locations further to the north are bigger than for southern locations closer to the equator, where the parallels above the Earth are relatively smaller. In the first diurnal rotation of the universe, it takes more time [for the Sun] to traverse the larger arcs above the Earth and reach the western horizon. For this reason the lengths of days and nights are different in the various habitations. Evidently, for habitations further to the north the longest days are proportionately larger and the nights smaller. But for the habitations at the equator, it follows from what we said that both days and nights have 12 hours each. As we advance further north along successive parallels of our ecumene, the elevation of the North Pole and arcs of the parallels above the local horizons become larger and the days are longer, with some of them exceeding 12 hours. They increase continuously as the habitations and their inclinations advance further to the north.

Παμπλείστων δὲ ὄντων τῶν ἀπὸ τοῦ ἰσημερινοῦ πρὸς τὸ βόρειον μέρος νοουμένων παραλλήλων, οἱ μὲν Πρόχειροι Κανόνες, ὡς ἔφημεν, μόνων ἑπτὰ καταγραφὰς ποιοῦνται τῶν διὰ τῶν ἑπτὰ εἰρημένων κλιμάτων ὁλοσχερέστερον πρὸς τὴν χρείαν συντάττοντες· καὶ καθ' ἡμιώριον τὰς παραυξήσεις τῶν ἡμερῶν ποιοῦνται ἐν τοῖς τοιούτοις κλίμασι, τὴν ἀρχὴν ὑποτιθέμενοι ἀπὸ τοῦ κατὰ Μερόην καὶ καταντῶντες εἰς τὸ κατὰ τὰς ἐκβολὰς Βορυσθένους. ἐν δὲ τῷ πρώτῳ τῆς Γεωγραφίας ὁ Πτολεμαῖος, ὡσαύτως καὶ ἐν τῷ δευτέρῳ τῆς Συντάξεως, ὡς ἔφημεν, κατὰ πλείονας καί, ὡς ἂν εἴποι τις, πυκνοτέρους τοὺς παραλλήλους διαιρῶν τὸ ἀπὸ τοῦ ἰσημερινοῦ πρὸς βορρᾶν ἄνω τεταρτημόριον. κἀντεῦθεν πολλοὺς τοὺς παραλλήλους ἀπαριθμούμενος τὰς ὑπ' αὐτοὺς οἰκήσεις δηλοποιεῖ, παραυξάνων τὰς ἐν αὐταῖς μεγίστας ἡμέρας ἐξαρχῆς κατ' ὀλίγον, ἤτοι τετάρτῳ ὥρας ἡμερινῆς. πρῶτον οὖν τιθεὶς τὸν τῶν παραλλήλων μέγιστον αὐτὸν τὸν ἰσημερινὸν καὶ τὴν κατ' αὐτὸν ἡμέραν ὡρῶν ιβ, ἑξῆς ἕτερον παράλληλον τίθησι δεύτερον διὰ Ταπροβάνης, τῆς μεγίστης νήσου τῆς ἐν τῷ μεγάλῳ Ἰνδικῷ πελάγει, ἐν ᾗ τὴν μεγίστην ἡμέραν ἰσημερινῶν ὡρῶν τίθησι ιβ καὶ τετάρτου. καὶ τούτῳ τρίτον ἑξῆς τίθησι παράλληλον τὸν διὰ τοῦ Ἀβαλίτου κόλπου ἐν τῷ αὐτῷ Ἰνδικῷ πελάγει, ἔνθα ἡ μεγίστη ἡμέρα ὡρῶν ἐστι ιβ ἡμίσεος. εἶτα τούτῳ ἐφεξῆς τέταρτον παράλληλον τίθησι τὸν διὰ τοῦ Ἀδολιτικοῦ κόλπου καὶ αὐτὸν ἐν τῷ αὐτῷ μεγάλῳ Ἰνδικῷ πελάγει καταγραφόμενον, ἔνθα τὴν μεγίστην ἡμέραν ὡρῶν τίθεται ιβ ἡμίσεος καὶ τετάρτου. τούτους τρεῖς παραλλήλους ἐν τῷ μέσῳ κατατάττει τοῦ τε ἰσημερινοῦ καὶ τοῦ κατὰ τὸ πρῶτον εἰρημένον κλίμα. εἶτα αὐτὸν τοῦτον κατατάττει πέμπτον παράλληλον τὸν κατὰ τὸ εἰρημένον κλίμα πρῶτον, τὸ διὰ Μερόης, ἔνθα τὴν μεγίστην ἡμέραν ὡρῶν ιγ ὑποτίθεται.

Εἶτα καθ' εἱρμὸν τῇ αὐτῇ τάξει χρώμενος, ἄλλους πλείστους ἐν μέσῳ τῶν ἑπτὰ κλιμάτων παραλλήλους καταγράφων, οὕτω δὴ καὶ ὑπὲρ τὸ ἕβδομον τὸ κατὰ τὰς ἐκβολὰς Βορυσθένους ἑξῆς πλείστους παραλλήλους ἀπαριθμούμενος, καὶ μέχρι πολλοῦ τὰς ὑπ' αὐτοὺς

149 Cf. Ptol. *Proch. Kan.* p. 101-128 Tihon **154-55** Cf. Ptol. *Geog.* 1.23; *Alm.* 2.6

156 εἴποι : εἴπη C **168** αὐτὸν C[pc] : αὐτοῦ VC[ac]

The parallels to the north of the equator are very numerous, but the *Handy Tables* record only seven, one for each of the seven zones, in order to simplify their use. For these habitations the days are recorded in half-hour increments. They begin at Meroë and end at the mouth of the Borysthenis. Ptolemy records them this way in the first book of *Geography*, as also in the second book of *Syntaxis*: he divides the quadrant starting at the equator into more parallels, being so to speak denser. After recording the many parallels he enumerates the places below them, increasing the largest days somewhat at the beginning, by a quarter of an hour. He first arranges the largest among the parallels, the equator, where the day has 12 hours; as a second parallel, the one through Taprobane, being the largest island in the great Indian Ocean and there he arranges the longest day to have 12 and a quarter hours. Third, he arranges the parallel through the Avalite Gulf in the same Indian Ocean, where the largest day is 12 and a half hours. After this, the fourth parallel is the one through the Adulitic Gulf, in the great Indian Ocean, where the longest day is set at 12 and three quarter hours. He arranges these three parallels to be between the equator and the first habitation. Next he arranges the fifth parallel, that of Meroë, where the longest day is 13 hours.

Following the same methods, he then records many parallels in between, also enumerating parallels beyond the seventh habitation to the north of the river Borysthenis. He gives in detail the location below them and the increase

οἰκήσεις δηλοποιῶν καὶ τὰς ἐπαυξήσεις τῶν ἡμερῶν τελευτῶν καταντᾷ μέχρις οὗ γένοιτ' ἂν ὁ βόρειος τῆς ὀρθῆς σφαίρας πόλος κατὰ κορυφὴν τῇ θέσει τῆς γῆς ἐκείνης καὶ ὁ ἰσημερινὸς ὁρίζοντος λάβῃ τάξιν τηνικαῦτα καὶ θέσιν, ὡς ἐξανάγκης ἕπεσθαι ἑξαμηνιαῖον καιρὸν συνεχῶς ἐκεῖ ὡς ἡμέραν μίαν εἶναι. καθ' ὃν δηλαδὴ καιρὸν ἑξαμηνιαῖον διϊὼν ὁ ἥλιος τὸ ἀπὸ τοῦ ἰσημερινοῦ βόρειον ἡμικύκλιον τοῦ ζῳδιακοῦ, ἤτοι τὰ ἓξ ζώδια, Κριόν, Ταῦρον, Διδύμους, Καρκίνον, Λέοντα καὶ Παρθένον, ἐξανάγκης ὑπεράνω συνεχῶς ἔσται τῆς ἐκεῖσε γῆς. καὶ αὖθις ἑξαμηνιαῖον καιρὸν συνεχῶς ἐκεῖ ὡς μίαν εἶναι νύκτα, καθ' ὃν δηλαδὴ ἑξαμηνιαῖον καιρὸν διϊὼν ὁ ἥλιος τὸ ἀπὸ τοῦ ἰσημερινοῦ νότιον αὖθις ἡμικύκλιον τοῦ ζῳδιακοῦ, ἤτοι τὰ λοιπὰ ἓξ ζώδια, Ζυγόν, Σκορπίον, Τοξότην, Αἰγοκέρωτα, Ὑδροχόον καὶ Ἰχθύας, ἐξανάγκης ὑποκάτω συνεχῶς ἔσται τῆς ἐκεῖσε γῆς.

Καὶ οὕτω μὲν ὁ Πτολεμαῖος τὸ ἀκόλουθον τῆς Γεωγραφίας καθιστορῶν τοὺς εἰρημένους παραλλήλους ἑξῆς ἐκτίθεται. ἐμοὶ δ' οὖν πρόσφορον δοκεῖ καὶ μάλα προσῆκον πρὸς τὴν προκειμένην ἡμῖν χρῆσιν μήτε τοῖς τῶν ἑπτὰ καλουμένων κλιμάτων μόνοις παραλλήλοις ἀρκεῖσθαι, ἐπειδὴ καὶ ἄλλαι οἰκήσεις ἐφ' ἑκάτερα τούτων εἰσί, καὶ μάλιστα ἐπὶ τὰ βόρεια οὐκ ὀλίγαι καὶ σφόδρα ἐγνωσμέναι ἡμῖν, μετὰ τὸ ἕβδομον δηλαδὴ κλίμα τὸ κατὰ τὰς ἐκβολὰς Βορυσθένους, μήτε εἰς τὰ ἀοίκητα παντάπασι καὶ ἄγνωστα παρεκτείνεσθαι καὶ προχωρεῖν τοῖς παραλλήλοις τούτοις, ἀλλὰ μέχρι τούτου καταλογίζεσθαι τοὺς παραλλήλους καὶ τὰς κατ' αὐτοὺς ἐπαυξήσεις τῶν ἡμερῶν, μέχρις οὗ τὰς ὑπ' αὐτοὺς οἰκήσεις ἱστοροῦμεν καὶ καταλαμβάνομεν ἐγνωσμένας, μέχρι δηλαδὴ τῶν βορειοτάτων τῆς μικρᾶς Βρεττανίας, ἣν δὴ καὶ ὁμολογουμένως νῦν γινώσκομεν οἰκουμένην. καθ' ἣν καὶ τὴν μεγίστην ἡμέραν ὡρῶν ὑποτίθεται ιθ ὁ Πτολεμαῖος ἐν τῷ δευτέρῳ τῆς Συντάξεως καὶ τὸν δι' αὐτῆς παράλληλον εἰκοστὸν ἕβδομον καταγράφων ἀπέχειν, φησί, τοῦ ἰσημερινοῦ μοίρας ξα. εἰ δὲ βούλεταί τις, ἔστω ἡμῖν ἔσχατος τῶν παραλλήλων ὁ διὰ Θούλης εἰκοστὸς ἔννατος καταγραφόμενος,

191 Cf. Ptol. *Geog.* 1.23 **204** Cf. Ptol. *Alm.* 2.6 (vol.1.1, p. 114.1-2 Heiberg)

181 τάξιν post τηνικαῦτα transp. C **202** Βρετανίας C

[in the length of the days], finally ending up at the North Pole of the *sphaera recta*, on the top of the Earth where the horizon is parallel to the equator. It now follows that at the North Pole there is one continuous day lasting for six months. This is the time the Sun needs to travel the semi-circle to the north of the equator, that is the six constellations of Aries, Taurus, Gemini, Cancer, Leo and Virgo, at which time it is always above that section on the Earth. Immediately afterwards there is a continuous night for six months. These are the six months that the Sun travels the semi-circle of the ecliptic to the south of the equator that is through the remaining constellations of Libra, Scorpio, Sagittarius, Capricorn, Aquarius and Pisces. At that time [the Sun] is continuously below that section of the Earth.

In this manner Ptolemy narrates and describes the parallels in his *Geography*. Personally, I do not think it is very useful, nor does it suffice to record only the parallels for seven habitations, since there are numerous other inhabited places on either side of them, especially in the north, and I am terribly aware that beyond the seventh habitation at the mouth of the Borysthenis we do not enter an uninhabited, unknown region. But we enumerate the parallels and the corresponding extensions of the days up to the region we consider as known, up to the inhabited regions of Little Brittania, which we readily accept as ecumene. In the second book of *Syntaxis* Ptolemy accepts the longest day to have 19 hours and the corresponding 27th parallel to be 61° from the equator. If one wishes to take as the very last parallel the 29th, passing through Thoule,

ἐν ᾧ τὴν μεγίστην ἡμέραν ὡρῶν ὁ Πτολεμαῖος ὑποτίθεται κ ἔν τε τῷ αὐτῷ δευτέρῳ βιβλίῳ τῆς Συντάξεως καὶ ἐν τῷ πρώτῳ τῆς Γεωγραφίας καὶ ἀπέχειν φησὶν τὸν τοιοῦτον παράλληλον τοῦ ἰσημερινοῦ μοίρας ξγ στ′′.

209 Cf. Ptol. *Alm.* 2.6 (vol.1.1, p. 114.9-10 Heiberg); *Geog.* 1.23.11-13 (p. 58 Müller)

208 ὁ Πτολεμαῖος post ὑποτίθεται transp. C

mentioned in the second book of *Syntaxis* and the first book of *Geography*, the largest day there has 20 hours and the parallel is 63° and 6″ from the equator.

13 περὶ τῆς τάξεως τῶν οὐρανίων σφαιρῶν καὶ ὅτι πᾶσαι ὑποκάτω εἰσὶ τῆς ἀπὸ ἀνατολῶν εἰς δυσμὰς κατὰ τὴν πρώτην ὡς εἴρηται φορὰν κινουμένης σφαίρας καὶ πᾶσαι ἐναντίως αὐτῇ ἀπὸ δυσμῶν εἰς ἀνατολὰς φέρονται

Τὰ μὲν δὴ τῶν παραλλήλων τῷ ἰσημερινῷ κύκλῳ καὶ τῶν ὑπ᾽ αὐτοὺς ἑκάστων οἰκήσεων, καὶ ὅπως ἐγκεκλιμένη πρὸς τὴν ὀρθὴν σφαῖραν ἡ καθ᾽ ἡμᾶς οἰκουμένη καταλαμβάνεται πρὸς τὰ βόρεια μέρη τῆς γῆς, καὶ ὅτι διὰ ταύτην τὴν αἰτίαν καὶ ὁ βόρειος τῆς ὀρθῆς σφαίρας πόλος κατ᾽ ὀλίγον ἐξαίρεται ὑπὲρ γῆν, καὶ κατ᾽ ἶσον ὁ νότιος πόλος ὑπὸ γῆν κρύπτεται ἀναλόγως, μᾶλλον δὲ κατ᾽ ἰσότητα τῆς ἀποστάσεως, ἣν ἔχει ὁ ἑκάστης οἰκήσεως παράλληλος πρὸς τὸν ἰσημερινόν, διὰ βραχέων καὶ κατ᾽ ἐπιτομὴν οὕτως εἴρηται, ἀναγκαῖά γε ὄντα κομιδῇ προκατειλῆφθαι πρὸς τὰς ἑξῆς χρείας, ὡς προϊὼν ὁ λόγος δείξει σαφῶς. ὅπερ δὲ προελέγομεν, ἄνευ μόνης τῆς ὑπερτάτης καὶ καθολικῆς καὶ ἀριδηλοτάτης καὶ ταχυτάτης περιφορᾶς τοῦ παντὸς οὐρανίου σώματος, τῆς ἀπὸ ἀνατολῶν εἰς δυσμάς, αἱ ἐντὸς καταλαμβανόμεναι ὀκτὼ σφαῖραι, αἱ καὶ τοὺς ἀστέρας ἅπαντας περιφέρουσαι ἀπὸ δυσμῶν εἰς ἀνατολὰς ποιοῦνται τὰς κινήσεις, ἐναντίας τῇ πρώτῃ, λοξὰς μέντοι καὶ περὶ τοὺς πόλους τοῦ ζωδιακοῦ, οὐ περὶ τοὺς πόλους τοῦ ἰσημερινοῦ, ἤτοι τῆς ὀρθῆς σφαίρας.

Ὅπως δὲ τάξεως ἔχουσιν αἱ τοιαῦται σφαῖραι ἤδη λέγομεν. πρώτη μὲν οὖν νοεῖται σφαιρικὴ περίοδος περὶ τοὺς πόλους τοῦ λοξοῦ καὶ ζωδιακοῦ κύκλου εἰς τὰ ἑπόμενα φερομένη, ὡς πολλάκις εἴρηται, ἡ τῶν ἀπλανῶν ἁπάντων ἀστέρων, οἳ καὶ πεπηγέναι δοκοῦσι καὶ ἐπὶ τοῦ αὐτοῦ ἵστασθαι, κινούμενοι δ᾽ ἀναμφιβόλως καταλαμβάνονται, βραδύτατα δὲ σφόδρα ὡς πρὸς τοὺς ἄλλους. μετὰ δὲ τὴν τῶν ἀπλανῶν σφαῖραν ἑξῆς ἡ τοῦ Κρόνου καταλαμβάνεται, καὶ ταχύτερον μὲν ἐν ταύτῃ καὶ πάνυ τοι ταχύτερον οὗτος ὁ ἀστὴρ περιοδεύει ἢ κατὰ τοὺς ἀπλανεῖς ἀστέρας, βραδύτερον δὲ ὅμως πολλῷ τῷ μέτρῳ, ἤπερ οἱ ἑξῆς

2 οἰκήσεων : ἐγκλίσεων C **10** μόνης post τῆς ὑπερτάτης transp. C **12** ἀπὸ : ἀπ᾽ C
22 τῶν bis exhib. C

13 The order of the celestial spheres: the first sphere that moves from East to West with all other spheres located below and rotating from West to East

Returning to the parallels, the inhabited regions below them as well as our ecumene are located in the northern hemisphere and are visualised to be inclined relative to *sphaera recta*; a consequence is that the North Pole of the *sphaera recta* appears to be a little elevated above the Earth and the South Pole is concealed by an equal amount, or rather by an amount proportional to the distance measured directly and in the shortest possible way between the parallel of the habitation and the equator. All these have been mentioned briefly, because it is necessary to understand them well in advance, for future use, as will become clear in the next chapters. As we mentioned, except for the highest, universal, very evident and fastest revolution of the stellar globe from east to west, the eight spheres inside contain all the stars, rotate from west to east and revolve in opposite sense to the rotation of the first sphere around the poles of the ecliptic and not around the poles of the equator (poles of the *sphaera recta*).

We shall now describe the classification of these spheres. First we conceive the circular period around the poles of the inclined circle, being also the poles of the ecliptic, toward the subsequent constellations. All stars are considered to be fixed on this sphere; we have no doubt that they rotate, even though extremely slowly relative to any other sphere. Next to the sphere of the fixed stars is the sphere of Saturn, which rotates much faster than that of the

ἕτεροι ἀστέρες. ἑξῆς δὲ ταύτῃ, ἡ περίοδός ἐστι τοῦ Διός, ταχυτέρα μὲν τῆς τοῦ Κρόνου, βραδυτέρα δὲ τῶν ἄλλων. καὶ μετὰ ταύτην ἔτι ἡ τοῦ Ἄρεος περίοδος, ταχυτέρα μὲν καὶ αὕτη τῶν πρὸ αὐτῆς, καὶ πολλῷ ταχυτέρα, βραδυτέρα δὲ τῶν ἄλλων. εἶτα ἑξῆς αἱ τρεῖς περίοδοι, ἥ τε τοῦ ἡλίου, ἡ τῆς Ἀφροδίτης καὶ ἡ τοῦ Ἑρμοῦ. κατελήφθησαν δὲ αὗται τοῖς ἀστρονόμοις ἰσοταχεῖς τῇ ὁμαλῇ κινήσει.

Ὅπως δὲ αἱ ὁμαλαὶ κινήσεις τῶν ἀστέρων καὶ αἱ ἀνώμαλοι καὶ ἄνισοι θεωροῦνται καὶ διὰ τίνας τὰς αἰτίας, ἐν τοῖς ἑξῆς κατὰ καιρὸν ἐροῦμεν, ἔνθα καὶ καθ᾽ ὁπόσον καιρὸν ἡ ἑνὸς ἑκάστου ὁμαλὴ περίοδος ἀποκαθίσταται διαληψόμεθα. τὰ γὰρ ὁμαλὰ αὐτῶν κινήματά εἰσι καὶ εὐόριστα, τῶν δὲ ἀνωμάλων κινήσεων αὐτῶν αἱ καταλήψεις πολὺ τὸ ἐργῶδες ἔχουσιν καὶ δυσόριστοί πώς εἰσι καταλαβεῖν, ὡς προϊὼν ὁ λόγος δηλώσει, κατὰ καιρὸν τὰ περὶ αὐτῶν ἑκάστων διευκρινῶν. ἐσχάτη δ᾽ οὖν ἁπασῶν τῶν περιόδων ἡ τῆς σελήνης καὶ ταχυτέρα πολλῷ τῷ μέτρῳ πασῶν.

27 ἔτι : ἐστὶν C

fixed stars, but much slower than the rotation of the other planets. Next is the rotation of Jupiter, being faster than that of Saturn and slower than the others. Next is the sphere of Mars which is much faster than all previous rotations, but slower than the others. Finally, there are three rotations, one for each of the following: Sun, Venus and Mercury. The astronomers established that they rotate with the same uniform speed.

We will describe at the appropriate time the explanation for the uniform and the anomalous and unequal rotations of the stars, as well as the periods during which the uniform motions are restored. The uniform rotations are easy to describe; however, understanding of the anomalous motions is a tedious and difficult task as will become evident when we proceed to elucidate them for each of the planets. The last rotation is that of the Moon which is much faster than all the other rotations.

14 ὅτι οἱ κύκλοι πάντες, ἐφ' ὧν φέρονται ὅ τε ἥλιος καὶ ἡ σελήνη καὶ οἱ πέντε πλανῆται, οὐχ ὁμόκεντροί εἰσι τῷ παντί, ἀλλ' ἔκκεντροι, ὡς εἶναι τὰ μὲν αὐτῶν ἀπόγεια μέρη, τὰ δὲ περίγεια

Καὶ ἡ μὲν τάξις τῶν εἰς τὰ ἑπόμενα περιόδων τῶν ἐναντίων τῇ πρώτῃ περιφορᾷ οὕτω δὴ καταλαμβάνεται. καταλαμβάνεται δὲ ἡ περίοδος τῶν εἰρημένων πέντε ἀστέρων τῶν πλανωμένων καὶ τοῦ ἡλίου καὶ τῆς σελήνης ἀποτελουμένη, οὐ καθ' ὁμοκέντρων κύκλων τῷ παντί, ἀλλὰ κατ' ἐκκέντρων. ὁ μὲν γὰρ ζωδιακὸς κύκλος νοεῖται ἀνωτάτω ὑπὲρ τὰς τοὺς ἀστέρας ἐχούσας σφαίρας, ὡς εἴρηται, λοξὸς διὰ τοῦ οὐρανίου σώματος, μέγιστος ὢν τῶν ἐν αὐτῷ κύκλων, ὡς καὶ τοῦτ' εἴρηται, καὶ διὰ τοῦτο πάντως καὶ τὸ αὐτὸ κέντρον ἔχων τῇ σφαίρᾳ. καὶ ὁμόκεντρος ὢν αὐτῇ καὶ περὶ τὸ τοῦ κόσμου παντὸς καὶ τοῦ οὐρανίου σώματος κέντρον, καὶ αὐτὸς νοεῖται τοῖς πᾶσιν αὐτοῦ μέρεσιν ἴσην ἔχων ἀπόστασιν ἀπ' αὐτοῦ τοῦ κέντρου, ἤτοι τῆς γῆς.

Τὴν γὰρ γῆν ἔφημεν ἀποδεδεῖχθαι τοῖς μαθηματικοῖς, διά τε τὸ μέγεθος τοῦ οὐρανοῦ καὶ τὴν ἀπ' αὐτῆς ἀπόστασιν καὶ τὸ αὐτῆς πρὸς τὸ οὐράνιον σῶμα ὀλιγομερὲς καὶ σχεδὸν παντάπασιν ἀνεπαίσθητον καὶ ἀλόγιστον, κέντρου καὶ σημείου λόγον πρὸς αὐτὸν ἔχειν τὸν οὐρανόν, κἀντεῦθεν καὶ πρὸς αὐτὸν ἄρα τὸν ἐν αὐτῷ μέγιστον εἰρημένον κύκλον, τὸν λοξὸν καὶ ζωδιακόν. τοιγαροῦν, ὅταν τὸ κέντρον αὐτοῦ θεωρῶμεν, τὴν γῆν αὐτὴν ὅλην λέγομεν καὶ νοοῦμεν κατὰ πᾶν μέρος αὐτῆς κέντρου λόγον πρὸς αὐτὸν ἔχουσαν, διὰ τὸν ἀπαράλληλον καὶ ἀπαραλόγιστον ὄγκον αὐτῆς ὀλιγομερέστατον, ὡς ἔφημεν, ὄντα πρὸς αὐτὸ τὸ παμμηκέστατον οὐράνιον σῶμα, καὶ τὴν ἑπομένην ἐξανάγκης, ἐπειδὴ σφαιρικός ἐστιν ὁ οὐρανὸς καὶ αὐτὴ ἡ γῆ, ἀπ' αὐτῆς ἀπόστασιν ἀλόγιστον καὶ ἄληπτον, ὡς εἰπεῖν. καὶ ἡμεῖς ἑστῶτες ἐπὶ τῆς γῆς, ἔνθα ἂν τῶν μερῶν αὐτῆς ἑστῶτες ὦμεν, πάντα τὰ φαινόμενα τοῦ ζωδιακοῦ μέρη ἐπίσης ἀφ' ἡμῶν ὡς ἀπὸ οἰκείου κέντρου ἀφεστῶτα ἔχομεν καὶ οὐδὲν μᾶλλον οὐδ' ἧττον τόδε ἢ τόδε, οὐδὲ τὸ μὲν τῶν τοῦ ζωδιακοῦ τμημάτων κατανοεῖν ἔχομεν ἡμῶν ἐγγύτερον καὶ προσγειότερον, τὸ δὲ ἀπογειότερον καὶ πορρωτέρω· ἐπὶ γὰρ τοῦ κέντρου αὐτοῦ, ὡς ἔφην,

14 All the circles which carry the Sun, the Moon and the five planets are not homocentric to the Universe, but eccentric, and how their apogees and perigees are arranged

This is the arrangement of the periods [for the spheres rotating] to the subsequent constellations, i.e. in opposite sense to the first revolution. The rotations of the five planets, the Sun and the Moon are understood to consist of circles, which are not homocentric to the universe but eccentric. The circle of the ecliptic is thought to be far above the spheres of the stars being a great inclined circle, as mentioned already. For this reason it has the same centre as the celestial sphere, being homocentric with the universe and with the centre of the celestial sphere, from which we understand that all its points are at equal distance from the centre, that is the Earth.

It has been proven by the mathematicians that the size of the Earth in comparison to the size and distance of the celestial sphere is very small, immeasurable and infinitesimal. It has the ratio of the centre [of a circle] and of a point relative to the celestial sphere and consequently to the major inclined circle of the ecliptic. Therefore, when we consider the centre, we refer to and mean every part of the Earth because of its negligible and infinitesimal volume relative to the incomparable size of the celestial body; this necessarily follows because the sky is spherical as also is the Earth itself, whose dimension is, so to speak, negligible. As we stand at whatever point on the Earth, we observe all points and phenomena on sections of the ecliptic from its inherent centre; all sections are equivalent to us and at equal distance, none of them being closer

ἑστῶτές ἐσμεν. καὶ ὁ μὲν λοξὸς καὶ ζωδιακὸς οὕτω δὴ καταλαμβάνεται, αἱ δὲ ὑπ᾿ αὐτὸν σφαῖραι καὶ οἱ κύκλοι ἐν αἷς περιοδεύουσι οἵ τε πλανώμενοι πέντε ἀστέρες καὶ ὁ ἥλιος καὶ ἡ σελήνη, ἅπαντες ἔκκεντροί εἰσι πρὸς αὐτὸν τὸν ζωδιακὸν καὶ κοινὸν οὐκ ἔχουσι κέντρον οὔτε πρὸς ἀλλήλους, ὡς προϊόντες ἐροῦμεν, οὔτε πρὸς αὐτὸν τὸν ζωδιακὸν οὐδὲ ὁμόκεντροί εἰσι καθόλου τῷ παντὶ κόσμῳ καὶ τῷ οὐρανίῳ σώματι.

Διὰ τοῦτο καὶ ἡμεῖς οἱ ἐπὶ τῆς γῆς ἑστῶτες ἐπὶ τοῦ κοσμικοῦ, ὡς ἔπος εἰπεῖν, κέντρου οὐ τὴν αὐτὴν καὶ ἴσην διάστασιν ἔχομεν πρὸς ἅπαντα τὰ τμήματα ἑνὸς ἑκάστου κύκλου τῶν τοιούτων ἀστέρων τῶν πέντε πλανωμένων καὶ τοῦ ἡλίου καὶ τῆς σελήνης, ἀλλὰ τινὰ μὲν τμήματα τῶν τοιούτων κύκλων, ἐφ᾿ ὧν, ὡς εἴρηται, φέρονται οἱ ἀστέρες, εἰσὶ πορρωτέρω ἡμῶν καὶ ἀπογειότερα, τὰ δὲ ἐγγύτερα καὶ προσγειότερα διὰ τὴν αὐτὴν αἰτίαν ἣν λέγομεν τῆς ἐκκεντρότητος. πρὸς μὲν γὰρ τὰ οἰκεῖα κέντρα οἱ τοιοῦτοι κύκλοι ἐξανάγκης διὰ πάντων τῶν οἰκείων τμημάτων ἴσην ἔχουσι τὴν ἀπόστασιν. πρὸς δὲ τὴν γῆν καὶ ἡμᾶς τοὺς ἐν αὐτῇ ἐν τῷ κοσμικῷ, ὡς ἔφημεν, κέντρῳ νοουμένους, οὐκ ἴσην πανταχόθεν ἔχουσι τὴν ἀπόστασιν διὰ τὴν αὐτῶν ἐκκεντρότητα, ἀλλὰ τινὰ μὲν τῶν αὐτῶν κύκλων μέρη ἐξανάγκης διὰ ταύτην τὴν αἰτίαν ἔγγιστα καταλαμβάνεται ἡμῶν, τινὰ δὲ πορρωτέρω.

or farther away. For we stand at its centre and observe the inclined of the ecliptic and the spheres and circles below it, with the Sun, the Moon and the five planets moving on eccentrics. They do not have a common centre with the ecliptic, nor with the celestial sphere, nor with each other.

For this reason, as we stand on the Earth, practically at the cosmic centre, we are not at equal distance from all points of the circles for each of the five planets or for the Sun or the Moon, but some are further away from us at the apogees and others are closer (at the perigees), because of the property we call eccentricity. By definition, all points on these circles are at equal distance from their own centre, but relative to the Earth and to us who stand at the cosmic centre they are not at equal distance due to the eccentricity. Consequently, some parts of the circles are closer and others farther away from us.

15 ὅτι τὰ τοιαῦτα ἀπόγεια καὶ περίγεια αὐτῶν οὐκ ἐπὶ τῶν αὐτῶν καταλαμβάνονται μορίων ἀεὶ τοῦ ζωδιακοῦ, ἀλλὰ διόλου μετατιθέμενα ἐν ἄλλοτε ἄλλοις

Καὶ οὕτω μὲν οἱ ἑπτὰ κύκλοι ἐφ' ὧν φέρονται οἱ πέντε ἀστέρες οἱ πλανώμενοι καὶ ὁ ἥλιος καὶ ἡ σελήνη οὐχ ὁμόκεντροι καταλαμβάνονται τῷ ζωδιακῷ, ἀλλ' ἔκκεντροι, ὡς εἶναι τάδε μὲν τὰ τμήματα αὐτῶν τὰ ὑποτείνοντα τάδε τὰ δωδεκατημόρια τοῦ ζωδιακοῦ, ἤτοι τὸν Κριὸν καθ' ὑπόθεσιν ἢ τὸν Λέοντα ἢ τὸν Τοξότην ἢ ἄλλο ὁτιοῦν, ἀπόγεια καὶ πορρωτέρω ἡμῶν, τὰ δὲ ὑποτείνοντα τὰ κατὰ διάμετρον τοῖς αὐτοῖς δωδεκατημορίοις εἶναι περίγεια καὶ ἐγγυτέρω ἡμῶν.

Κατελήφθησαν δὲ τοῖς μαθηματικοῖς διαφόροις τρόποις ἀποδείξεων, περὶ οὓς οὐκ ἂν ἔγκαιρον εἴη διατρίβειν, τὰ τοιαῦτα ἀπόγεια καὶ περίγεια τῶν πέντε πλανωμένων καὶ τῆς σελήνης, οὐκ ἐπὶ τοῦ αὐτοῦ τμήματος καὶ σημείου τοῦ ζωδιακοῦ ἀεὶ ἱστάμενα καὶ συμπεριφερόμενα, ἀλλὰ μεταβαίνοντα καὶ μετατιθέμενα ἐν ὡρισμένοις χρόνοις ὅσον ἑξῆς ἐφ' ἑκάστου εἰρήσεται, ὡς νῦν μὲν εἶναι τὸ ἀπόγειον, φέρε εἰπεῖν, τοῦ Κρόνου εἰς πρώτην μοῖραν τοῦ Κριοῦ καὶ τὸ περίγειον ἐξανάγκης κατὰ διάμετρον εἰς τὴν πρώτην μοῖραν τοῦ Ζυγοῦ, νῦν δὲ εὑρίσκεσθαι προχωρήσαντα εἰς τὰ ἑπόμενα τοῦ ζωδιακοῦ, οἷον εἰπεῖν εἰς δεκάτην μοῖραν ἢ εἰκοστὴν ἢ εἰκοστὴν ὀγδόην τοῦ Κριοῦ τὸ ἀπόγειον καὶ ἐκ διαμέτρου εἰς τὰς αὐτὰς τοῦ Ζυγοῦ τὸ περίγειον· ὡσαύτως καὶ τοῦ Διὸς καὶ τοῦ Ἄρεος καὶ τῆς Ἀφροδίτης καὶ τοῦ Ἑρμοῦ, ἔτι γὲ μὴν καὶ τῆς σελήνης.

Μετατίθεται μὲν γὰρ καὶ αὐτῆς τό τε ἀπόγειον καὶ τὸ περίγειον τοῦ ἐκκέντρου, πλὴν οὐκ εἰς τὰ ἑπόμενα, καθώσπερ ἔφημεν, καὶ τὰ τῶν ἄλλων, ἀλλ' εἰς τὰ ἡγούμενα. κοινὸν μὲν γὰρ ἔχει σὺν αὐτοῖς ὅτι μετατίθενται καὶ αὐτῆς τὰ τοιαῦτα μέρη τοῦ ἐκκέντρου αὐτῆς, τὸ ἀπόγειον καὶ τὸ περίγειον, ἤτοι ὁ ὅλος ἔκκεντρος. ἀλλ' ἐπ' ἐκείνων μὲν ἡ μετάθεσις τῶν ἐκκέντρων αὐτῶν εἰς τὰ ἑπόμενα καταλαμβάνεται τοῦ ζωδιακοῦ καὶ οἱονεὶ σύνδρομός ἐστιν ἡ τοιαύτη προχώρησις τῶν

22 καθώσπερ : καθὼς C[ac]

15 The apogees and perigees of the planets and the Moon are not always at the same place of the ecliptic but they move

Thus the seven circles which carry the five planets, the Sun and the Moon are not homocentric to the ecliptic but eccentric, with the positions of their apogees stretching further away at some locations of the ecliptic, let us say in Aries or Leo or Sagittarius or at whatever other place. Their perigees are closer to us, being located diametrically opposite on the ecliptic.

They have been determined with mathematical methods and proofs but this is not yet the appropriate time to discuss them. The apogees and perigees of the five planets and the Moon are not always at the same place of the ecliptic, but they move and relocate themselves at specific time intervals which we will describe later. Let us say the apogee of Saturn is at the moment in the first degree of Aries and the perigee by necessity diametrically opposite at the first degree of Libra. At another time we find that [the apogee] has advanced to the tenth or twentieth or the twenty eighth degree of Aries, with the perigee at the diametrically opposite degree of Libra. The same thing happens with Jupiter, Mars, Venus and Mercury, as well as with the Moon.

The Moon's apogee and perigee in the eccentric also moves, with the difference that the perigee of the Moon does not move forward to the subsequent but backwards to the preceding constellations. In addition, a common property [of the planets] is that all parts of the eccentric move together with the apogee, the perigee and all the eccentric.

ἀπογείων τῶν ἐκκέντρων αὐτῶν τῇ κατὰ φύσιν κινήσει αὐτῶν, ὅτι δὴ καὶ αὕτη, καθὼς προείπομεν, εἰς τὰ ἑπόμενά ἐστι καὶ τὸν τρόπον τοῦτον προκοπήν τινα δίδωσι καὶ προσθήκην τῇ κατ᾽ αὐτοὺς κινήσει τῶν ἀστέρων. ἐπὶ δὲ τῆς σελήνης τοὐναντίον ἔχει ἢ κατ᾽ αὐτοὺς τοὺς ἄλλους ἀστέρας, καὶ ἐναντίως καταλαμβάνεται μετατίθεσθαι τὸ εἰρημένον ἀπόγειον τοῦ ἐκκέντρου αὐτῆς, τῇ κατ᾽ αὐτήν, ὡς εἰπεῖν, φυσικῇ κινήσει καὶ περιόδῳ εἰς τὰ ἑπόμενα· εἰς γὰρ τὰ ἡγούμενα τοῦ ζωδιακοῦ τὸ τοιοῦτον ἀπόγειον αὐτῆς μεταχωρεῖ καὶ μετατίθεται.

Ἀλλὰ περὶ μὲν τούτων ὅπως ἔχουσι καὶ πῶς κατελήφθησαν τὰ μὲν τῶν ἄλλων πέντε πλανωμένων ἀστέρων ἀπόγεια εἰς τὰ ἑπόμενα συγκινούμενα τῇ κατ᾽ αὐτοὺς οἰκείᾳ κινήσει—καὶ ὁπόσον ἐν τοῖς ὡρισμένοις χρόνοις ταῦτα μετατίθενται καὶ προχωροῦσιν εἰς τὰ ἑπόμενα καὶ κατὰ ποίων μερῶν τοῦ ζωδιακοῦ τὰ τοιαῦτα ἀπόγεια τῶν ἐκκέντρων αὐτῶν τῶν ἀστέρων καὶ τὰ κατὰ διάμετρον αὐτοῖς πάντως περίγεια ἐν τοῖς καθ᾽ ἡμᾶς τούτοις χρόνοις εὑρίσκεσθαι ἐπιλογιζόμεθα—τὰ δὲ τοῦ ἐκκέντρου τῆς σελήνης ἀπόγεια καὶ περίγεια εἰς τἀναντία μετατιθέμενα, ἤτοι εἰς τὰ ἡγούμενα, ἐν τοῖς ἑξῆς κατὰ μέρος ἐροῦμεν πρὸς τὴν ἀπαιτοῦσαν ἀκολουθίαν καὶ χρῆσιν ἀπαντῶντες.

The precession for all of them is to the subsequent [constellations], as if running together with their natural rotation; it is forward and this way it adds to and increases the speeds of the stars. [The precession] of the Moon is in the reverse direction to that of the other stars with its apogee moving in the opposite sense to its natural rotation. The apogee moves and relocates itself to the preceding sings of the zodiac.

[Concerning the nature of the precessions], how we understood that the apogees of the five planets move forward and in the same sense as their inherent motion — how much they advance at a given time and in which positions of the ecliptic the apogees of the eccentrics and the diametrically opposite perigees are located in our own times will be calculated —, as well as the precession of the apogee and perigee of the Moon, moving in opposite sense [to the proceeding constellations], will be discussed separately with due attention to useful applications.

16 ὅτι τὸ τῆς ἡλιακῆς σφαίρας μόνης ἀπόγειον καὶ περίγειον ἐπὶ τῶν αὐτῶν μορίων τοῦ ζωδιακοῦ εἰσιν ἀεὶ ἀκίνητα, καὶ ὅτι μόνη ἡ πάροδος τοῦ ἡλίου ἀεὶ ἐπὶ τοῦ μέσου ἐστὶ τοῦ ζωδιακοῦ, αἱ δὲ τῶν ἄλλων πάροδοι παρεκκλίνουσιν ἐφ' ἑκάτερα τοῦ μέσου καὶ βόρεια καὶ νότια

Τὸ μέντοι τοῦ ἡλιακοῦ ἐκκέντρου ἀπόγειον καὶ περίγειον οὐχ ὡσαύτως τοῖς ἄλλοις κινούμενα καὶ μετατιθέμενα καὶ αὐτὰ καταλαμβάνεται ἢ εἰς τὰ ἑπόμενα ἢ εἰς τὰ ἡγούμενα, ἀλλὰ μένοντα καὶ ἀεὶ ὥσπερ ἑστῶτα καὶ ἀμετάθετα ἐπὶ τῶν αὐτῶν τοῦ ζωδιακοῦ τμημάτων. κατελήφθη δὲ τὸ μὲν ἀπόγειον τοῦ ἐκκέντρου τοῦ ἡλίου δι' ἀποδείξεων θαυμασίων καὶ ἀναντιρρήτων παρὰ τοῦ Πτολεμαίου ἀείποτε εὑρισκόμενον ἐν Διδύμων μοίραις ε λʹ, τὸ δὲ περίγειον κατὰ διάμετρον πάντως ἐν Τοξότου μοίραις ε λʹ. καὶ ἔστι τοῦτο ἐπὶ τοῦ ἡλίου ἰδιότροπον καὶ θαυμάσιον καὶ ξένον καὶ παρηλλαγμένον ἢ κατὰ τοὺς ἄλλους, μᾶλλον δὲ ἀνενδεές, ὡς εἰπεῖν, καὶ βασιλικὸν ὥσπερ καὶ ἄτρεπτον, καθὼς δὴ καὶ ἀλλάττα περὶ αὐτὸν θεωρεῖται διαφορώτερα καί, ὡς ἂν εἴποι τις, γεννικώτερα καὶ μεγαλοπρεπέστερα καὶ ἁπλούστερα ἢ κατὰ τοὺς ἄλλους.

Οἷον οἱ μὲν πέντε ἀστέρες οἱ πλανώμενοι καὶ ἡ σελήνη οὐ παροδεύουσιν ἀεὶ ἐπὶ τοῦ μέσου τοῦ ζωδιακοῦ, ἀλλὰ καὶ ἐφ' ἑκάτερα αὐτοῦ γίνονται καὶ βόρεια καὶ νότια. ὁ δὲ ἥλιος αὐτὸς τὰ μέσα παροδεύει τοῦ ζωδιακοῦ διαπαντὸς κατ' οὐδέτερα παρεκκλίνων, ἀλλ' ὡς ἀληθῶς ὁδῷ μέσῃ καὶ ὄντως βασιλικῇ χρώμενος καὶ διϊὼν ἁπλοϊκῶς οὕτως καὶ περιπολευόμενος ἅπαντα τὸν αἰῶνα. ἔτι γὲ μὴν αὐτὸς καὶ μέσος ἐστὶ καὶ ἡ κατ' αὐτὸν κίνησις τῶν ὅλων σὺν αὐτῷ οὐρανίων ἑπτὰ σωμάτων, ἃ κατὰ τὸ οὐράνιον βάθος, ὡς προέφημεν, ἑξῆς ἀλλήλων καὶ ὑπ' ἄλληλα περιοδεύει ὑπὸ τὸν ζωδιακὸν δι' ἐκκέντρων πρὸς αὐτὸν κύκλων. τρία γάρ, ἐν οἷς περὶ τῆς τάξεως αὐτῶν καὶ τοῦ βαθμοῦ

5-6 Cf. Ptol. *Alm.* 3.4 (vol. 1.1, p. 237 Heiberg), 3.7 (vol. 1.1, p. 255.21-256.1 Heiberg)

5 post ἡλίου add. ἀεὶ ἐπὶ τοῦ μέσου ἐστὶ τοῦ ζωδιακοῦ. αἱ δὲ τῶν ἄλλων πάροδοι παρεκκλίνουσιν ἐφ' ἑκάτερα τοῦ μέσου καὶ βόρεια καὶ νότια V in mg. (manus secunda) **7** ἐν : τοῦ V **10** ἀλλάττα scripsi : ἄλλάττα V C **11** εἴποι : εἴπῃ C **14** τοῦ om. C

16 Only the apogee and the perigee of the solar sphere are always at the same place on the zodiac, and only the trajectory of the Sun is always in the middle of the zodiac. The trajectories of the others deviate from the centre, being at some times to the north and at other times to the south

And so the apogee and perigee of the Sun's eccentric does not move, nor is it displaced to the subsequent or preceding constellations like the rest. It remains stationary and unchanged at the same place of the ecliptic as has been demonstrated by Ptolemy, with remarkable and reliable proofs that the apogee of the eccentric is always located in the sign of Gemini at 5° 30′ and the perigee at the diametrically opposite position in Sagittarius at 5° 30′. This is considered to be a special, remarkable and unusual property of the Sun, distinctly different from the stars, just as many things about the Sun are very different and, so to speak, nobler, more magnificent and simpler than for the other stars.

Thus, the five planets and the Moon do not always rotate through the middle of the ecliptic, but deviate on both sides, sometimes to the north and other times to the south. The Sun however always rotates through the middle of the ecliptic without any deviation. It follows the middle and noble path eternally. And indeed the Sun follows a median path among the seven celestial bodies, which in the celestial depth traverse their eccentric circles close to the zodiac. Three among the celestial bodies, according to order and degree,

προολίγου ἐλέγομεν, τὰ ὑπεράνω αὐτοῦ, Κρόνος, Ζεύς, Ἄρης, καὶ τρία τὰ ὑποκάτω, Ἀφροδίτη, Ἑρμῆς, σελήνη. καὶ δόξα τοιαύτη παρὰ τοῖς πλείοσι τῶν ἀστρονόμων κρατεῖ, ᾗ καὶ Πλάτων καὶ Ἀριστοτέλης συντίθεται, ἐπ᾽ ἀμφότερα τὰ μέρη ἐπίσης διανενεμημένα παρ᾽ αὐτοῦ καὶ τοῖς ἓξ σώμασι δεσποτικῶς ἐκ τοῦ μέσου, ὡς εἰπεῖν, καὶ κραταιῶς τὸν φωτισμὸν διαδίδοσθαι.

Ἔτι γὲ μὴν πρὸς τούτῳ ἐπί τε τῶν ἄλλων ἓξ τούτων οὐρανίων σωμάτων καὶ τῶν κατ᾽ αὐτὰ περιόδων νοεῖται καὶ ὁμαλὴ κίνησις καὶ ἀνώμαλος καὶ ἐπ᾽ αὐτοῦ τοῦ ἡλίου. ἀλλ᾽ ἐπὶ μὲν τοῦ ἡλίου μονότροπός ἐστι ἡ ἀνώμαλος αὐτοῦ κίνησις καὶ ἁπλοϊκή, ὥσπερ καὶ τἄλλα τὰ κατ᾽ αὐτὸν ἔφημεν ἁπλοϊκά, ἐπὶ δὲ τῶν ἄλλων οὐ καταλαμβάνεται μία καὶ ἁπλῆ ἡ ἀνωμαλία τῆς κινήσεως αὐτῶν, ἀλλὰ ποικίλη καὶ πολύτροπος καὶ διαφορὰς ἔχουσα, ὡς κατὰ καιρὸν περὶ τούτων διαληψόμεθα.

25 Πλάτων καὶ Ἀριστοτέλης : Ἀριστοτέλης καὶ Πλάτων C[ac] **27** ἐκ τοῦ μέσου post ὡς εἰπεῖν transp. C

are above the Sun: Saturn, Jupiter and Mars. Three others are below: Venus, Mercury and the Moon. This opinion is held by most astronomers, with Plato and Aristotle among them, that the six celestial bodies are distributed three on either side of the Sun, which occupies a central position, powerfully dispersing its light.

Furthermore, it is understood that the motions and periods of the other six celestial bodies and that of the Sun contain uniform and anomalous components, with the exception that, for the Sun, the anomalous motion is achieved in one step [i.e. with one epicycle] and is simple like its other properties. For the other celestial bodies, the anomaly in the rotations is not one and simple, but more elaborate, diverse and different as we shall explain at the appropriate place.

17 ὅτι ἐπὶ τοῦ ἡλίου καὶ τῆς σελήνης καὶ τῶν πέντε πλανωμένων ἀστέρων θεωροῦνται ὁμαλαὶ καὶ ἀνώμαλοι κινήσεις, καὶ ὅτι κατὰ δύο τρόπους αἱ ἀνώμαλοι καταλαμβάνονται τῶν ἀστέρων κινήσεις, κατά τε τὸ ἐπὶ ἐκκέντρων, ὡς εἴρηται, φέρεσθαι καὶ κατὰ τὸ ἐπὶ ἐπικύκλων ἰδίᾳ πάλιν αὐτοὺς παροδεύειν τοὺς ἀστέρας

Ὅπως μέντοι καταλαμβάνονται αἵ τε ὁμαλαὶ κινήσεις τῶν ἀστέρων καὶ αἱ ἀνώμαλοι διευκρινῆσαι λοιπὸν ἀναγκαῖον. κατὰ δύο τοίνυν τρόπους, καθόλου γε ἐρεῖν καὶ ὁλοσχερέστερον, ἡ τοιαύτη ἀνωμαλία τῆς κινήσεως τῶν ἀστέρων καταλαμβάνεται γίνεσθαι, ἤτοι διὰ τὴν ἐκκεντρότητα τῶν κύκλων ἐφ᾽ ὧν φέρονται ὡς πρὸς τὸν κύκλον τὸν ζωδιακόν, μᾶλλον δὲ τὸν ὁμόκεντρον αὐτῷ τῷ ζωδιακῷ ἐφ᾽ ἑκάστης ἐπισκέψεως καὶ θεωρίας αὐτῶν τῶν ἀστέρων· ὅτι δηλαδὴ ἡ περίοδος ἑνὸς ἑκάστου αὐτῶν ἐπὶ κύκλων γίνεται, ὧν τὸ κέντρον ἑκάστου διάφορον ἔχει τὴν θέσιν πρὸς τὸ κέντρον τοῦ ὅλου οὐρανίου σώματος καὶ τοῦ ζωδιακοῦ πάντως, ἤγουν αὐτῆς τῆς γῆς καὶ ἡμῶν τῶν ἐπὶ γῆς θεωρούντων, ἢ ὅτι ἐπὶ ἐπικύκλων καταλαμβάνονται φερόμενοι αὐτοὶ οἱ ἀστέρες.

Καὶ οἱ μὲν ἐπίκυκλοι οὗτοι φέρονται ὁμαλῶς ἐπ᾽ αὐτῶν τῶν κύκλων· τὰ γὰρ κέντρα τῶν τοιούτων ἐπικύκλων ἐφάπτονται ἀεὶ τῶν τοιούτων κύκλων, αὐτοὶ δὲ οἱ ἀστέρες καταλαμβάνονται, ὡς λέγομεν, παροδεύοντες καὶ φερόμενοι ἐπ᾽ αὐτῶν τῶν ἐπικύκλων. καὶ λοιπὸν ἄλλοτε ἄλλας τὰς θέσεις ἐπὶ τῶν ἐπικύκλων ἔχοντες ποιοῦσι διὰ τούτων διαφορὰς τῶν τόπων αὐτῶν ἐπὶ τῶν κύκλων αὐτῶν, ἀλλὰ δὴ καὶ τοῦ ζωδιακοῦ, τῶν μὲν κέντρων τῶν ἐπικύκλων, ἤτοι τῶν ἐπικύκλων αὐτῶν, ὁμαλῶς καὶ ἴσως ἐν τοῖς ἴσοις χρόνοις, ὡς ἔφημεν, παροδευόντων εἰς τὰ ἑπόμενα τοῦ κύκλου ἐφ᾽ οὗ φέρονται, τῶν δὲ ἀστέρων αὐτῶν ἐν τῇ ἐπ᾽ αὐτῶν τῶν ἐπικύκλων παρόδῳ ἄλλοτε ἄλλας ἐχόντων τὰς θέσεις, νῦν μὲν προλαμβανούσας τὰ κέντρα τῶν ἐπικύκλων, νῦν δὲ ὄπισθεν αὐτῶν

17 post ἐπὶ add. τῶν κύκλων αὐτῶν, ἀλλὰ δὴ καὶ V^{ac} C **21** post τοῦ add. ἐκκέντρου C^{ac} V^{ac}

17 On the motions of the Sun, the Moon and the five planets - some of them are considered to be regular and others irregular. The irregular motions of the stars are understood in two ways: they are carried along on the eccentric circles, as we have mentioned already, or they move on their epicycles

Since the rotations of the stars have uniform and anomalous components it is necessary to specify them. Generally speaking, the anomaly of the motion is understood in two ways, namely through the eccentricity of the circles on which [the stars] are carried along relative to the ecliptic, or to say it better relative to a circle homocentric to the ecliptic. Each periodic rotation takes place either on circles whose centres are different from the centre of the stellar globe and of the ecliptic — consequently different from the Earth and from us who observe them — or the stars are understood to rotate on epicycles.

The epicycles rotate uniformly on the own circles, with their centres remaining attached into these circles, while the stars themselves are understood to rotate and be carried on the epicycles. Consequently, they occupy various positions on the epicycles and also assume different positions relative to the ecliptic. The centres of the epicycles move smoothly and uniformly, and as we said, they cover equal intervals in equal times as they proceed to the subsequent parts of the circle on which they rotate. The stars on their revolutions on the epicycles assume different positions: sometimes they run ahead and other times lag behind the centres of the epicycles. It follows from this that

καταλαμβανομένας· ὡς ἐντεῦθεν πάντως ἀκολουθεῖν ἄλλην εἶναι τὴν φαινομένην ἐποχὴν καὶ τὸν τόπον τοῦ ἀστέρος αὐτοῦ ἐπί τε τοῦ κύκλου αὐτοῦ καὶ τοῦ ζωδιακοῦ, καὶ ἄλλην τὴν τοῦ κέντρου τοῦ ἐπικύκλου αὐτοῦ, καθ' ἣν νοεῖται ἡ ὁμαλὴ κίνησις τοῦ ἀστέρος. καὶ περὶ τούτου μέν, ἤτοι τῆς κατὰ τοὺς ἐπικύκλους ὑποθέσεως, ποιησόμεθα τὸν λόγον ὕστερον ἐντελέστερον διευκρινοῦντες ὅπως ἐντεῦθεν ἡ ἀνωμαλία τῆς κινήσεως τῶν ἀστέρων ἀποτελεῖται· νῦν δέ γε πρότερον περὶ τῆς διὰ τὴν ἐκκεντρότητα ἀνωμαλίας ἐροῦμεν.

Οἱ ἀστέρες ἐπὶ τῶν οἰκείων κύκλων τῶν ἐκκέντρων κινούμενοι ἀεὶ τὰ αὐτὰ κινήματα ἐν τοῖς ἴσοις χρόνοις ποιοῦνται καὶ ἴσα, οἷον τὴν αὐτὴν περιφέρειαν καὶ ἴσην παροδεύουσιν ἐπὶ τῶν οἰκείων ἐκκέντρων, ἐν ἑκάστῳ ἰσημερινῆς ὥρας χρόνῳ καὶ ἐν ἑκάστῳ ἡμέρας καὶ ἔτους χρόνῳ, καὶ οὐ νῦν μὲν πλέον, νῦν δὲ ἔλαττον. ἐπεὶ δὲ τὰ τοιαῦτα κινήματα αὐτῶν καταλογίζονται ἡμῖν τοῖς ὁρῶσιν ἀπὸ τοῦ κοσμικοῦ τοῦδε κέντρου αὐτῆς τῆς γῆς, οὐκ ἐπὶ τῶν κύκλων αὐτῶν τῶν ἐκκέντρων, ἀλλ' ἐπὶ τοῦ ζωδιακοῦ, καὶ τοὺς τόπους τῶν ἀστέρων ἐπ' αὐτοῦ τοῦ ζωδιακοῦ λέγομεν καὶ ἀποφαινόμεθα, οἷον φέρε εἰπεῖν, ὡς νῦν μέν ἐστιν ὁ Κρόνος ἐν Ὑδροχόου μοίρᾳ δεκάτῃ καὶ ὁ Ζεὺς ἐν Λέοντος εἰκοστῇ, νῦν δὲ ὁ Κρόνος ἐν Κριοῦ πρώτῃ, ὁ δὲ Ζεὺς ἐν Τοξότου πέμπτῃ ἢ ἄλλαις δή τισιν, ἐξανάγκης διαφοραὶ τῶν κινημάτων αὐτῶν τῶν ἀστέρων ἐντεῦθεν καταλαμβάνονται. καὶ ἀκολουθεῖ διὰ τὸν τρόπον τοῦτον μὴ πάντοτε τὰ κινήματα αὐτῶν εἶναι ἴσα ὡς πρὸς τὸν ζωδιακόν, ἀλλὰ καὶ πλείονα καὶ ἐλάττονα, διὰ τὸ νῦν μὲν αὐτοὺς εἶναι ἐν τοῖς ἀπογείοις τῶν ἐκκέντρων αὐτῶν, νῦν δὲ ἐν τοῖς περιγείοις, νῦν δὲ ἐν τοῖς μέσοις τόποις.

Ποιησόμεθα δὲ σαφέστερον τὸν λόγον ἐπὶ ὑποδείγματος καὶ καταγραφῆς. ἔστω ὁμόκεντρος τῷ παντὶ καὶ τῷ ζωδιακῷ κύκλῳ ὁ αηβγχδ, κέντρον δὲ αὐτοῦ τὸ θ. ἕτερος δὲ κύκλος ἴσος αὐτῷ, ἔκκεντρος δὲ πρὸς αὐτόν, ὁ κλβζφδ, ἐφ' οὗ φέρεται τῶν ἀστέρων τις (ἔστω ὁ Ζεύς), κέντρον δὲ αὐτοῦ τὸ μ. οὗτοι δὴ οἱ κύκλοι ἔκκεντροι πρὸς ἀλλήλους ὄντες καὶ τέμνοντες ἀλλήλους κοινοὺς ἔχουσι συνδέσμους τὸ β καὶ τὸ δ, ὡς ἐντεῦθεν τοῦ ἑνὸς κύκλου τοῦ ἐκκέντρου τὴν μὲν εἶναι περιφέρειαν ὡς πρὸς ἡμᾶς τοὺς ἐπὶ τοῦ κέντρου ὄντας τοῦ ὁμοκέντρου τῷ ζωδιακῷ τῆς γῆς, ἤτοι τοῦ θ ἐγγυτέρω καὶ προσγειοτέραν, ἤτοι

the apparent longitude and position of the star is different in its own circle and on the ecliptic and different from the centre of the epicycle which defines the uniform rotation of the star. The subject of the epicycles will be discussed later in greater detail, clarifying how the anomaly of the motion is created. Now we shall describe the anomaly created by the eccentricity.

The stars moving on their eccentric circles traverse equal intervals at equal times, i.e. they travel equal arcs in each equinoctial hour, or in each day, or in each year and never faster or slower. But they appear irregular, since we observe them from the cosmic centre on the Earth, being also centre of the ecliptic, and not from the centre of the eccentric circle. We say and describe their positions on the ecliptic, for example by saying that Saturn is at this moment 10° in the constellation of Aquarius and Jupiter at 20° in Leo; now Saturn is in the first degree of Aries and Jupiter 5° of Sagittarius, or that they are found at a different position of their motion. In this manner their revolution on the ecliptic is not always equal, but larger or smaller because sometimes they are at the apogees of their eccentric circles, sometimes at the perigees and other times in between.

We can make this argument more evident with an example and a diagram. Let the circle αηβγχδ with its centre at θ be homocentric to the universe and the ecliptic. Another circle, the κλβζφδ, with centre at μ of the same size but eccentric, is where the star (let us say Jupiter) is transported. These two circles are eccentric relative to each other and intersect at the nodes β and δ. It follows now that the arc βλκδ of the eccentric circle is closer to us, who are located on the Earth, and at the centre of the homocentric to the

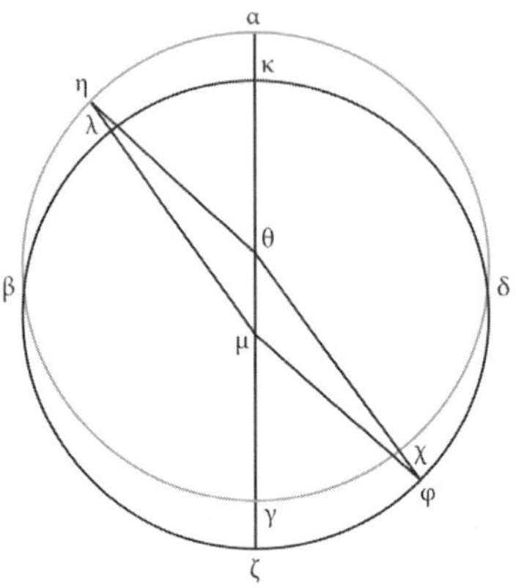

τὴν βλκδ, τὴν δὲ εἶναι πορρωτέρω καὶ ἀπογειοτέραν πρὸς ἡμᾶς, ἤτοι τὴν βζφδ. ὁ μὲν οὖν ἀστὴρ ὁ Ζεύς, ὡς ἔφημεν, τὸν οἰκεῖον ἔκκεντρον περίεισι τὸν κλβζφδ ὁμαλῶς, ἤτοι ἐν ἴσοις χρόνοις ἶσα τοῦ τοιούτου περιφερείας διέξεισι. θεωρούμενος δὲ ἐπὶ τοῦ ὁμοκέντρου τῷ ζωδιακῷ, ἐφ᾽ οὗ λαμβάνομεν τὰς ἐποχὰς καὶ τοὺς τόπους αὐτοῦ, οὐκ ἀείποτε ἐν τοῖς αὐτοῖς καὶ ἴσοις χρόνοις τὰς ἴσας περιφερείας τοῦ αὐτοῦ ὁμοκέντρου τῷ ζωδιακῷ καταλογίζεται περιοδεύων, οὐδὲ ἐν τοῖς αὐτοῖς χρόνοις τὰς ἴσας καταλογίζεται κινεῖσθαι ἐπί τε τοῦ οἰκείου ἐκκέντρου κύκλου καὶ τοῦ ὁμοκέντρου τῷ ζωδιακῷ, καίτοι γε ἴσων ὑποτιθεμένων τῶν τοιούτων κύκλων, ἀλλὰ ἀνίσους. ἐν μὲν γὰρ τοῖς ἀπογείοις τοῦ οἰκείου ἐκκέντρου κινούμενος πλείονα ἐπιλαμβάνει περιφέρειαν τοῦ οἰκείου ἐκκέντρου κύκλου ἢ τοῦ ζωδιακοῦ, ἐν δὲ τοῖς περιγείοις τοῦ οἰκείου ἐκκέντρου κύκλου περιϊὼν τὸ ἐναντίον ἐν τοῖς αὐτοῖς χρόνοις πλείονα περιφέρειαν ἐπιλαμβάνει ἐπὶ τοῦ ζωδιακοῦ ἤπερ ἐπὶ τοῦ ἰδίου ἐκκέντρου κύκλου.

Κεκινήσθω γὰρ ὁ ἀστὴρ ἀπὸ τοῦ περιγείου σημείου τοῦ ἐκκέντρου αὐτοῦ τοῦ κ ἐν ἡμέραις ξ ἐπὶ τοῦ ἰδίου ἐκκέντρου τὴν κλ οὖσαν ὡς ἐν ὑποθέσει μοιρῶν ι. καὶ κείσθω εὐθεῖα διήκουσα ἀπὸ τοῦ κέντρου τοῦ παντός, ἤτοι τῶν ὀμμάτων ἡμῶν κατὰ τὴν ἀρχὴν τῆς κινήσεως, ἡ θκ καταντῶσα εἰς τὸν ὁμόκεντρον τῷ ζωδιακῷ καὶ τῷ παντὶ κατὰ τὸ α σημεῖον, ἑτέρα δὲ εὐθεῖα ἀπὸ τοῦ αὐτοῦ κέντρου, ἤτοι τῶν ὀμμάτων ἡμῶν, ἡ θλ καταντῶσα εἰς τὸ η σημεῖον ἐπὶ τοῦ ὁμοκέντρου τῷ παντί, ἤτοι τῷ ζωδιακῷ. καταλογίζεται οὖν κεκινῆσθαι ἐν τῷ τοιούτῳ χρόνῳ ὁ ἀστὴρ ἐπὶ τοῦ ὁμοκέντρου τῷ ζωδιακῷ τὴν αη περιφέρειαν, ἤτοι τὴν ὑπὸ αθη γωνίαν, ἐπὶ δὲ τοῦ οἰκείου ἐκκέντρου τὴν κλ περιφέρειαν, ἤτοι τὴν ὑπὸ κμλ γωνίαν. καὶ ἔστιν ἡ κλ περιφέρεια ἐλάττων τῆς αη διὰ τὸ τριακοστὸν δεύτερον θεώρημα τοῦ πρώτου Στοιχείου τοῦ Εὐκλείδου,

81-84 κατὰ τὴν ἀρχὴν ... ὀμμάτων ἡμῶν om. C^{ac} (homoeotel.)

zodiac, i.e. it is closer to θ. The other arc βζφδ is further away from us. As we said, Jupiter travels its own eccentric κλβζφδ, and at equal times traverses equal segments of the circumference. Looking at it on the homocentric to the ecliptic where the longitudes and positions are defined, it never appears to traverse equal arcs at equal times; nor is Jupiter observed to travel in equal time intervals equal arcs of the eccentric and the homocentric, in spite of the fact that these circles are assumed to be equal. When it moves at the apogee of its eccentric, it appears to cover larger arcs on the eccentric than on the zodiac; at the perigee it appears to cover larger arcs on the zodiac than on the eccentric.

Let the star travel in 60 days from point κ on the perigee of its eccentric, the arc κλ corresponding to 10° degrees. Let us draw a line from the centre of the universe, where we are, to the beginning of the movement, the line θκ crossing the homocentric to the ecliptic at point α. Another line θλ is drawn from the same centre, i.e. from our position, crossing the ecliptic at the point of observation, and ends up at the point η. We consider that in this time interval the star travelled on the ecliptic the arc αη, that is the angle αθη; however, on the eccentric it travelled the arc κλ, that is the angle κμλ. Then the arc κλ is smaller than the arc αη according to the thirty-second theorem in the first

ἐν ᾧ φησί· ʽπαντὸς τριγώνου μιᾶς τῶν πλευρῶν προσεκβληθείσης ἡ ἐκτὸς γωνία ἴση ἐστὶ δυσὶ ταῖς ἐντὸς καὶ ἀπεναντίας᾽.

Καὶ τοίνυν ἐνταῦθα τριγώνου τοῦ μηθ ἡ ἐκτὸς γωνία ἡ ὑπὸ ηθκ δυσὶ ταῖς ἐντὸς καὶ ἀπεναντίας τῇ τε ὑπὸ θημ καὶ τῇ ὑπὸ ημθ ἴση ἐστίν, ἀλλ᾽ ὑπὸ ημθ, τουτέστιν ἡ ὑπὸ λμθκ, ὑποτείνει περιφέρειαν τοῦ ἐκκέντρου ὡς ὑπόκειται, ἐφ᾽ οὗ φέρεται ὁ ἀστὴρ μοιρῶν δέκα. ἔσται οὖν καὶ ἡ γωνία τῶν αὐτῶν μοιρῶν δέκα, ἐπειδὴ καὶ ἴση ἐστὶν ἡ γωνία τῇ περιφερείᾳ, ἣν ὑποτείνει καὶ ἐφ᾽ ἧς βέβηκεν· ὥστε ἡ ὑπὸ ηθα μείζων ἔσται τῆς ὑπὸ ημκ τῇ ὑπὸ θημ, ἤτοι ἡ ἀνώμαλος καὶ ἐπὶ τοῦ ζωδιακοῦ τῆς ὁμαλῆς κινήσεως τῆς ἐπὶ τοῦ ἐκκέντρου τοῦ ἀστέρος καταλογιζομένης ἐν τοῖς περιγείοις τοῦ ἐκκέντρου παροδεύοντος τοῦ ἀστέρος. καθ᾽ ὅσον δέ ἐστι μείζων ἐφ᾽ ἑκάστων τμημάτων τῶν ἀπὸ τοῦ περιγείου τοῦ ἐκκέντρου ἡ ἀνώμαλος τῆς ὁμαλῆς, ἤγουν ὅση ἐστὶν ἡ διαφορὰ αὐτῶν ἡ ὑπὸ μηθ γωνία καὶ ὅσων μοιρῶν ἢ λεπτῶν ἢ καὶ μοιρῶν καὶ λεπτῶν ἐστιν αὐτὴ ἡ γωνία, γραμμικαῖς ἀποδείξεσιν ἀνευρὼν ὁ Πτολεμαῖος ἔπειτα καὶ κατὰ μέρος κανονικῶς ἐκτίθεται ἐφ᾽ ἑκάστων τμημάτων, οἷον ὅταν ἀπέχῃ ὁ ἀστὴρ ἕκαστος, ὡς νῦν ἐπὶ ταύτης τῆς ὑποθέσεως ὁ Ζεύς, ἀπὸ τοῦ περιγείου τοῦ ἐκκέντρου αὐτοῦ, ἤτοι τοῦ κ, ἢ δέκα μοίρας ὡς ὑπεθέμεθα ἢ εἴκοσιν ἢ τριάκοντα ἢ ὁσασδήποτε. τὸ ἀνάπαλιν δέ, ὡς προείρηται, ἐν τοῖς ἀπογείοις τοῦ οἰκείου ἐκκέντρου παροδεύων ὁ ἀστὴρ πλείονα περιφέρειαν ἐπιλαμβάνει αὐτοῦ τοῦ ἐκκέντρου ἢ τοῦ ὁμοκέντρου τῷ ζωδιακῷ ἐν τοῖς αὐτοῖς χρόνοις.

90 Eucl. *Elem.* 1.32a **91** ἴση ... ἀπεναντίας : δυσὶ ταῖς ἐντὸς καὶ ἀπεναντίον ἴση ἐστίν Eucl. **92-103** add. in mg. C

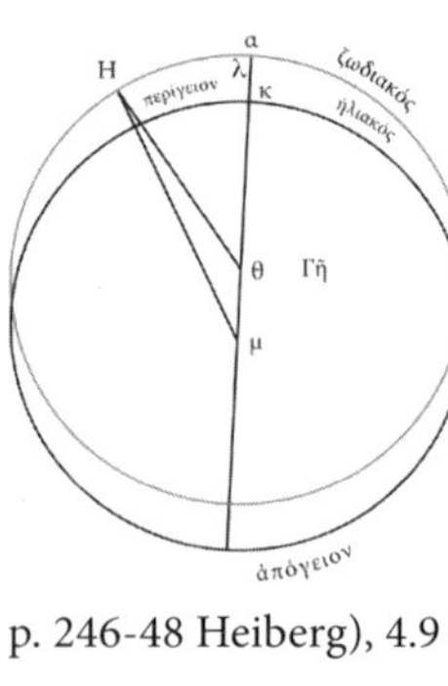

104 Cf. Ptol. *Alm.* 3.5 (vol. 1.1, p. 246-48 Heiberg), 4.9 (vol. 1.1, p. 335 Heiberg)

book of the *Elements* of Euclid, where it is stated that, “for any triangle, when one side is extended, the external angle is equal to the sum of the two internal angles on the opposite side”.

For our case the external angle ηθκ of the triangle μηθ is equal to the sum of the two internal and opposite angles θημ plus ημθ. The angle ημθ, which is the same as λμθκ, subtends on the eccentric an arc equal to 10°, which is the arc the star travelled on the circumference. It follows now that the angle ηθα is larger than ημκ by the angle θημ. As the star travels close to the perigee, the anomalous motion on the ecliptic [is larger] than the uniform on the eccentric, since it is larger for every segment, and their difference is the angle μηθ, which is expressed in either degrees or minutes or degrees and minutes. Using geometrical methods Ptolemy computed this angle and recorded the values for each segment and for each star, in our case Jupiter, how far each star is from its perigee of the eccentric, that is from point κ; if it is 10°, 20°, 30° or whatever. Alternatively, as we mentioned already, in the apogees a star travels in equal time larger arcs on the eccentric than on the ecliptic.

Ἐν γὰρ τῇ αὐτῇ πάλιν καταγραφῇ κεκινήσθω ὁ ἀστὴρ ἀπὸ τοῦ ζ, ὃ ὑπεθέμεθα ἀπόγειον εἶναι τοῦ ἐκκέντρου αὐτοῦ, ἐν ταῖς ἴσαις ἡμέραις ταῖς ἀνωτέρω εἰρημέναις ἑξήκοντα τὴν ζφ περιφέρειαν ἴσην πάντως τῇ κλ δέκα μοιρῶν, ἐπειδὴ καὶ ἔφημεν ἐν τοῖς ἴσοις χρόνοις τὰ ἶσα κινεῖσθαι καὶ ὁμαλῶς κινεῖσθαι τοὺς ἀστέρας ἐπὶ τῶν οἰκείων ἐκκέντρων, ταύτην τὴν περιφέρειαν ὑποτείνει γωνία πάλιν, ἡ ὑπὸ φμζ, ἥτις καὶ αὐτὴ τῶν ἴσων ἐστὶ πάντως δέκα. ὅθεν δὴ καὶ τριγώνου τοῦ θφμ διὰ τὸ αὐτὸ τὸ ἀνωτέρω εἰρημένον τριακοστὸν δεύτερον θεώρημα τοῦ πρώτου Στοιχείου δυσὶ ταῖς ἐντὸς καὶ ἀπεναντίας τῇ ὑπὸ φμγ, ἤτοι τῇ ὑπὸ χθγ καὶ τῇ ὑπὸ μφθ, ἴση ἐστίν. ἔστιν ἄρα ἡ ὑπὸ ζμφ τῆς ὁμαλῆς παρόδου τοῦ ἀστέρος γωνία ἐνταῦθα ἐπὶ τοῦ ἀπογείου τοῦ ἐκκέντρου μείζων τῆς ὑπὸ γθχφ γωνίας τῆς ἀνωμάλου παρόδου τοῦ ἀστέρος, τῆς καταλογιζομένης δηλονότι ἐπ᾽ αὐτοῦ τοῦ ὁμοκέντρου τῷ ζωδιακῷ τῇ ὑπὸ μφθ γωνίᾳ· ἥντινα δὴ τὴν ὑπὸ μφθ γωνίαν τὴν διαφορὰν τῆς ὁμαλῆς καὶ ἀνωμάλου παρόδου τοῦ ἀστέρος ἀνευρίσκει καὶ αὐτὴν καὶ κατὰ μέρος κανονικῶς ὁ Πτολεμαῖος ἐκτίθεται ἐφ᾽ ἑκάστων τμημάτων ὡσαύτως, ὧν ἀπέχει ἀπὸ τοῦ ἀπογείου αὐτοῦ ὁ ἀστὴρ εἴτε δέκα εἴτε εἴκοσιν εἴτε τριάκοντα εἴτε ὅσων δήποτε. ἀποδείκνυται μέν γε παρ᾽ αὐτοῦ γραμμικαῖς ταῖς ἀποδείξεσι θαυμασιώτατα καὶ ἀναντιρρήτως ἡ τοιαύτη διαφορά, ἡ παρὰ τὸν ἔκκεντρον τῆς ὁμαλῆς καὶ ἀνωμάλου κινήσεως ἑκάστου ἀστέρος ἴση οὖσα ἐν τοῖς ἴσοις τμήμασιν, οἷς ἀφίσταται ὁμαλῶς ὁ ἀστὴρ κινούμενος ἐπὶ τοῦ οἰκείου ἐκκέντρου ἀπό τε τοῦ ἀπογείου τοῦ ἐκκέντρου αὐτοῦ καὶ ἀπὸ τοῦ περιγείου τοῦ αὐτοῦ ἐκκέντρου. ἐν μὲν γὰρ τοῖς ἴσοις χρόνοις προέφημεν ὁμαλῶς καὶ ἴσως κινεῖται ἐπ᾽ αὐτοῦ τοῦ ἐκκέντρου καὶ τὰς αὐτὰς ἐπιλαμβάνει περιφερείας αὐτοῦ, εἴτε ἐπὶ τοῦ περιγείου καταλογιζόμεθα αὐτὸν κινεῖσθαι εἴτε ἐπὶ τοῦ ἀπογείου καὶ ἐπὶ τῶν ἴσων παρόδων ἐπ᾽ ἀμφότερα εἴτε τριάκοντα τμημάτων εἴτε τεσσαράκοντα εἴτε εἴκοσιν εἴτε ὅσων δήποτέ εἰσιν.

Ἡ αὐτὴ καὶ ἴση διαφορὰ ἐξανάγκης ἀκολουθεῖν ἀποδείκνυται τῶν τε ὁμαλῶν καὶ ἀνωμάλων κινήσεων ἑκάστου ἀστέρος, ἤτοι τῶν ἐπὶ τοῦ ἐκκέντρου αὐτοῦ θεωρουμένων καὶ ἐπὶ τοῦ ὁμοκέντρου τῷ

127 Cf. Ptol. *Alm.* 3.5 (vol. 1.1, p. 241-46 Heiberg), 3.6, 4.10, 11.11

120 φθμ V C : correxi

Again on the same diagram, let the star move from point ζ, which is the apogee of the eccentric, and in the same 60 days mentioned above cover the arc ζφ being equal to κλ, i.e. 10°, since the star moves uniformly on its eccentric. This arc subtends the angle φμζ, being equal to 10°. Again for the triangle, θφμ according to the thirty-second theorem of the first book of the *Elements*, "the angle φμγ is equal to the sum of the two opposite angles χθγ and μφθ. Therefore, the angle ζμφ for the uniform rotation of the star at the neighbourhood of the eccentric apogee is larger than the angle γθχφ of the anomalous rotation on the ecliptic by the angle μφθ. Ptolemy calculated correctly the angle μφθ for the difference of the uniform motion of the star from the anomalous and tabulated the results as a function of the distance that the star travelled from its apogee, being ten, twenty, thirty degrees and so on. He calculated the difference between the uniform (on the eccentric) and the anomalous (on the ecliptic) for each star by a method very remarkable and free of objections, and found them to be equal for locations in the motion of each star which are at equal distance from the apogees and perigees of its eccentric. As we mentioned, the star moves uniformly on its eccentric covering equal arcs at equal time intervals, independently of whether we consider its motion on the perigee or apogee or in variable intervals of thirty, forty, twenty, or as many as we wish, degrees.

It is still necessary to demonstrate that this difference agrees with the uniform and anomalous movements of each star, which are the rotations on the eccentric and the homocentric to the ecliptic. These changes must agree

ζωδιακῷ, ἴση μέν γε οὖσα ἡ τοιαύτη διαφορὰ διὰ τὴν ἐκκεντρότητα, ὡς λέγομεν, ἐπί τε τῶν ἀπογείων καὶ τῶν περιγείων τῶν ἐκκέντρων. ἐπὶ μὲν τῶν ἀπογείων ἔστιν ἀφαιρετική· ἀφαιρεῖται γὰρ ἡ τοιαύτη διαφορά, ὅση δήποτέ ἐστιν ἑκάστοτε ἀπὸ τοῦ εὑρεθέντος ἀριθμοῦ τῆς ὁμαλῆς παρόδου τοῦ ἀστέρος καὶ ὁ καταλειπόμενος ἀριθμὸς ἀναφαίνεται εἶναι τῆς ἀνωμάλου παρόδου τοῦ ἀστέρος, ἥτις δὴ καὶ ἀκριβὴς καλεῖται, ὡς ἐπὶ τοῦ ζωδιακοῦ καταλογιζομένη. ἐν δὲ τοῖς περιγείοις τῶν ἐκκέντρων ἡ τοιαύτη διαφορά ἐστι προσθετική· προστίθεται γὰρ ὁ ἀριθμὸς αὐτῆς τῷ τῆς ὁμαλῆς τοῦ ἀστέρος παρόδου ἀριθμῷ, κἀντεῦθεν ἀναφαίνεται ἡ ἀνώμαλος αὐτοῦ καὶ ἀκριβὴς πάροδος ἡ ἐπὶ τοῦ ζωδιακοῦ. ἀποδείκνυται μέν γε καὶ τοῦτο γραμμικαῖς ὡσαύτως ἀναντιρρήτως ταῖς δείξεσι τῷ Πτολεμαίῳ, ὡς τηνικαῦτά εἰσιν αἱ μείζονες διαφοραὶ εἴτε προσθετικαὶ εἴτε ἀφαιρετικαὶ τῶν ἀνωμάλων καὶ ὁμαλῶν κινήσεων τῶν ἀστέρων, ὅταν ἀπέχωσιν ταῖς παρόδοις αὐτῶν τεταρτημοριαίας ἔγγιστα ἀποστάσεις ἀπὸ τῶν ἀπογείων καὶ περιγείων τῶν ἐκκέντρων αὐτῶν, ὅταν δηλονότι περὶ τὰ β καὶ δ σημεῖα καταλαμβάνωνται. καὶ τὰ μὲν περὶ τῆς διὰ τὴν ἐκκεντρότητα τῶν κύκλων ἀνωμαλίας τοῦτον ἔχειν τὸν τρόπον καὶ ἱκανῶς ἐμοὶ δοκεῖν εἴρηται καὶ διηυκρίνηται.

Ἵνα δὲ καὶ τὸν δεύτερον τρόπον τῆς ἀνωμαλίας τὸν διὰ τὴν ἐπὶ τοῦ ἐπικύκλου περιφορὰν τοῦ ἀστέρος κατίδωμεν, ἔσται δευτέρα αὕτη καταγραφή. ἔστω γὰρ ὁ κύκλος ἐφ᾽ οὗ φέρεται ὁ ἀστὴρ ὁ αβγδ, οὗ τὸ κέντρον ἐστὶ τὸ ξ, καὶ καταγεγράφθω ἐπ᾽ αὐτοῦ ἐπίκυκλος βραχὺς ὁ ζθκλ, περιφερόμενος τὸν εἰρημένον κύκλον. καὶ τὸ κέντρον τούτου ἔστω διαπαντὸς ἐπὶ τῆς περιφερείας τοῦ τοιούτου κύκλου, τοῦ αβγδ. ὑποκείσθω τοίνυν ὁ τοιοῦτος ἐπίκυκλος εἶναι ἐπὶ τοῦ α σημείου τοῦ μεγίστου κύκλου, ὁ δὲ ἀστὴρ εἶναι ἐπὶ τοῦ ζ σημείου τοῦ ἐπικύκλου, τοῦ ἀπογείου δηλονότι τοῦ ἐπικύκλου. ὁ μὲν οὖν ἐπίκυκλος κεκινήσθω ἀπὸ τοῦ α σημείου πρὸς τὰ ἑπόμενα, τὴν αβ περιφέρειαν τεταρτημόριον οὖσαν τοῦ ὅλου μεγίστου κύκλου· ἐν ᾧ δὲ ὁ ἐπίκυκλος κινεῖται ἀπὸ τοῦ α ἐπὶ τὸ β, κινείσθω καὶ ὁ ἀστὴρ ἐπὶ τοῦ ἐπικύκλου ἀπὸ τοῦ ζ ἀπογείου

154 Cf. Ptol. *Alm.* 3.4 (vol. 1.1, p. 238-39 Heiberg), 11.10 (vol. 1.2, p. 433 Heiberg)

145-46 πότε ἀφαιρετικὴ καὶ πότε προσθετικὴ ἡ τοιαύτη διαφορά Sch in mg. C **155** προσθετικαὶ εἴτε ἀφαιρετικαὶ : ἀφαιρετικαὶ εἴτε προσθετικαὶ C[ac] **161** τὸν[2] om. C

with those produced by the eccentricity at the apogees and perigees. In the apogees the effect is subtractive, i.e. we subtract the difference from the accepted uniform motion and the resulting number is quoted as the anomalous and precise motion on the ecliptic. In the perigees of the eccentric the difference is additive; that is, the number is added to the uniform rotation and the resulting number is quoted as the anomalous and precise motion on the ecliptic. This is proven with geometric methods and beyond doubt by Ptolemy. As a result they produce the maximum additive and subtractive differences between the anomalous and uniform motions when the stars are approximately located one quadrant away from the apogees and perigees i.e. at the points β and δ. This is the way that the anomaly is created, due to the eccentricity, and I believe I described and clarified it in a satisfactory manner.

We describe next the second method of producing the anomalous rotation of the star by using epicycles. Let us denote by αβγδ the circle on which the star rotates with the centre at ξ. On this we draw a small circle ζθκλ, which moves around the mentioned circle with its centre always attached on the circumference of αβγδ. Let the small circle be at the point α of the deferent and the star at point ζ of the epicycle, that is at the apogee of the epicycle. Then, as the epicycle moves from point α to the subsequent points of (the circumference) through the arc αβ, being one quadrant of the circle, the star

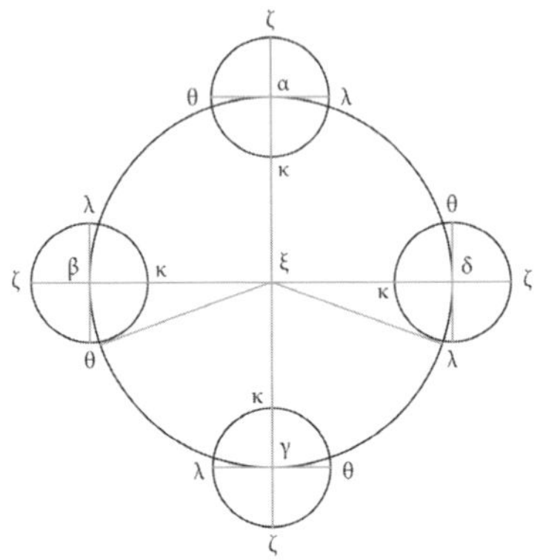

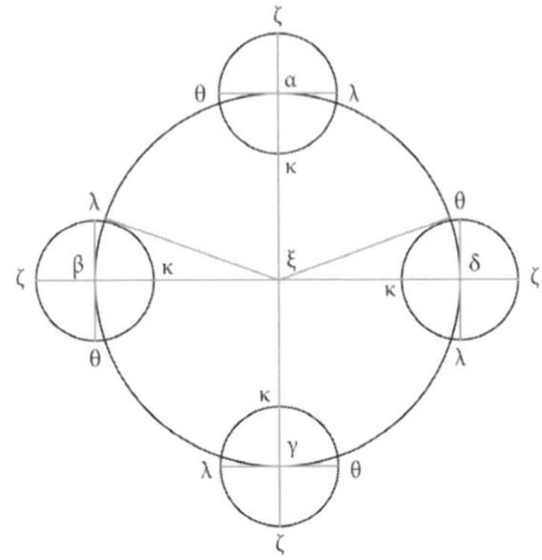

αὐτοῦ ἢ ὡς πρὸς τὰ ἑπόμενα τοῦ ἐπικύκλου τὴν ζθ περιφέρειαν ἢ ὡς πρὸς τὰ ἡγούμενα τὴν ζλ. ἔστω δ' οὖν πρῶτον ὡς πρὸς τὰ ἑπόμενα τὴν ζθ περιφέρειαν. ἡ μὲν οὖν ὁμαλὴ κίνησις τοῦ ἀστέρος ἐστὶν ἡ τοῦ ἐπικύκλου, ἐφ' οὗ φέρεται προχώρησις, ἤτοι τοῦ κέντρου τοῦ ἐπικύκλου. τὸ γὰρ κέντρον ἐστὶ τὸ μέσον τοῦ κύκλου καὶ ἡ προχώρησις καὶ κίνησις τοῦ κύκλου ὀφείλει καταλογίζεσθαι εἰς τὸ κέντρον αὐτοῦ. ἡ μὲν οὖν ὁμαλή, ὡς ἔφην, κίνησις τοῦ ἀστέρος ἐστὶν ἡ τοῦ κέντρου τοῦ ἐπικύκλου αὐτοῦ προχώρησις. καὶ ἔστιν ἐνταῦθα ἡ αβ περιφέρεια, ἡ δὲ ἀνώμαλος καὶ φαινομένη διὰ τὴν ἐπὶ τοῦ ἐπικύκλου κίνησιν τοῦ ἀστέρος γινομένη ἐστὶν ἡ αθ· ἀπὸ τοῦ ζ γὰρ εἰς τὸ θ σημεῖον προέκοψεν ὁ ἀστὴρ διὰ τὴν ἰδίαν αὐτοῦ τὴν ἐπὶ τοῦ ἐπικύκλου κίνησιν.

Τὸ μὲν γὰρ κέντρον τοῦ ἐπικύκλου, ὡς εἴρηται, ἐκινήθη ἀπὸ τοῦ α ἐπὶ τὸ β, συγκινουμένου πάντως καὶ συμπεριφερομένου ἐξανάγκης καὶ τοῦ ἐπ' αὐτοῦ ἀστέρος. ἐπεὶ δὲ ἐν ᾧ τὸ κέντρον τοῦ ἐπικύκλου ἀπὸ τοῦ α ἐπὶ τὸ β κινεῖται, κινεῖται καὶ ὁ ἀστὴρ ἐπὶ τοῦ ἐπικύκλου καὶ γίνεται ἐπὶ τοῦ θ σημείου. ὁ ἀστὴρ προστίθησι τῇ κινήσει τοῦ ἐπικύκλου, τουτέστι τῇ ὑπὸ αξβ γωνίᾳ, τὴν ὑπὸ βξθ γωνίαν, ἤτοι τὴν περιφέρειαν τὴν βθ καὶ διὰ τοῦτο γίνεται ἡ ἀνώμαλος κίνησις ἐνταῦθα πλείων τῆς ὁμαλῆς. καὶ εὑρίσκεται τῷ Πτολεμαίῳ διὰ γραμμικῶν ἀποδείξεων ὁπόση ἐστὶν ἡ γωνία ἑκάστοτε ἡ ὑπὸ βξθ ἐφ' ἑνὸς ἑκάστου τῶν ἀστέρων καὶ κανονικῶς ἐκτίθεται ὑπ' αὐτοῦ κατὰ μέρος. πάλιν νοείσθω κεκινημένος ὁ ἐπίκυκλος, ἤτοι τὸ κέντρον αὐτοῦ ἀπὸ τοῦ β σημείου ἐπὶ τὸ γ. ἐν ᾧ

182 ζ C^{pc} : β V C^{ac} **191** Cf. Ptol. *Alm*. 3.5 (vol. 1.1, p. 243-44 Heiberg), 3.6, 4.10, 11.11

also rotates on the epicycle away from the apogee ζ either forward through the arc ζθ or backward through ζλ. Let us consider first the forward motion through the arc ζθ. The uniform motion refers to the rotation of the centre of the epicycle, being the middle point of a circle, in this case the arc αβ. The anomalous and apparent motion is the rotation αθ, created by the rotation of the star on the epicycle, with the star moving from ζ to θ.

In the first step, the centre of the epicycle moved from α to β, moving with it everything and carrying along the star. As the centre of the epicycle moves from α to β, the star also moves on the epicycle and arrives at the point θ. The star adds to the rotation of the centre of the epicycle αξβ the angle βξθ, producing the arc βθ; for this reason the anomalous rotation is larger than the uniform. Ptolemy found with geometrical methods how much larger the anomalous motion is in relation to the regular, i.e. how big is the angle βξθ for each star, and tabulated them for each star at each position. Again, let us consider the centre of the epicycle moving from point β to γ

δὲ κινεῖται ὁ ἐπίκυκλος ἀπὸ τοῦ β ἐπὶ τὸ γ, κεκινήσθω καὶ ὁ ἀστὴρ ἐπὶ τοῦ ἐπικύκλου ἀπὸ τοῦ θ ἐπὶ τὸ κ. κἀνταῦθα συμβαίνει τὴν αὐτὴν εἶναι κίνησιν καὶ ἀποκατάστασιν καὶ κατευθεῖαν τήν τε τοῦ κέντρου τοῦ ἐπικύκλου καὶ τὴν τοῦ ἀστέρος ἐπὶ τοῦ ἐπικύκλου, καὶ οὐδεμία γίνεται αὐτῶν διαφορά. ἡ γὰρ αὐτὴ εὐθεῖα καὶ ἀπὸ τοῦ ξ δίεισι καὶ διὰ τοῦ ἀστέρος ἐπὶ τοῦ κ καὶ διὰ τοῦ κέντρου τοῦ ἐπικύκλου, τουτέστι τοῦ γ, καὶ διαφορά τις οὐκ ἔστι τῆς ἀνωμάλου παρὰ τὴν ὁμαλὴν κίνησιν.

Πάλιν κεκινήσθω τὸ κέντρον τοῦ ἐπικύκλου ἀπὸ τοῦ γ σημείου ἐπὶ τὸ δ, ἐν τοσούτῳ δὲ καὶ ὁ ἀστὴρ ἐπὶ τοῦ ἐπικύκλου, ἀπὸ τοῦ κ ἐπὶ τὸ λ. ἡ μὲν οὖν ὁμαλὴ πάλιν κίνησίς ἐστιν ἡ ὑπὸ γξδ γωνία, ἤτοι ἡ γδ περιφέρεια, ἣν ὑποτείνει ἡ τοιαύτη γωνία. ἡ δὲ ἀνώμαλος, ἥτις γίνεται διὰ τὴν εἰρημένην αἰτίαν, ἤτοι τὴν ἐπὶ τοῦ ἐπικύκλου κίνησιν τοῦ ἀστέρος, ἐστὶν ἡ ὑπὸ γξλ· καὶ ἔστιν αὕτη ἐλάττων τῆς ὁμαλῆς τῇ ὑπὸ δξλ γωνίᾳ, ἤτοι τῇ λδ περιφερείᾳ. καὶ πάλιν κινουμένου καὶ παροδεύοντος τοῦ κέντρου τοῦ ἐπικύκλου ἀπὸ τοῦ δ ἐπὶ τὸ α, ἐν τοσούτῳ δὲ καὶ τοῦ ἀστέρος ἐπὶ τοῦ ἐπικύκλου γινομένου ἀπὸ τοῦ λ ἐπὶ τὸ ζ, ἡ αὐτὴ ἔσται κίνησις, ἥ τε ὁμαλὴ καὶ ἡ ἀνώμαλος. καὶ ὁ ἀστὴρ καταλαμβάνεται ἐπὶ τῆς αὐτῆς καὶ μιᾶς εὐθείας διϊούσης διά τε τοῦ α καὶ τοῦ ζ, ἤτοι τῆς ξαζ, καὶ ἀποκαθίσταται καὶ ὁ ἐπίκυκλος καὶ ὁ ἀστὴρ εἰς τὴν πρώτην θέσιν. καὶ οὕτως μὲν γίνεται ἡ διαφορὰ τῆς ὁμαλῆς καὶ ἀνωμάλου κινήσεως καὶ διὰ τὴν αἰτίαν τῆς περιφορᾶς τοῦ ἀστέρος ἐπὶ τοῦ ἐπικύκλου, ὅταν κινῆται ὁ ἀστὴρ εἰς τὰ ἑπόμενα τοῦ ἐπικύκλου, καθὼς ἔχει ἡ μία καταγραφή.

191-217 add. in mg. C

ὅμοιον τῷ προτέρῳ

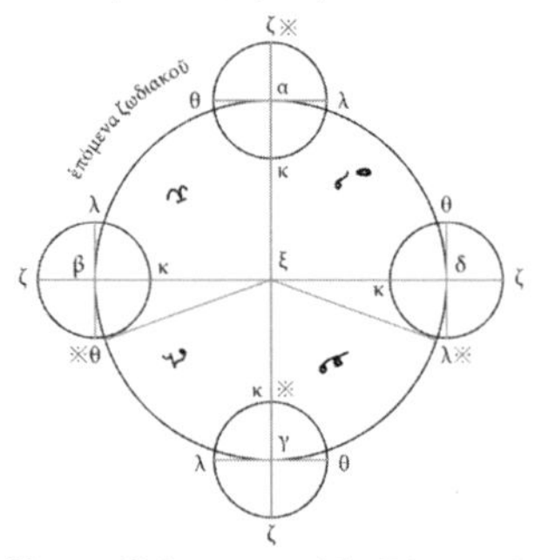

ὅμοιον τῷ δευτέρῳ

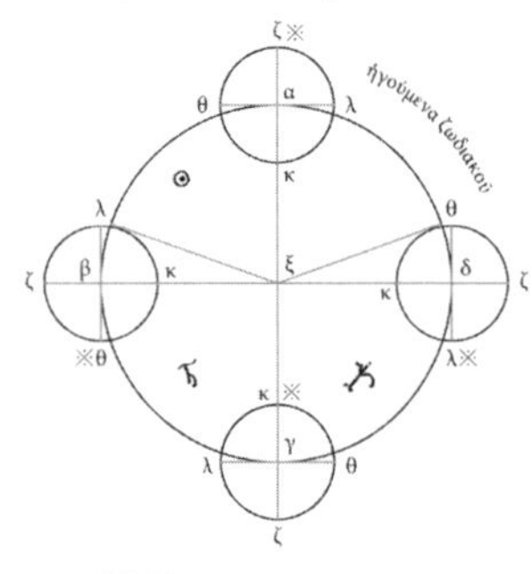

199 δίεισι : διέρχεται add. C^{sl} **202** σημείου : σημεῖον V C^{ac}

as the star moves on the epicycle from point θ to κ. In this position the rotation of the centre of the epicycle and of the star on the epicycle restores the star on the straight line passing from ξ through the star at κ and through the centre of the epicycle γ. Thus there is no difference between anomalous and uniform rotations.

Then, let again the centre of the epicycle move from point γ to δ, and at the same time let the star move on the epicycle from κ to λ. The uniform motion is through the arc γδ, which subtends the angle γξδ. The anomalous motion is produced by the same arguments, from the rotation of the star on the epicycle corresponding to the angle γξλ. This motion is less than the uniform by the angle δξλ, that is by the arc λδ. Then again as the centre of the epicycle moves and travels from δ to α, at the same time the star moves on the epicycle from λ to ζ; these are the movements: the uniform and the anomalous. Finally the star is located on the straight line passing through α and ζ, that is the line ξαζ. In this way the epicycle and the star return to their initial positions. This is how the difference between the uniform and the anomalous motion is generated from the forward rotation of the star on the epicycle, according to one [of the two] descriptions.

Τὰ αὐτὰ δὲ ἀντιστρόφως συμβαίνει καὶ ὅταν εἰς τὰ ἡγούμενα τοῦ ἐπικύκλου κινῆται ὁ ἀστήρ, καθάπερ ἔχει ἡ ἑτέρα καταγραφή. κεκινήσθω μὲν γὰρ τὸ κέντρον τοῦ ἐπικύκλου ἐπὶ τοῦ μείζονος κύκλου τὴν αβ περιφέρειαν. ἐν ᾧ δὲ κινεῖται τὸ κέντρον τοῦ ἐπικύκλου τὴν αβ περιφέρειαν, κεκινήσθω καὶ ὁ ἀστὴρ ἐπὶ τοῦ ἐπικύκλου ἀπὸ τοῦ ζ ἀπογείου αὐτοῦ πρὸς τὰ ἡγούμενα τὴν ζλ περιφέρειαν. ἡ μὲν οὖν ὁμαλὴ κίνησις τοῦ ἀστέρος ἐστὶν ἡ αβ περιφέρεια, ἣν κινεῖται τὸ κέντρον τοῦ ἐπικύκλου, τοῦ συμπεριφέροντος αὐτὸν τὸν ἀστέρα. ἡ δὲ ἀνώμαλος γίνεται ἡ αλ, ἐλλιπὴς οὖσα τῆς ὁμαλῆς τῇ αβ, τῇ ὑπὸ βξλ γωνίᾳ, ἤτοι τῇ βλ περιφερείᾳ τῇ ὑποτεινομένῃ ὑπὸ τῆς τοιαύτης γωνίας. πάλιν κεκινήσθω ὁ ἐπίκυκλος ἀπὸ τοῦ β ἐπὶ τὸ γ, ἐν τοσούτῳ δὲ κεκινήσθω καὶ ὁ ἀστὴρ ἐπὶ τοῦ ἐπικύκλου, ἀπὸ τοῦ λ ἐπὶ τὸ κ. καὶ γίνεται τηνικαῦτα ἡ αὐτὴ ὁμαλὴ καὶ ἀνώμαλος ἀποκατάστασις, καὶ μία εὐθεῖα κατ᾽ ἰσότητα διήκει ἀπὸ τοῦ ξ σημείου διά τε τοῦ κέντρου τοῦ ἐπικύκλου, ἤτοι τοῦ γ καὶ τοῦ κ ἀστέρος. πάλιν κεκινήσθω ὁ ἐπίκυκλος ἀπὸ τοῦ γ ἐπὶ τὸ δ, ἐν τοσούτῳ δὲ καὶ ὁ ἀστὴρ ἀπὸ τοῦ κ ἐπὶ τὸ θ. ἡ μὲν οὖν ὁμαλὴ πάροδός ἐστι ἡ τοῦ ἐπικύκλου ἡ γδ, ἡ δὲ ἀνώμαλος ἡ γθ, ὑπερβάλλουσα τῆς γδ τῇ ὑπὸ δξθ γωνίᾳ, ἤτοι τῇ ὑποτεινομένῃ ὑπ᾽ αὐτῆς τῆς γωνίας περιφερείας, τῇ δθ. καὶ πάλιν ἀποκαταστάντος τοῦ ἐπικύκλου ἐπὶ τοῦ α σημείου ὡς πρότερον καὶ τοῦ ἀστέρος ἐπὶ τοῦ ζ, ἡ αὐτὴ ἔσται ὁμαλὴ καὶ ἀνώμαλος ἀποκατάστασις καὶ ἡ αὐτὴ εὐθεῖα κατ᾽ ἰσότητα διϊοῦσα ἀπὸ τοῦ ξ κέντρου τοῦ μείζονος κύκλου διά τε τοῦ α κέντρου τοῦ ἐπικύκλου καὶ τοῦ ζ ἀστέρος.

Οὕτω καὶ κατ᾽ ἀμφοτέρους τοὺς τρόπους τουτέστι κινουμένου τοῦ ἀστέρος ἐπὶ τοῦ ἐπικύκλου, εἴτε εἰς τὰ ἡγούμενα εἴτε εἰς τὰ ἑπόμενα, ὡσαύτως συμβαίνουσιν αἱ διαφοραὶ τῆς ὁμαλῆς κινήσεως καὶ τῆς ἀνωμάλου, νῦν μὲν κατὰ πρόσθεσιν, νῦν δὲ κατὰ ἔλλειψιν, καὶ αἱ ἐπὶ τοῦ αὐτοῦ ἀποκαταστάσεις. καὶ νῦν μὲν ἡμῖν τὸ ὑπόδειγμα ὁλοσχερέστερον ἐκτέθειται ἐπισκεπτομένοις τήν τε κίνησιν τοῦ ἐπικύκλου καὶ τὴν κίνησιν τοῦ ἀστέρος ὡς ἐπὶ τεταρτημοριαίας περιφερείας καὶ ὡς ἰσοταχῶς ἀποκαθισταμένων τῶν τε ἐπικύκλων εἰς τὰ τεταρτημόρια τῶν μειζόνων κύκλων καὶ τῶν ἀστέρων εἰς τὰ

218 ὃ μάλιστα ὁ Πτολεμαῖος ἀποδέχεται add. C^{sl} **219** κινεῖται V^{ac} C^{ac}

The same also happens in the reverse situation, [which in fact is prefered by Ptolemy], when the star rotates backwards on the epicycle, and for this case we present another explanation. Let the centre at the epicycle move on the deferent through the arc αβ. While the centre of the epicycle moves on the arc αβ, the star rotates on the epicycle from its apogee ζ toward the preceding points through the arc ζλ. The uniform motion of the star is the arc αβ, on which travels the centre of the epicycle carrying the star with it. The anomalous motion is the arc αλ which is less than the uniform motion αβ by the angle βξλ. Again, let the epicycle move from β to γ and the star rotate on the epicycle from λ to κ (this way the uniform motion and anomalous rotation takes place) and exactly a straight line passes from the point ξ through the centre of the epicycle γ and the star at κ. Let again the epicycle move from γ to point δ, in which time the star moves from κ to θ. The uniform movement is that of the epicycle across γδ and the anomalous is the arc γθ, being larger than the γδ by the angle δξθ, in other words by the arc δθ of the circumference. Then takes place the restoration of the epicycle back to the original point α, and the star returns to the point ζ. This is the restoration of the uniform and anomalous motion, and precisely a straight line starting at the centre of the larger circle passes through the centre of the epicycle α and meets the star at ζ.

In this manner we selected the rotation of the star on the epicycle in two ways, being either forward or backward, and we generated the differences between the uniform and anomalous motions, [with the anomaly] being sometimes additive, at other times subtractive, and sometimes restoring the uniform motion. We presented examples in great detail by selecting for the rotation of the epicycle and of the star intervals of 90°, with the epicycles rotating with uniform velocity of 90° on the large circle and with the stars also

τεταρτημόρια τῶν ἐπικύκλων. ἔξεστι δὲ συλλογίζεσθαι τὰ παραπλήσια καὶ ἀνάλογα συμβαίνειν, κἂν εἴ τις καὶ ἐπὶ ἡττόνων περιφερειῶν ἢ κατὰ τεταρτημόρια θεωροίη τὰς τοιαύτας παρόδους ἢ κἂν εἴ τις οὐκ ἰσοταχεῖς τὰς κινήσεις τῶν τε ἐπικύκλων καὶ τῶν ἀστέρων ἐπὶ τῶν ἐπικύκλων ὑποτιθείη, ἀλλὰ τὴν μὲν ταχυτέραν, τὴν δὲ βραδυτέραν, ὥσπερ καὶ ἐροῦμεν ἐν τοῖς ἑξῆς καὶ κατὰ μέρος περὶ ἑνὸς ἑκάστου τῶν ἀστέρων λεχθῆναι μέλλουσι. νῦν δὲ εἰς τοσοῦτο μόνον ἐχρησάμεθα τῷ ὑποδείγματι, εἰς ὅσον ὑποτυπῶσαι καὶ δηλοποιῆσαι ὁλοσχερέστερον ὡς εἴρηται, ὅπως ἡ παρὰ τὴν ἐπὶ τοῦ ἐπικύκλου κίνησιν τοῦ ἀστέρος καταλαμβάνεται διαφορὰ τῶν κινήσεων καὶ ἀνωμαλία.

256 λεχθῆναι : ῥηθῆναι C

rotating 90° on the epicycles. From these, one can contemplate that something very similar and analogous happens for arcs smaller than quadrants, or even when one assumes that the rotations of the epicycles and of the stars on the epicycles are not uniform but sometimes faster and other times slower, as we shall describe further in the discussion for each of the stars separately. For now we only used an example in order to demonstrate and clarify in detail how the rotation of a star on an epicycle generates differences of rotations and the anomaly.

18 ὅτι τῆς σελήνης καὶ τῶν ἄλλων ἀστέρων καὶ κατ' ἀμφοτέρους τοὺς τρόπους κινουμένων ἀνώμαλον κίνησιν, μόνος ὁ ἥλιος καθ' ἕνα τούτων μόνον τρόπον ἀνωμάλως κινεῖται καὶ ἁπλῆν μόνην ἔχει τὴν αἰτίαν τῆς ἀνωμάλου κινήσεως

Ἀλλ᾽ ἐπαναληπτέον τὸν λόγον εἰς ὅπερ ἦν ἡ πρόθεσις, ὅτι διπλοῦς μὲν ὁ τρόπος τῆς ἀνωμάλου κινήσεως τῶν ἀστέρων κατὰ δύο τὰς αἰτίας καταλαμβανόμενος, ὡς ὑπεδείξαμεν, τήν τε ἐκκεντρότητα τῶν κύκλων ἐφ᾽ ὧν φέρονται καὶ τὸ ἐπὶ ἐπικύκλων αὐτοὺς τοὺς ἀστέρας κινεῖσθαι. ἔστι δὲ ὁ διπλοῦς τρόπος οὗτος τῆς ἀνωμαλίας, ἐπί τε τῶν πέντε πλανωμένων καὶ ἐπὶ τῆς σελήνης. διπλῆ γὰρ καὶ σύνθετος ἐπ᾽ ἀμφοτέρων ἡ κατ᾽ αὐτοὺς ἀνωμαλία καταλαμβάνεται, ὡς ἐροῦμεν ἑξῆς ἐν τοῖς περὶ ἑκάστων λόγοις. κατ᾽ ἀμφότερα γὰρ μεμιγμένως συμβαίνει ἐπ᾽ αὐτῶν τὰ τῆς ἀνωμάλου κινήσεως, κατά τε τὴν ἐκκεντρότητα καὶ κατὰ τὴν ἐπὶ τοῦ ἐπικύκλου κίνησιν αὐτῶν τῶν ἀστέρων.

Ἐπὶ δὲ τοῦ ἡλίου, ὥσπερ καὶ τἄλλα τὰ κατ᾽ αὐτὸν καθάπερ προολίγου ἐλέγομεν, παρηλλαγμένα καὶ διάφορα καί, ὡς ἄν τις εἴπῃ, ἑνοειδέστερα καὶ ἀποίκιλα, οὕτως ἔχει καὶ τὰ περὶ τῆς αὐτοῦ κινήσεως. οὐ γὰρ διπλῆ ἐν αὐτῇ θεωρεῖται ἡ ἀνωμαλία, ὥσπερ ἐν ταῖς τῶν ἄλλων, ἀλλ᾽ ἁπλῆ τε καὶ μία· καὶ οὐ μία μόνον, ἀλλὰ καὶ βραχεῖά τίς ἐστιν ἡ διαφορὰ τῆς ἀνωμάλου κινήσεως τοῦ ἡλίου πρὸς τὴν ὁμαλὴν αὐτοῦ, καὶ ὅταν ἡ μεγίστη διαφορὰ αὐτῶν καταλαμβάνηται. ἐπ᾽ ἐκείνων δὲ τῶν ἄλλων ἀστέρων, οὐ μόνον διπλῆ ὡς ἔφημεν, ἀλλὰ καὶ πολλή τις καὶ πολυμήκης ἐστὶν ἡ διαφορὰ τῆς ὁμαλῆς κινήσεως πρὸς τὴν ἀνώμαλον, καὶ δῆλον ἔσται τοῦτο ἐν τοῖς ἔμπροσθεν λεχθησομένοις κατὰ τοὺς οἰκείους τόπους. νῦν δέ, ὅπερ ἐλέγομεν, ἡ ἀνώμαλος διαφορὰ τῆς ἡλιακῆς κινήσεως οὐκ ἔστιν ὥσπερ καὶ ἐπὶ τῶν ἄλλων διπλῆ, ἀλλ᾽ ἁπλῆ καταλαμβάνεται καὶ μία, καθ᾽ ἕνα δὴ πάντως μόνον τρόπον τῶν

5 τρόπος post οὗτος transp. C **6** ἐπ᾽ : ἀπ᾽ V **12** εἴπῃ : εἴποι C[pc] **22** ἡλιακῆς κινήσεως : ὁμαλῆς κινήσεως τοῦ ἡλίου C (τοῦ ἡλίου s.l.) **23** μόνον post τρόπον transp. C

18 The Moon and the other planets have anomalous motions described by one of the two methods. Only the Sun moves, according to one of the models, and has a simple explanation

It was our intention to repeat the theory because, as we demonstrated, there are two methods and arguments with which one understands the anomalous motion of the stars: one with the eccentricity of the circles on which the stars are carried and one with the epicycles on which they rotate. For the five planets and the Moon, the anomaly is double and we understand it with the mixture of two movements, as we shall explain when we describe each of them. In both cases, the motion is a composition of both the eccentricity and the rotation of the star on the epicycle.

As we said earlier for other properties of the Sun, its motion is unique and different, and one could say that it is easier to understand it without many variations. Its anomaly is not double, as for the other stars, but single and simple; in addition to having only one anomaly, the difference between the anomalous and the uniform motion at the time of its maximum is very short. In contrast, the other stars have a double anomaly that is large, and the difference between the anomalous from the uniform motion is long, as is evident by what we say later at the appropriate places. For now, we are saying that the difference caused by the anomaly in the Sun's motion is not double as for the other planets, but simple and single, caused either by the eccentricity or by the rotation on the epicycle.

εἰρημένων, εἴτε τὸν διὰ τὴν ἐκκεντρότητα εἴτε τὸν διὰ τὴν ἐπὶ τοῦ ἐπικύκλου φοράν.

Ποιούμενος γὰρ τὴν περὶ τούτων ἐξέτασιν ὁ Πτολεμαῖος ἀνευρίσκει γραμμικαῖς δείξεσιν ὅτι, κἄν τε ἔκκεντρον μόνον νοήσωμεν πρὸς τὸν ζωδιακὸν τὸν κύκλον, ἐφ᾽ οὗ φέρεται ὁ ἥλιος αὐτὸς καθ᾽ ἑαυτὸν οὐκ ἐπὶ ἐπικύκλου, κἄν τε ὁμόκεντρον μὲν τῷ ζωδιακῷ τὸν κύκλον, ἐφ᾽ οὗ φέρεται ὁ ἥλιος, ἐπ᾽ αὐτοῦ δὲ κινούμενον ἐπίκυκλον καὶ ἐπ᾽ αὐτοῦ τοῦ ἐπικύκλου τὸν ἥλιον, τὰ αὐτὰ συμβαίνει καὶ ἡ αὐτὴ τῆς ἀνωμάλου τοῦ ἡλίου κινήσεως πρὸς τὴν ὁμαλὴν διαφορά. εἴπερ μόνον ὑποθεῖτό τις τὴν διάστασιν, ἣν ἔχει τὸ κέντρον τοῦ ἐκκέντρου, ἐφ᾽ οὗ θεωρεῖν ἐστι παροδεύοντα τὸν ἥλιον, πρὸς τὸ κέντρον τοῦ ὁμοκέντρου τῷ ζωδιακῷ, ἴσην τῇ ἀπὸ τοῦ κέντρου τοῦ ἐπικύκλου γραμμῇ πρὸς τὴν περιφέρειαν αὐτὴν τοῦ ἐπικύκλου, ἔτι δὲ καὶ τὴν ἐπὶ τοῦ ἐπικύκλου πάροδον τοῦ ἡλίου πρὸς τὰ ἡγούμενα καὶ ἰσοταχῆ τὴν ἀποκατάστασιν αὐτοῦ τῇ ἀποκαταστάσει αὐτοῦ τοῦ ἐπικύκλου.

Οἷον ὑποκείσθω ὁ ἔκκεντρος κύκλος, ἐφ᾽ οὗ νοεῖν ἐστι περιφερόμενον τὸν ἥλιον, ὁ αλβγδ· κέντρον δὲ αὐτοῦ ἔστω τὸ κ. καὶ ἕτερος ἴσος αὐτῷ ὁμόκεντρος τῷ ζωδιακῷ ὁ ζχβθδ, οὗ κέντρον ἔστω τὸ μ. καὶ τεμνέτωσαν ἀλλήλους κατὰ τὰ κοινὰ σημεῖα, τὰ βδ. διάστασις δὲ ἔστω ἀπὸ τοῦ κ κέντρου ἐπὶ τὸ μ κέντρον καθ᾽ ὑπόθεσιν τμημάτων δύο ἢ τριῶν, ὁποίων ἐστὶν ὁ κύκλος τξ καὶ ἡ διάμετρος αὐτοῦ ρκ, ἢ μᾶλλον κατ᾽ ἀλήθειαν τοσαύτη, ὅσην μετ᾽ ὀλίγον ἐροῦμεν τὴν μεγίστην εἶναι διαφορὰν τῆς ὁμαλῆς τοῦ ἡλίου κινήσεως πρὸς τὴν ἀνώμαλον.

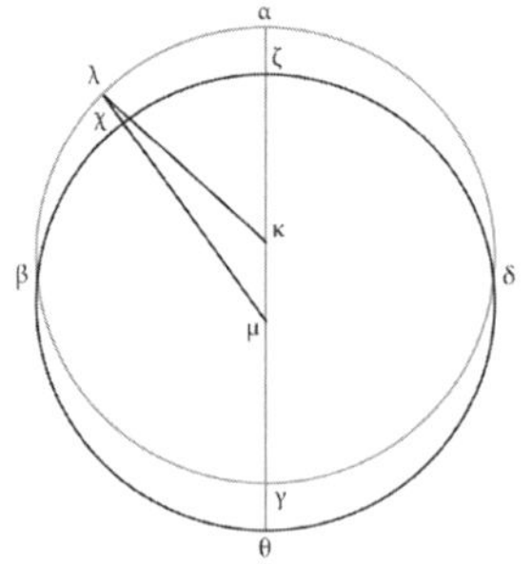

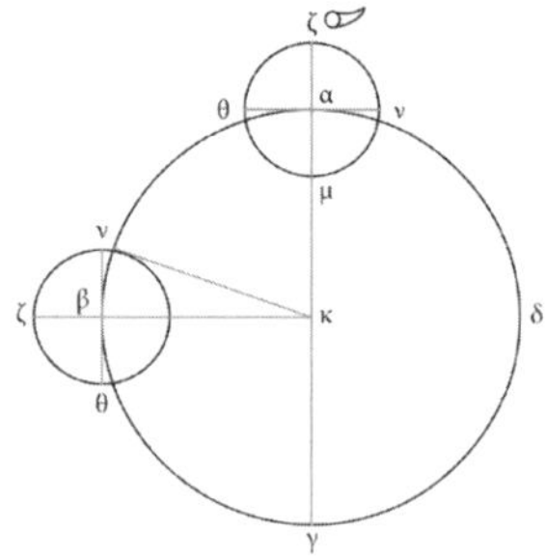

26 Cf. Ptol. *Alm.* 3.3 (vol. 1.1, p. 221-232 Heiberg)

32 τοῦ ἡλίου post κινήσεως transp. C **32** διαφορὰν C

Ptolemy, who investigated the subject, discovered and proved geometrically that if we conceive only an eccentric circle relative to the ecliptic, with the Sun carried on this circle and not on an epicycle, or if on the other hand we conceive a circle homocentric to the ecliptic with an epicycle and the Sun rotating on it, we obtain the same result and the same difference between the anomalous and uniform motion. This happens provided in the two cases one assumes that the distance from the [centre] of the eccentric (on which the Sun travels) to the centre of the ecliptic is equal to the radius of the epicycle. In addition, the angular velocity for rotation of the Sun on the epicycle toward the rear is equal to the rotation, which restores the epicycle to its original position on the homocentric.

Let us visualize the eccentric circle the αλβγδ, on which the Sun rotates with its centre κ. Let us consider a second circle of equal size, homocentric to the ecliptic, the ζχβθδ, with centre at μ and intersecting each other at the common points βδ. We assume the distance from the centre κ to μ to be two or three units with the circle being 360, and its diameter 120 units, or more accurately as large as the maximum difference of the uniform from the anomalous motion.

Πάλιν ὑποκείσθω κύκλος ὁμόκεντρος τῷ ζωδιακῷ ὁ αβγδ περὶ κέντρον τὸ κ, ἐπὶ τούτου δὲ φερέσθω ὁ ἐπίκυκλος ὁ ζθμν, τὸ κέντρον ἔχων ἐπὶ τοῦ εἰρημένου κύκλου, τοῦ αβγδ. νοείσθω τὸ κέντρον αὐτοῦ ἐπὶ τοῦ α καὶ ἡ ἀπὸ τοῦ κέντρου αὐτοῦ τοῦ κατὰ τὸ α ἐπὶ τὴν περιφέρειαν αὐτοῦ κατὰ τὸ θ σημεῖον ἢ τὸ ν ἢ ἕτερον ὁτιοῦν ἔστω ἴση τῇ μεταξὺ τῶν εἰρημένων δύο σημείων ἐπὶ τῆς ἄλλης ὑποθέσεως τῆς κατὰ τὸν ἔκκεντρον, τοῦ κ δηλαδὴ καὶ τοῦ μ. κεκινήσθω δὲ ὁ ἐπίκυκλος ἀπὸ τοῦ α ἐπὶ τὸ β, ὁ δὲ ἥλιος ἐπὶ τοῦ ἐπικύκλου ἰσοταχῶς τεταρτημοριαίαν ὡσαύτως διάστασιν εἰς τὰ ἡγούμενα ἀπὸ τοῦ ζ ἐπὶ τὸ ν. εἴτε οὖν καθὼς ἔχει ἡ μία καταγραφὴ ἐπὶ ἐκκέντρου πρὸς τὸν ζωδιακὸν φέρεται ὁ ἥλιος χωρὶς ἐπικύκλου εἴτε ἐπὶ ὁμοκέντρου τῷ ζωδιακῷ, ἐπὶ ἐπικύκλου δὲ τοσούτου μεγέθους ὄντος, ὅσου λέγομεν τὰ αὐτὰ συμβήσεται καὶ ἡ αὐτὴ ἀκολουθήσει διαφορὰ τῆς ὁμαλῆς τοῦ ἡλίου κινήσεως πρὸς τὴν ἀνώμαλον. καὶ τοῦτο δῆλον αὐτόθεν. τῆς μὲν γὰρ ὑπὸ ακλ γωνίας, ὅταν δηλονότι ὁ ἥλιος νοῆται ἐπὶ ἐκκέντρου μόνον κινούμενος, οὐ μὴν δὲ καὶ ἐπὶ ἐπικύκλου, ὡς ἐπὶ τῆς μιᾶς καταγραφῆς, τῆς γοῦν γωνίας ταύτης ἔστιν ἐλλάτων ἡ ὑπὸ ζμχ, ὡς ἀνωτέρω εἴρηται διὰ τὸ τριακοστὸν δεύτερον θεώρημα τοῦ πρώτου Στοιχείου, τῇ ὑπὸ κλμ γωνίᾳ. καὶ ἔστιν αὕτη ἡ διαφορὰ τῆς ὁμαλῆς καὶ ἀνωμάλου κινήσεως, ἤτοι ἡ ἐπὶ τῆς μκ νοουμένη περιφέρεια. πάλιν ὅταν ἐπὶ ὁμοκέντρου κινῆται ὁ ἐπίκυκλος τοῦ ἡλίου καὶ ὁ ἥλιος ἐπὶ τοῦ ἐπικύκλου, ὡς ἐν τῇ ἑτέρᾳ καταγραφῇ, ἔστι μὲν ἡ ὁμαλὴ τεταρτημοριαία κίνησις ἡ ὑπὸ ακβ, ἡ δὲ ἀνώμαλος ἡ ὑπὸ ακν. ἐπὶ γὰρ τοῦ ν ἐστὶν ὁ ἥλιος διὰ τὴν ἐπὶ τοῦ ἐπικύκλου τεταρτημοριαίαν ἰσοταχῆ πάροδον αὐτοῦ εἰς τὰ ἡγούμενα, ὡς ὑπεθέμεθα, ἀπὸ τοῦ ζ ἐπὶ τὸ ν. καὶ ἔστιν ἡ διαφορὰ τῆς ὁμαλῆς καὶ ἀνωμάλου κινήσεως ἡ ὑπὸ νκβ γωνία, ἤτοι ἡ νοουμένη περιφέρεια ἐπὶ τῆς βν εὐθείας, ἥτις ἐστὶν ἴση, ὡς ὑπόκειται, τῇ ἐπὶ τῆς μκ περιφερείᾳ, τῆς μεταξὺ δηλονότι τῶν δύο κέντρων, τοῦ τε ἐκκέντρου κύκλου καὶ τοῦ ὁμοκέντρου τῷ ζωδιακῷ, ὡς ἐπὶ τῆς πρώτης καταγραφῆς. οὕτω δὴ τὸ αὐτὸ μὲν συμβαίνει καθ᾽ ἑκατέραν ὑπόθεσιν διάφορον τῆς ὁμαλῆς κινήσεως πρὸς τὴν ἀνώμαλον ἐπὶ τοῦ ἡλίου.

76 ἑκατέραν : ἑτέραν C^{ac}

Let us again visualise a circle homocentric to the ecliptic, the αβγδ, whose centre is κ. This circle carries the epicycle ζθμν with its centre on the [circumference of] αβγδ. Let us consider the radius from point α to point θ on the circumference, or to point ν, or to any other point, to be equal [to the distance] between the two points κ and μ of the other hypothesis. Let the epicycle move from α to β, and the Sun move uniformly and in the opposite sense one quadrant from ζ to ν. In this case, the rotation of the Sun on the eccentric [circle] relative to the zodiac (without an epicycle), or the rotation on an epicycle of the appropriate size, produces the same difference between the uniform and the anomalous motion. This becomes evident as follows: when the Sun moves on the eccentric (not on the epicycle) it subtends the angle ακλ; [relative to this] the angle ζμχ is smaller by the angle κλμ, as it follows from the 32nd theorem of the first book of the *Elements*. This is the difference between the uniform and anomalous motion for the circumferences corresponding to [centres μκ]. Alternatively, when (according to the second description) the epicycle of the Sun moves on the homocentric by one quadrant and the Sun rotates on the epicycle, the uniform motion is ακβ and the anomalous the ακν. The Sun finds itself at the point ν due to the uniform backward movement on the epicycle from point ζ to ν. The difference between the uniform and anomalous motions is the angle νκβ, i.e. the circumference corresponding to the straight line βν, which is equal to the corresponding distance μκ, between the two centres of the eccentric and homocentric to the ecliptic of the first description. Thus, in each hypothesis the same result follows for the deviation of uniform from the anomalous rotation of the Sun.

Καὶ ἐπεὶ διὰ τῶν τηρήσεων ἓν δὴ καὶ ἁπλοῦν, ὡς ἔφημεν, ἀεὶ ὡσαύτως ἔχον κατελήφθη ἐπιμελῶς τὸν νοῦν προσχοῦσι καὶ ἀκριβῶς ἐπισκεψαμένοις τοῖς μαθηματικοῖς ἐπὶ τοῦ ἡλίου τὸ διάφορον τῆς ὁμαλῆς κινήσεως πρὸς τὴν ἀνώμαλον, καὶ εἷς ἐστι πάντως τρόπος ἐκ τῶν εἰρημένων δύο ὁ τοῦτο ποιῶν, δυνατόν ἐστι πάντως καθ' ὁπότερον ἄν τις βούλοιτο τῶν δύο τρόπων ἐπισυμβαίνειν τὰ τῆς τοιαύτης ἡλιακῆς ἀνωμαλίας καὶ τῇ ἀληθείᾳ ἀμφίβολόν ἐστιν. ἐπίσης γάρ, ὡς ἔφημεν, δυνατόν ἐστι τὴν τοιαύτην ἀνωμαλίαν ἐπὶ τοῦ ἡλίου καταλαμβάνεσθαι καὶ δι' ἐκκεντρότητα καὶ δι' ἐπίκυκλον, δοκεῖ δ' οὖν ὅμως Πτολεμαίῳ βέλτιον εἶναι δι' ἐκκεντρότητα μᾶλλον ἀποφαίνεσθαι τὴν τῆς ἡλιακῆς κινήσεως ἐπισυμβαίνειν ἀνωμαλίαν ἢ διὰ τὸν ἐπίκυκλον. αὐτὸς γάρ φησιν ὁ Πτολεμαῖος καὶ Πυθαγόρειοι πάντες καὶ μαθηματικοὶ προσῆκον εἶναι τὸν φιλόσοφον ἐν τοῖς περὶ τῶν οὕτω μεγίστων καὶ τιμίων λόγοις ταῖς ἁπλουστέραις μᾶλλον χρῆσθαι τῶν ὑποθέσεων. ἁπλούστερος δὲ ὁ τρόπος ὁ διὰ τὴν ἐκκεντρότητα ἤπερ ὁ διὰ τὸν ἐπίκυκλον. ἐπὶ μὲν γὰρ τοῦ τρόπου τῆς ἐκκεντρότητος μία νοεῖται κίνησις, ἡ τοῦ ἡλίου ἐπὶ τοῦ ἐκκέντρου κύκλου αὐτοῦ· ἐπὶ δὲ τοῦ κατ' ἐπίκυκλον τρόπου δύο τὲ ἐξανάγκης ὑποτίθενται κινήσεις, μία μὲν ἡ τοῦ ἐπικύκλου ἐπὶ τοῦ ὁμοκέντρου τῷ ζωδιακῷ, δευτέρα δὲ ἡ τοῦ ἡλίου αὐτοῦ ἐπὶ τοῦ ἐπικύκλου. καὶ πρὸς τούτοις ἔτι ὑποτίθεσθαι ἀναγκαῖον ἰσοταχεῖς τε εἶναι ταύτας τὰς κινήσεις καὶ ἐπὶ τὰ ἐναντία, τὴν μὲν εἰς τὰ ἑπόμενα, τὴν δὲ εἰς τὰ ἡγούμενα. ὅθεν δὴ καὶ μᾶλλον ἐκράτησεν ἡ δόξα παρὰ τοῖς ἀστρονόμοις αὕτη, μὴ ἐπὶ ἐπικύκλου κινεῖσθαι μόνον τὸν ἥλιον κατὰ τοὺς ἄλλους, ἀλλ' ἐπὶ ἐκκέντρου μόνον κύκλου καὶ αὐτὸν κατὰ τοὺς ἄλλους.

86 Cf. Ptol. *Alm.* 3.4 (vol. 1.1, p. 232.14-17 Heiberg) **89** Cf. Ptol. *Alm.* 3.1 (vol. 1.1, p. 201.18-20 Heiberg), 3.4 (vol. 1.1, p. 232.14-17 Heiberg), 13.2 (vol. 1.2, p. 532.19-21 Heiberg)

Those who contemplate and deal accurately with the mathematics of the Sun established from these observations one simple difference between the uniform motion and the anomalous, occurring always in the same way. Its [physical] origin is due to only one of the two models discussed so far, even though it is possible if one wishes, to produce the solar anomaly in two ways, which brings in an ambiguity: it is equally possible to understand it in terms of the eccentricity or the epicycles. Ptolemy, however, is of the opinion that it is better to conclude that the anomaly of the Sun originates from the eccentricity rather than from the epicycle. As stated by Ptolemy, by all the Pythagoreans, and the mathematicians, it is preferable for the philosopher in the greatest and most honourable theories to choose the simplest hypotheses. The simplest model is the eccentric model over the epicycle, because in the eccentric we conceive one motion for the Sun on the eccentric circle. For the model with epicycles it is necessary to introduce two movements: one for the motion of the epicycle on the homocentric and a second for the motion of the Sun on the epicycle. In addition, it is necessary to postulate that the rotations are uniform but in opposite directions, one forward and the other backward. From this the following opinion established itself among the astronomers, namely the Sun does not move on an epicycle as happens for other heavenly bodies but moves on an eccentric circle, in contrast to the other [celestial] bodies.

19 ὅτι οἱ ἀστρονόμοι αἰγυπτιακοῖς ἔτεσιν ἀπαριθμοῦνται τὰς τῶν ἀστέρων κινήσεις καὶ οὐχ ἑλληνικοῖς εἴτουν ῥωμαϊκοῖς, καὶ τίς ἡ τῶν αἰγυπτίων ἐτῶν πρὸς ταῦτα διαφορά

Ἀλλ᾽ ἐπειδὴ τὰς ὑποθέσεις καὶ τοὺς τρόπους εἴπομεν τῶν ὁμαλῶν καὶ ἀνωμάλων κινήσεων τῶν ἀστέρων, καὶ ὅτι ἐπὶ τῆς τοῦ ἡλίου κινήσεως εἷς ἐστι καὶ ἁπλοῦς ὁ τρόπος τῆς ἀνωμάλου κινήσεως αὐτοῦ, καὶ ἐξανάγκης μὲν εἷς τῶν εἰρημένων δύο νικᾷ δὲ μᾶλλον καὶ κρατεῖ, ὁ δι᾽ ἐκκεντρότητα ἢ ὁ δι᾽ ἐπίκυκλον, φέρε λοιπὸν ἀρξόμεθα τῆς περὶ τῆς ἑκάστου κινήσεως τῶν πέντε πλανωμένων καὶ τοῦ ἡλίου καὶ τῆς σελήνης διδασκαλίας καὶ διασαφήσεως, καὶ πρῶτόν γε κατὰ τὸ εἰκὸς ὡς περὶ ἁπλουστέρας καὶ τιμιωτέρας τῆς τοῦ ἡλίου κινήσεως, εἶθ᾽ ἑξῆς καὶ περὶ τῆς τῶν ἄλλων. ἐπεὶ δ᾽ ἐν χρόνῳ πᾶσα κίνησίς ἐστιν ἐξανάγκης καὶ κατὰ χρονικὰ διαστήματα μέλλομεν ἐκθέσθαι καὶ ἐπιλογίσασθαι τὰς κατὰ μέρος κινήσεις τοῦ ἡλίου, ὡς ἐν τοσῷδε χρόνῳ τοσόνδε κινεῖται διάστημα, ἤτοι ἐν ἡμέρᾳ τοσοῦτον καὶ ἐν νυκτὶ τοσοῦτον καὶ δι᾽ ἔτους τοσοῦτον, ἀναγκαῖόν ἐστι προδιαλαβεῖν σαφηνείας ἕνεκεν καὶ ἀσυγχύτου διδασκαλίας, ὁποίοις χρῶνται οἱ ἀστρονόμοι τοῖς χρονικοῖς ἀριθμοῖς καὶ διαστήμασι, καὶ ὅτι οὐχ ἑλληνικοῖς ἔτεσιν οὐδὲ ῥωμαϊκοῖς, ἀλλ᾽ αἰγυπτιακοῖς, καὶ τίς ἡ τούτων διαφορὰ πρὸς τὰ ἑλληνικὰ καὶ ῥωμαϊκά.

Καὶ μετὰ τὴν τούτων διασάφησιν, ἀναγκαίαν οὖσαν καὶ ἀπαραίτητον, ὡς ἔστι δῆλον, αὐτόθεν εἶθ᾽ οὕτως χωρῆσαι πρὸς τὴν διδασκαλίαν καὶ τὰς κανονικὰς κατὰ τοὺς ὡρισμένους χρόνους ἐκθέσεις τῆς ἡλιακῆς κινήσεως. ἰστέον δὴ ὅτι οἱ ἀστρονόμοι τοῖς αἰγυπτιακοῖς χρῶνται ἔτεσιν ἐν πάσαις ταῖς ψηφοφορίαις καὶ ἐπιλογισμοῖς τῶν κινήσεων τῶν ἀστέρων. ἐμοὶ δοκεῖν ὅτι ἐξ ἀρχῆς εἰς τοῦτο τὸ μάθημα πλέον τῶν ἄλλων ἐσχόλασαν Αἰγύπτιοι καὶ Χαλδαῖοι. εἰκὸς μὲν οὖν καὶ Χαλδαίους ἐν πάσαις ταῖς αὐτῶν ψηφοφορίαις τοῖς κατ᾽ αὐτοὺς

1 τὰς ὑποθέσεις καὶ om. C **5** ἀρξόμεθα : ἀρξώμεθα C (cf. 5.337) **8** ὡς περὶ : ὥσπερ V[ac]

19 The astronomers compute the motions of the stars according to the Egyptian and not the Graeco-Roman calendar. Description of the difference between the Egyptian calendar and the others

We discussed the hypotheses and methods used for describing the uniform and anomalous rotations of the stars, and pointed out that the Sun has one simple rotation, where one of the two models prevails: either the eccentric or the epicyclical. Let us now begin with the exposition and explanation of the rotations for each of the five planets, of the Sun and of the Moon. It is appropriate to begin with the simplest and most important rotation, that of the Sun, and then discuss the others. Since every motion takes place in time and we will present and calculate the specific rotations of the Sun in intervals of time, as the distance travelled in a time interval, i.e. in a day, in a night and in a year, it becomes necessary to specify from the very beginning and for the sake of clarity which units and intervals of time are used by the astronomers, because the astronomers do not use the Graeco-Roman years but the Egyptian, which differ from each other.

After this necessary and required clarification — which is evident —, we move in our instruction with the exposition of the uniform rotation for the Sun into specific time intervals. It is known that astronomers use in all calculations and deliberations Egyptian years. It is my opinion that from the very beginning the Egyptians and the Assyrians more than any people studied this field; thus it is very likely that the Assyrians used in all their calculations

νομίμοις χρόνοις οἷστισιν ἄρα χρῆσθαι· ἡμεῖς δ' οὖν, Ἕλληνές τε καὶ Ῥωμαῖοι, τὴν ἐπιστήμην ταύτην παρ' Αἰγυπτίων διαδεξάμενοι καὶ τὴν κατ' ἐκείνους τῶν χρόνων τάξιν ἐξ αὐτῶν παραλαβόντες καὶ τὰς κατ' αὐτὴν τὴν τάξιν τῶν χρόνων κανονικὰς ἐκθέσεις δηλοποιούσας ἐν ἀριθμοῖς κατὰ μέρος τὰς κινήσεις τῶν ἀστέρων, ὡς διεδεξάμεθα ταύτας καὶ παρελάβομεν, ἀκόλουθον κρίνομεν οὕτω δὴ συντηρεῖν, διὰ τὸ ἀσφαλὲς τῆς παραδόσεως τῆς παρὰ τῶν ἐπιστημόνων ἐκείνων, καὶ τοὺς ἐπιλογισμοὺς καὶ τὰς ψηφοφορίας κατὰ τὴν χρονικὴν ἐκείνων παράδοσιν ποιεῖσθαι.

Δοκεῖ δὲ καὶ ἄλλως εὐόριστον εἶναι μᾶλλον καὶ ῥᾷον εἰς κρίσιν τὸ τοὺς ἐπιλογισμοὺς τῶν κινήσεων ποιεῖσθαι καθ' ἡμέρας ὁλοκλήρους καὶ τελείας ἢ καθ' ἡμέρας ὁσασδήποτε καὶ μέρος ἐπ' αὐταῖς ὁτιοῦν ἡμέρας. αἱ ἐτήσιοι γοῦν περίοδοι καὶ οἱ χρόνοι Αἰγυπτίοις μὲν συμπεραίνονται εἰς τελείας ἡμέρας, Ἕλλησι δὲ καὶ Ῥωμαίοις εἰς ἡμέρας τὲ καὶ μέρος ἢ μέρη μιᾶς ἡμέρας. κατὰ τοῦτο γὰρ διαφέρει τὸ αἰγύπτιον ἔτος τοῦ ἑλληνικοῦ καὶ ῥωμαϊκοῦ, ὅτι Ῥωμαῖοι μὲν καὶ Ἕλληνες συμπεραίνουσι καὶ καταλογίζονται τὸν ἐτήσιον χρόνον δι' ἡμερῶν τξε καὶ τετάρτου, ὅτι δὴ καὶ διὰ τοσούτου τὴν οἰκείαν κυκλικὴν περίοδον ὁ ἥλιος ἀπαρτίζει καὶ ἀφ' οὗ σημείου κεκίνηται ἀποκαθίσταται εἰς αὐτό· Αἰγύπτιοι δὲ τὸν ἐτήσιον συμπεραίνουσι χρόνον δι' ἡμερῶν τξε ὁλοκλήρων μόνων καὶ τὸ λοιπὸν τέταρτον μόριον τῆς ἡμέρας, ὃ συναπαρτίζει τὴν τῆς ἡλιακῆς κινήσεως ἀποκατάστασιν, παρεῶσι καὶ ἀρχὴν τίθενται ἑτέρου ἔτους· ὡς ἐντεῦθεν ἀκολουθεῖν ἐξανάγκης, κατὰ τέτταρα ἔτη συντιθεμένων τῶν τοιούτων τεταρτημορίων ἡμέραν ὁλόκληρον ἀπαρτίζεσθαι καὶ προλαμβάνειν τὸ ἑλληνικὸν ἔτος ἐν ἡμέρᾳ μιᾷ.

Οἷον συμπληρωθέντων τεττάρων ἐτῶν ἑλληνικῶν καὶ ῥωμαϊκῶν αὐτοῖς μὲν Ἕλλησι καὶ Ῥωμαίοις, οὕτως ἄρχεται τὸ πέμπτον ἔτος. Αἰγύπτιοι δὲ συμπληρωθέντων τῶν τοιούτων τεττάρων ἑλληνικῶν τε καὶ ῥωμαϊκῶν ἐτῶν οὐκ ἀρχὴν τίθενται τὴν μετὰ τὴν συμπλήρωσιν πρώτην ἡμέραν ἑτέρου πέμπτου ἔτους, ἀλλὰ δευτέραν ταύτην ἡμέραν τοῦ ἑξῆς πέμπτου ἔτους τίθενται, προλαμβάνοντες καὶ προτιθέντες

28 αὐτῶν : ἐκείνων C **36** τῶν κινήσεων post ποιεῖσθαι transp. C **51** ἐτῶν post ἑλληνικῶν transp. C

the standard units of time. We Greeks and Romans inherited this science from the Egyptians and accepted their calendar. We used their calendar in order to present numerically the motions of the stars in our teachings, and subsequently we decided to keep it in order to maintain with certainty and for the sake of continuity the procedures, the numerical values and the traditions of these scientists.

Furthermore, it is my opinion that the various properties of the rotations are better defined and more easily distinguished when we carry the calculations in full and complete days rather than in a number of days plus a fraction of a day. The annual times and periods of the Egyptians are completed in full days, for the Greeks and the Romans in days and fractions of a day because the Romans and Greeks construct and count the year in 365 plus one-quarter days. In this time interval the Sun completes its own revolution and returns to the starting point. The Egyptians adopted the convention that the year consists of only 365 full days, with the remaining one-quarter day, which completes a full rotation, being carried over to the beginning of the next year. A necessary consequence of this is that every four years, the added one-quarter days comprise a whole day and is ahead of the Greek year by one day.

For the Greeks and Romans, whenever four of their years are completed then immediately after that date begins the fifth year. For the Egyptians, however, when the four Graeco-Roman years are completed they do not have the first day of a new year, but the second day of the fifth year. They are ahead

ἡμέραν πρώτην τοῦ τοιούτου ἑξῆς πέμπτου ἔτους τὴν συντιθεμένην ἐκ τῶν εἰρημένων τεταρτημορίων τῶν παρελθόντων τεττάρων ἐτῶν. καὶ οὐ συγκαταλογίζονται, ὡς ἡμεῖς, κατὰ μέρος αὐτὴν δὴ τὴν ἡμέραν τοῖς παρελθοῦσιν ἔτεσιν, ἀλλ᾽ ὡς ἀρχὴν καὶ πρώτην τοῦ ἑξῆς πέμπτου ἔτους τίθενται. καὶ οὕτω δὴ ἐκ τῆς προσθήκης ταύτης κατὰ τέτταρα ἔτη κατ᾽ αὐτούς, ἤτοι αἰγυπτιακά, προστιθέντες ἡμέρας καὶ μῆνα ἢ μῆνας ἑξῆς ἀπαρτίζουσιν, οἷς καὶ προλαμβάνουσι τὰς ἀρχὰς τῶν καθ᾽ ἡμᾶς ἐτῶν, ὡς ἄρχεσθαι τὸ ἔτος αὐτοῖς πρότερον καὶ προλαμβάνειν τοῦ καθ᾽ ἡμᾶς ἔτους ἐν ἑνὶ μηνὶ ἢ δυσὶν ἢ καὶ πλείοσι, καὶ διὰ αυξ ἐτῶν τοιούτων αἰγυπτιακῶν ἀπαρτίζεσθαι τέλειον αὐτοῖς τοιοῦτον πάλιν ἔτος, ἤτοι τξε ἡμερῶν, καὶ πάλιν κοινὴν ἀρχὴν ποιεῖσθαι ἔτους μετὰ Ἑλλήνων καὶ Ῥωμαίων.

Ἐπεὶ γοῦν τοῖς μὲν Αἰγυπτίοις οὕτως εἰς ὁλοκλήρους καὶ τελείας ἡμέρας τὸ ἔτος ἀπαρτίζεται, ἤτοι τὰς τξε ἡμέρας, καὶ ἀεὶ ὡσαύτως τοῦτ᾽ ἔχει καὶ ἴσον ἐστὶν αὐτοῖς τὸ ἔτος, Ἕλλησι δὲ καὶ Ῥωμαῖοις οὐχ οὕτως ἔχει, ἀλλὰ νῦν μὲν ἀρξάμενον τὸ ἔτος ἀπὸ τῆσδέ τινος τῆς ὥρας καὶ καταντῆσαν εἰς τελείας ἡμέρας κατ᾽ αὐτὴν τὴν ὥραν χρῄζει καὶ προσθήκης τεταρτημορίου ἡμέρας· νῦν δὲ περιοδεῦσαν καὶ καταντῆσαν πάλιν εἰς τὴν αὐτὴν ὥραν τὴν ἐξαρχῆς χρῄζει καὶ προσθήκης ἡμίσεος μέρους ἡμέρας· καὶ αὖθις περιοδεῦσαν καὶ ἀποκαταστὰν εἰς τὴν αὐτὴν πρώτην ἐξαρχῆς ὥραν χρῄζει καὶ προσθήκης ἑτέρας εἰς ἀποκατάστασιν ἡμίσεος καὶ ἔτι τετάρτου μορίων ἡμέρας· καὶ αὖθις τετράκις περιοδεῦσαν τὸ ἔτος εἰς τὴν ἐξαρχῆς ἐκείνην καὶ πρώτην ὥραν χρῄζει ἐπ᾽ αὐτῇ εἰς ἀπαρτισμὸν καὶ ἔτι ἡμέρας ὁλοκλήρου. καὶ ἔστιν οὕτω διόλου ἡ ἀποκατάστασις ἄνισος τῶν ῥωμαϊκῶν καὶ ἑλληνικῶν ἐτῶν καὶ πολὺ τὸ διάφορον ἔχουσα, εὐεπιλογιστότερόν ἐστι τῇ ἐπιστήμῃ τὸ ἀεὶ ὡσαύτως ἔχον καὶ ὡρισμένον. καὶ κατὰ τὰς αὐτὰς καὶ ἴσας καὶ ὡρισμένας ἀποκαταστάσεις ῥᾷον ἐκτίθενται πάντως καὶ ἁπλούστερον αἱ ἐπιτελούμεναι κατ᾽ αὐτὰς ἐνέργειαι δή τινες καὶ κινήσεις, καὶ ἡ περὶ τῶν τοιούτων κανονικὴ ἔκθεσις καὶ ἀπαρίθμησις προχειροτέρα καὶ ἀπονωτέρα καὶ σαφεστέρα.

65 post ἐτῶν add. τῶν C **66** τέλειον post αὐτοῖς transp. C **71** ἐστὶν post αὐτοῖς transp. C **82** post ἔχουσα add καὶ C[pc] (manus secunda)

by adding the day that resulted from the one-quarter days of the previous years, which brings them to the end of the first day of the fifth year. They do not count the fractions of the days in the previous four years like us, but they add one day in the fifth year. In this way the Egyptians, by supplementing a full day every four years, complete a month or months, which are found at the beginning of our years. Their years begin earlier and are ahead from us by one, two, or more months. In 1460 Egyptian years the excess amounts to a complete year of 365 days, at which time the beginning of their year coincides with the beginning of the Graeco-Roman year.

To sum up, the Egyptian year contains 365 full and complete days and is always the same and all their years are equal. The Greeks and the Romans do not count time this way, but when the year starts at some specific hour and arrives at a complete day, at that moment it is necessary to add one-quarter of a day. Then, advancing and arriving at the same starting hour, it is necessary to add half a day; again advancing and arriving at the starting hour, it is necessary to add half and one-quarter days. Once more, when it travels through four years, arriving at the starting hour, it becomes necessary to add a whole additional day. This way we restore the Graeco-Roman years, which are unequal and contain many differences. However, in science it is better to compute always with the same and precise rules, and following them, to achieve the same, equal and precise return to the starting point. Finally, presenting the development in time and their orbits with the same rules makes the exposition and the counting simpler, clearer and requires less effort.

Κατὰ τούτους τοὺς λόγους ἐμοὶ δοκεῖν προείλοντο οἱ ἀστρονόμοι τοὺς ἐπιλογισμοὺς τῶν κινήσεων τῶν ἀστέρων καὶ τὰς κανονικὰς ἐκθέσεις ποιήσασθαι κατὰ χρονικὰ διαστήματα καὶ ἔτη αἰγύπτια· εἰ δὲ καὶ κατ᾽ ἄλλους τινὰς τρόπους, πολυπραγμονεῖν νῦν οὐκ ἀναγκαῖον· ὃ δ᾽ οὖν ἐλέγομεν, τοῦτ᾽ ἐστίν, ὅτι οἱ ἀστρονόμοι ἐν ταῖς ψηφοφορίαις καὶ τοῖς ἐπιλογισμοῖς τῶν κινήσεων τῶν ἀστέρων ἔτεσιν αἰγυπτιακοῖς χρῶνται καὶ οὐχ ἑλληνικοῖς ἢ ῥωμαϊκοῖς καὶ τοῦτο σύνηθες τοῖς ἐξαρχῆς περὶ τὴν ἀστρονομικὴν ἐπιστήμην σπουδάσασι. καὶ ἡμεῖς γοῦν οὕτως ἀνάγκην ἔχομεν χρήσασθαι τοῖς προλαβοῦσιν ὡς εἰκὸς ἑπόμενοι καὶ προμηθευόμενοι τοῦ ἀσφαλοῦς τῇ παρούσῃ Στοιχειώσει τῆς ἐπιστήμης ταύτης καὶ διδασκαλίᾳ τῶν κινήσεων καὶ τῶν τόπων, καθ᾽ οὓς εὑρίσκονται ἑκάστοτε οἵ τε πέντε πλάνητες καὶ ἡ σελήνη καὶ ὁ ἥλιος.

For these reasons astronomers chose to describe and present their calculations for the rotations of the stars in terms of Egyptian time intervals and years. We do not consider it necessary to elaborate any more on the other methods, but only to say that astronomers use in their calculations the Egyptian years and not the Graeco-Roman. From the very beginning this has been the practice for those who study astronomy. For us, who live many generations later, it is natural to take over their tables for the locations and motions of the Sun, Moon and the five planets. This makes the presentation in this Treatise more reliable.

20 πῶς ἐμεθοδεύθη τοῖς ἀστρονόμοις ἡ τῶν ὁμαλῶν κινήσεων κατὰ χρόνους ὡρισμένους ἀπαρίθμησις, καὶ ὅτι ἀναγκαῖον ταύταις ὑποτεθῆναι τροπικὴν ἀρχήν, ἀφ᾽ ἧς ἔσται ἡ προχώρησις καὶ ἡ ἀπαρίθμησις, ἔτι γε μὴν καὶ χρόνου τινὸς ἀρχήν, ἀφ᾽ οὗ ἔσται ὡσαύτως ὁ ἐπιλογισμὸς τῶν ὁμαλῶν κινήσεων ἡλίου καὶ σελήνης καὶ τῶν εἰρημένων ἄλλων ἀστέρων

Ἐπεὶ γοῦν ἡ πρόθεσις ἡμῖν νῦν ἐστι περὶ τοῦ ἡλίου καὶ πῶς οἷόν τέ ἐστι καθ᾽ οἱονδήτινα καιρὸν δεδομένον, εἴτε τυχὸν νῦν ἐφ᾽ ἡμῶν εἴτε καὶ κατὰ τοὺς παρελθόντας χρόνους εἴτ᾽ ἴσως καὶ κατὰ τοὺς μέλλοντας, ἀνευρίσκειν ἀκριβῶς καὶ ἐπιλογίζεσθαι τὴν ἐποχὴν αὐτοῦ, ἤτοι καθ᾽ οἵου τόπου καὶ τμήματός ἐστι τοῦ ζωδιακοῦ, ἀναγκαῖον πρῶτον κανονικῶς ἐκθέσεις ποιήσασθαι τῶν χρονικῶν αὐτοῦ κινημάτων τῶν ὁμαλῶν. τοῦτο γὰρ χρὴ πρότερον προϋποστήσασθαι καὶ διευκρινῆσαι, ἤτοι τὰ διαστήματα τῆς ὁμαλῆς αὐτοῦ κινήσεως, ἐξαρχῆς προϋποκειμένου πάντως τοῦ καὶ τὸν ἥλιον καὶ τὴν σελήνην καὶ τοὺς πέντε πλανωμένους κατὰ τὸ εἰκὸς τῇ τῶν ὄντων εὐταξίᾳ, ὡς πολλάκις εἴρηται, ὁμαλῶς φέρεσθαι καὶ ἐν ἴσοις χρόνου τμήμασιν ἶσα παροδεύειν τμήματα ἐπὶ τῶν οἰκείων κύκλων, οὓς ἅπαντας ἐκκέντρους εἶναι πρὸς τὸν ζωδιακὸν προέφημεν. μετὰ δὲ τὴν τοιαύτην παράδοσιν καὶ κανονικὴν ἔκθεσιν τῶν ὁμαλῶν τοῦ ἡλίου κατὰ χρόνους ὡρισμένους κινήσεων, ἔπειθ᾽ οὕτως καὶ τὰς ἀνωμάλους αὐτοῦ κινήσεις καὶ τὰς ἑκάστοτε ἐποχὰς αὐτοῦ καὶ τόπους ἀληθεῖς καὶ ἀκριβεῖς λεγομένους παρὰ τῶν ἀστρονόμων μεθοδεύσομεν κατὰ τὴν τῶν παλαιοτέρων ἐπιστημόνων ἀσφαλεστάτην εὕρεσιν, καὶ κανονικῶς καὶ τὰ περὶ τούτων ἐκθησόμεθα.

Διῄρηνται τοίνυν παρὰ τῶν ἀστρονόμων αἱ ὁμαλαὶ τοῦ ἡλίου κινήσεις κατὰ χρόνους ὡρισμένους τὸν τρόπον τοῦτον. ἐπεὶ γὰρ ὁ ἥλιος ἐν τξε ἡμέραις καὶ τετάρτῳ μορίῳ ἡμέρας περιοδεύει τὸν οἰκεῖον κύκλον καὶ ἀποκαθίσταται ἀπὸ τοῦ αὐτοῦ σημείου εἰς τὸ αὐτό, ὁ δὲ

tit. post ἀναγκαῖον add. καὶ C^{sl} **tit.** τροπικὴν : τοπικὴν C **8** αὐτοῦ post κινήσεως transp. C

20 How did astronomers come to the idea to calculate the uniform movement in certain time periods, and why it is necessary to postulate a beginning, from which we count the subsequent movements. Indeed, we specify a time from which we begin to calculate the uniform rotations of the Sun, the Moon and the other planets

It is our intension now to deal with the Sun and show how it is possible to accurately compute for a given time (for example, now or in the past or future) its longitude, i.e. its position at a section of the ecliptic. To this end, it is necessary first to present in a proper manner its regular motions. However, before doing this, it is necessary to establish and specify the intervals of its uniform motion, taking as starting principle that reasonable order governs all beings, including the Sun, the Moon and the five planets. As we have mentioned many times, everything moves uniformly and travels equal intervals in equal times on their circles, which are eccentric to the ecliptic. After this exposition and precise presentation of the uniform rotations of the Sun as a function of time, we shall in addition present its anomalous motion and longitudes for specific times, which astronomers call true and precise. For this we shall follow the methods of ancient scientists, which are most reliable for presenting the results accurately.

Astronomers divide the uniform motions of the Sun according to time intervals in the following way. In 365 and a quarter days the Sun travels its own circle, returning back to the point where it started,

κύκλος τέμνεται εἰς τξ μοίρας, ὡς πολλάκις ἔφημεν, ἀναλύοντες τὰς τοιαύτας μοίρας εἰς τὰς εἰρημένας ἡμέρας εὑρίσκουσιν οἱ ἀστρονόμοι μιᾷ ἑκάστῃ ἡμέρᾳ ἀνήκειν, κατὰ ὁμαλὴν τῆς αὐτῆς παρόδου διαίρεσιν, κινεῖσθαι τὸν ἥλιον μιᾶς μοίρας ἑξηκοστὰ νθ. ἐπεὶ δὲ ἑκάστη ἡμέρα εἰς εἰκοσιτέσσαρας ὥρας ἰσημερινὰς διαιρεῖται, τέμνουσι τὸ εἰρημένον ἡμερήσιον κίνημα τοῦ ἡλίου εἰς κδ τμήματα, καὶ τὸ κδ τούτου αὐτό ἐστι τὸ ὡρισμένον ὁμαλὸν κίνημα τοῦ ἡλίου. πάλιν τριακοντάκις ποιήσαντες τὸ εἰρημένον ἡμερήσιον ὁμαλὸν κίνημα τοῦ ἡλίου, τὸν συντιθέμενον ἐντεῦθεν ἀριθμὸν αὐτόν φασιν εἶναι ἡλιακὸν κίνημα μηνός. καὶ τοῦτο δωδεκάκις ποιήσαντες ἔτι τὲ προσθέντες τῷ συντιθεμένῳ ἐντεῦθεν ἀριθμῷ καὶ πέντε ἡμερῶν κινήματα κατὰ τὸ ἀνωτέρω εἰρημένον ἡμερήσιον κίνημα, ἐντεῦθεν ἀπαρτίζουσι τοῦ αἰγυπτιακοῦ ἔτους, ἤτοι τῶν τξε ἡμερῶν, τὸν ἀριθμὸν τῆς κινήσεως τοῦ ἡλίου. καὶ οὕτω δὴ λοιπὸν ἀναφαίνεται ὁπόσον ὁ ἥλιος κινεῖται καθ᾽ ἕκαστον ὡρισμένον χρόνον, ἤτοι κατὰ ὥραν, κατὰ ἡμέραν, κατὰ μῆνα καὶ κατὰ ἔτος αἰγυπτιακόν.

Εἶθ᾽ οὕτω κανόνας ἐκτίθενται τῶν τοιούτων χρονικῶν τοῦ ἡλίου κινήσεων· καὶ προηγουμένως ἐκτίθενται κανόνα τῶν ὡρῶν, πόσον κινεῖται ὁ ἥλιος ἐν μιᾷ ὥρᾳ, πόσον ἐν δευτέρᾳ, πόσον ἐν τρίτῃ καὶ ἑξῆς μέχρι τῆς εἰκοστῆς τετάρτης. εἶτα συντιθέντες τὰ τῶν κδ ὡρῶν κινήματα, ἡμερήσιον ταῦτα ὁμοῦ ποιοῦνται κίνημα. καὶ ἑξῆς τιθέασιν ἕτερον κανόνα δεύτερον, τῶν ἡμερῶν, ἤτοι πόσον ἐν μιᾷ ἡμέρᾳ ὁ ἥλιος κινεῖται, πόσον ἐν δευτέρᾳ, πόσον ἐν τρίτῃ καὶ ἑξῆς μέχρι τῆς τριακοστῆς. εἶτα συντιθέντες τὸ τῶν λ ἡμερῶν κίνημα τοῦ ἡλίου, μηνὸς ἑνὸς κίνημα τοῦτο ἀποφαίνονται, ἐπειδὴ καὶ Ἕλληνες καὶ Αἰγύπτιοι τὸν μῆνα τριακονθήμερον τίθενται. καὶ δὴ λοιπὸν ἐπὶ τούτοις ἕτερον κανόνα τῶν μηνῶν ἐκτίθενται, οἷον πόσον ὁ ἥλιος κινεῖται ἐν ἑνὶ μηνί, πόσον ἐν δευτέρῳ, πόσον ἐν τρίτῳ καὶ ἑξῆς μέχρι τῶν δώδεκα μηνῶν. καὶ ἐπεὶ οἱ ιβ μῆνες ἀνὰ λ ἡμέρας ἔχοντες οὐδὲν ἀπαρτίζουσι τὸ αἰγυπτιακὸν ἔτος, ἀλλὰ χρῄζουσι καὶ ἡμερῶν πέντε, ἐπὶ τούτοις ἐκτίθενται καὶ τῶν πέντε ἡμερῶν τὸ κίνημα. καὶ ὁμοῦ συντιθέντες τὰ τῶν ιβ μηνῶν κινήματα καὶ τὰ τῶν ε ἡμερῶν, ὃν καὶ ἐλάχιστον μῆνα

26 νθ deest V (rasura)

and each circle is divided in 360 degrees. Astronomers divide the degrees by the aforementioned days and thus find that the Sun rotates each day 59′ of a degree. They divide the daily motion of the Sun into 24 parts, with each one-twenty-fourth being the hourly uniform motion the Sun. Then again, taking thirty times the daily uniform motion, they obtain a number called the monthly motion of the Sun. Then, multiplying the result by twelve and adding to the product the motion of five days, already determined above, they form the Egyptian year, that is the motion of the Sun in 365 days. This way it becomes evident how much the Sun moves at specific time intervals that are [classified] in an hour or a month or in an Egyptian year.

[The results] are presented in tables for the uniform motion of the Sun and for the aforementioned time intervals. First are the tables for hours: how much does the Sun move in one hour, in the second hour, in the third hour up to 24 hours, which when added together comprise the daily motion. Then they place another Table for the days. This gives how much the Sun moves in one day, in two, in three up to the thirtieth day, which all together comprise the rotation in one month. Since the Greeks and the Egyptians have thirty-day months, it is necessary to present them in a Table of months. And in fact they place another table that gives how much the Sun moves in the first month, in the second month, in the third and so on up to twelve months. Now, since the 12 months of 30 days each do not complete an Egyptian year, being shorter by five days, they also present the motion for five days. Then, combining the rotation for 12 months plus that of five days, which astronomers call minimal

καὶ ἐπαγόμενον οἱ ἀστρονόμοι καλοῦσι, τὸν ἀπὸ τούτων ὁμοῦ πάντων συναγόμενον ἀριθμὸν ἔτους αἰγυπτιακοῦ κίνησιν ἡλιακὴν τίθενται. καὶ τὰ μὲν ἡλιακὰ κινήματα οὕτω κατὰ τοὺς ὡρισμένους εἰρημένους χρόνους ἐπιλογίζονται οἱ ἀστρονόμοι.

Ἐπεὶ δὲ πᾶσα κίνησις ἀπὸ χρόνου τινὸς λαμβάνει ἀρχήν, ὡσαύτως δὲ καὶ ἀπὸ τόπου ὡρισμένου, ἵν᾽ ἐκεῖθεν καταλογιζομένου τοῦ ἑξῆς καὶ συνεχοῦς χρόνου συγκαταλογίζηται τῷ ἀριθμῷ αὐτοῦ καὶ τῷ ὡρισμένῳ διαστήματι ἡ κατὰ μῆκος προχώρησις τῆς κινήσεως ἀπὸ τοῦ δοθέντος τῆς ἀρχῆς τόπου, ἐδέησεν ἐξανάγκης τοῖς ἀστρονόμοις ὅρον τινὰ καὶ ἀρχὴν ὑποθέσθαι καὶ χρόνου καὶ τόπου τῆς τῶν ἀστέρων κινήσεως· ἵν᾽ ἐκεῖθεν ἀρχόμενος ὁστισοῦν καὶ προειλημμένον καὶ δεδομένον ἔχων ὡς ἄρ᾽ ἔν τινι τῷδε τῷ χρόνῳ τόνδε μὲν τὸν τόπον καὶ τὸ τμῆμα τοῦ ζωδιακοῦ ἐπεῖχεν ὁ ἥλιος, τόνδε δὲ ἡ σελήνη, τόνδε δὲ ἕκαστος τῶν ε πλανωμένων, λοιπὸν εἰς ὡρισμένους καθεξῆς ἔπειθ᾽ ἑκάστοτε χρόνους ῥᾷστα δὴ ἐπιλογίζηταί τε καὶ ἀνευρίσκῃ τὰς προχωρήσεις καὶ ἀποκαταστάσεις τῶν κινήσεων αὐτῶν. καὶ ἔχοι ἂν ἀποφαίνεσθαι ἐν οἷς ἕκαστος τόποις ἐστὶ κατὰ τὴν κανονικὴν ἔκθεσιν καὶ παράδοσιν ἐν τοῖς ὡρισμένοις χρόνοις τῶν ὁμαλῶν αὐτῶν κινήσεων, ἐπειδὴ καὶ παντάπασιν ἀνεπιχείρητον ἂν εἴη τὸ ψηφοφορεῖν καὶ ἀνευρίσκειν τοὺς ἑκάστων ἀστέρων τόπους νῦν ἑκάστοτε καὶ τὰς αὐτῶν ἐποχὰς ἐν τῷ ζωδιακῷ, κἂν εἴ τις εὖ μάλα καὶ ἀσφαλέστατα τὰς ὁμαλὰς αὐτῶν κινήσεις εἰδὼς εἴη. εἰ μή τις ὑποθεῖτο ὅθεν δεῖ ψηφοφορεῖν αὐτὸν τὰς τοιαύτας κινήσεις αὐτῶν καὶ ἀπὸ ποίας ἀρχῆς καὶ χρόνου καὶ τόπου, τί γὰρ ἂν μᾶλλον ἐν τούτοις ἀνύτειν ἔχῃ τοῦ μηδ᾽ ὁπωσοῦν εἰδότος τὰ ποσὰ τῶν χρονικῶν κινήσεων τῶν ἀστέρων ὁ περὶ τούτων ἐπιστήμην ἀκριβεστάτην ἔχων, ὡς ἄρα ἐν ὥρᾳ μὲν ὁ δεῖνα ἀστὴρ τοσοῦτον κινεῖται, ἐν ἡμέρᾳ δὲ τοσοῦτον, ἐν ἔτει δὲ τοσοῦτον, ἐν τοσούτοις δὲ ἔτεσιν ἀναλόγως τοσοῦτον; τί δ᾽ ἂν μᾶλλον ἀποφαίνεσθαι δύναιτο, ὡς νῦν ὁ δεῖνα ἀστὴρ ἐν τῷδε τῷ τμήματι τοῦ ζωδιακοῦ εὑρίσκεται, ὁ δεῖνα δὲ ἐν τῷδε καὶ ἄλλος ἐν ἄλλῳ ἢ καὶ ἐν τῷ αὐτῷ, ἂν μὴ προωρισμένους καὶ προεγνωσμένους ἔχοι ἀσφαλῶς καὶ βεβαίως τοὺς τόπους ἑκάστων τῶν ἀστέρων ἐν τῷδέ τινι τῷ προλαβόντι χρόνῳ; τούτου γε μὴν

74 τοὺς : τῶν C[ac]   **78** ἔχῃ : ἔχοι C

month or *epagomenon*, they obtain the rotation in one Egyptian year. This way astronomers compute the motions of the Sun for specific intervals of time.

Now, since every motion begins at some point in time and space, and from then on we compute continuously its trajectory, it became necessary and also desirable for astronomers to define a bench-mark as the origin of time and place for the motion of the stars, in order that, beginning at the specific initial time and space and longitude for the Sun, the Moon and each of the five planets, one can easily compute their positions at future times. Thus, one expresses with appropriate computation of their uniform rotation the position for each of them. Even though we may know accurately the uniform rotations, it is impossible to write down every longitude and every location of the stars. If one does not specify the initial time and place of the calculation, what more did he accomplish than the one who does not know the quantitative movements of the stars as a function of time? That is, how much a star moves in an hour, so much in a day and so much in a year or by analogy in so many years. How can he determine and state that a certain star is on this section of the ecliptic, the other at another or the same location, if he did not determine and know accurately and with certainty the position of the star at the predetermined

προεγνωσμένου καλῶς, ῥᾳδίως ἔπειτ᾽ ἂν ἔχοι περαίνειν ὁ τῶν ἑκάστου κινήσεων ἐπιστήμων καὶ συλλογίζεσθαί τε καὶ ἀποφαίνεσθαι, ὡς ἄρα ἐν τοῖσδέ τισιν τοῖς ἐκεῖθεν ὡρισμένοις χρόνοις προεχώρησαν καὶ εὑρίσκονται ἕκαστος τῶν ἀστέρων ἐν τοῖσδέ τισι τμήμασι καὶ τόποις τοῦ ζωδιακοῦ.

Τοιγαροῦν ὁ μὲν Πτολεμαῖος τὰς ἀρχὰς ταύτας ὑποτίθεται τῆς τῶν ἀστέρων κινήσεως, ὡς ἤδη φράσαντες ἔφημεν, κατὰ τὸ πρῶτον ἔτος τῆς Ναβονασάρου βασιλείας, ὁ δὲ Θέων ἐν τοῖς Προχείροις Κανόσι κατὰ τὸ πρῶτον ἔτος Φιλίππου, μετὰ τὴν τελευτὴν εὐθὺς Ἀλεξάνδρου τοῦ μεγάλου Μακεδόνων βασιλέως. ἐν τούτοις γὰρ τοῖς χρόνοις ἑκάτερος, ἀκριβέσι ψηφοφορίαις καὶ ἐπιλογισμοῖς εὑρηκότες τοὺς ἐπὶ τοῦ ζωδιακοῦ τόπους, ἐν οἷς εὑρίσκοντο τηνικαῦτα οἱ ἀστέρες, τὰς ἀρχὰς λοιπὸν τῶν ψηφοφοριῶν τῶν κινήσεων τῶν ἀστέρων ἐκεῖθεν ὑποτίθενται ῥᾳδίως τοῖς βουλομένοις ἀνευρίσκειν τὰς ἑκάστοτε ἐποχὰς καὶ τόπους τῶν ἀστέρων κατὰ τὰς παραδεδομένας αὐτῶν κανονικὰς ἐκθέσεις τῶν ὁμαλῶν ἑκάστου τῶν ἀστέρων κινήσεων. ἔξεστι γὰρ ἐκεῖθεν λαβόντα τοσούσδε χρόνους καὶ ἔτι πρὸς μηνῶν καὶ ἡμερῶν ἀριθμὸν καὶ ὡρῶν ἴσως ἐπιλογίζεσθαι πόσαι ἀνήκουσι τῇ τεταγμένῃ καὶ ὁμαλῇ κινήσει ἑνὸς ἑκάστου τῶν ἀστέρων μοῖραι τοῖς τοσοῖσδε χρόνοις καὶ ὁμοῦ ταύτας συντιθέντα ταῖς τῆς ἀρχῆς προϋποτεθειμέναις μοίραις ἑνὶ ἑκάστῳ τῶν ἀστέρων, ἂν ὑπερβαίνωσι κυκλικὴν περίοδον ἢ καὶ κυκλικὰς τυχὸν ἑτέρας οὐκ ὀλίγας περιόδους, ἀφαιρεῖν τὸν ἀριθμὸν τῶν τοιούτων κυκλικῶν περιόδων, ὅσος ἂν εἴη, καταλογιζόμενον καθ᾽ ἑκάστην περίοδον τξ μοίρας. εἶτα τὸν ἀπολιμπανόμενον ἀριθμὸν τῶν ἡμερῶν κατέχειν καί, ὅσος ἂν εἴη, τοῦτον τὸν ἀριθμὸν ἀποφαίνεσθαι προκεχωρηκέναι ἕκαστον ἀστέρα, περὶ οὗ ὁ λόγος ἀπὸ τῆς προϋποκειμένης καὶ δεδομένης τοπικῆς ἀρχῆς αὐτοῦ.

Προσεπενοήθη δὲ τοῖς ἀστρονόμοις καὶ τόδε χρήσιμον ὡς ἀληθῶς εἰς τὸ τῶν ψηφοφοριῶν εὐμεταχείριστον. ἐπεὶ γὰρ εἰκὸς ἦν μετὰ τὴν ὑποτεθεῖσαν καὶ δεδομένην ἀρχὴν χρονικὴν τῆς ἑκάστου

92 Cf. Ptol. *Alm.* 3.7 (vol. 1.1, p. 254 Heiberg) **94-95** Cf. Ptol., *Proch. Kan.* vol. 1, p. 1 Halma; Theon, *PC* p. 200 Tihon

98 ηὑρίσκοντο C

time? If he knows them precisely, the scientist easily completes the computation for each rotation by reasoning and presenting the distance that each star travelled in that time [interval] and the sections of the zodiac where it is located.

In this approach, Ptolemy selects as starting point for the motion of stars the first year of the reign of Nabonassar, as we stated already. On the other hand, Theon in his *Handy Tables* sets the original mark at the first year of the reign of Philip, immediately after the death of Alexander the Great, King of the Macedonians. For these, each author finds with precise calculations and arguments the positions on the ecliptic, where the stars were located at that time. These are assumed to be the starting points for the motions of the stars, and from then on one can proceed to calculate precisely and easily the locations and longitudes for subsequent times by using the given and tabulated uniform motion of each star. From then on, taking the years, months, days and hours he computes how many degrees correspond to the normal and uniform rotation of each star and adds them to the original [position] of each star. Whenever the calculation exceeds a complete revolution or several revolutions, they subtract complete revolutions (how many they may be) attributing to each period 360°. Then, for the remaining number of days, he determines the degrees that the star advanced from the assumed predetermined starting position. Astronomers devised this convention, being indeed useful and convenient in computations, because, as is usually the case after the starting time of each star,

τῶν ἀστέρων κινήσεως, ἔπειθ' ὕστερον χρόνων ἴσως πολλῶν παραδραμόντων ψηφοφορίας γίνεσθαι πρὸς εὕρεσιν ποῦ τῶν τοῦ ζωδιακοῦ μερῶν καὶ τμημάτων ἢ ὁ ἥλιος καταλαμβάνεται ἢ ἡ σελήνη ἤ τις τῶν πέντε πλανωμένων. καὶ λοιπὸν ἦν ἐργῶδες καὶ ἐπίπονον καθ' ἕκαστον ἔτος ἐπιλογίζεσθαι τὰς κινήσεις ἑκάστου τῶν τοιούτων ἀστέρων. εὐμεθόδως ἔταξαν καὶ συνέθεντο ἵνα, εἰ πλήθη χρόνων ὦσι τὰ διὰ μέσου τῆς ὑποτεθείσης καὶ δεδομένης χρονικῆς ἀρχῆς τῶν κινήσεων μέχρις αὐτοῦ τοῦ καιροῦ ἑκάστου, καθ' ὃν ζητεῖται ἡ εὕρεσις τῆς ἐποχῆς καὶ τοῦ τόπου πρὸς τὸν ζωδιακὸν οὑτινοσοῦν ἀστέρος, μερίζωσι τὰ ἐν μέσῳ τοιαῦτα πλήθη τῶν ἐτῶν κατὰ ἀριθμὸν κε ἐτῶν, ὡς ὁ Θέων ἐν τοῖς Προχείροις, ὡς δ' ὁ Πτολεμαῖος ἐν τῇ Συντάξει, κατὰ ἀριθμὸν ιη ἐτῶν. καὶ ἐπιλογισάμενοι μετὰ ἀκριβείας ὁπόσας μοίρας κινεῖται ἕκαστος τῶν ἀστέρων κατὰ εἰκοσιπενταετηρίδα ἢ ὀκτωκαιδεκαετηρίδα μετὰ κύκλου ἢ κύκλων ἴσως περίοδον, τὰς ἐναπολιμπανομένας ἔταξαν ἀνήκειν τῇ εἰκοσιπενταετηρίδι, εἰ ἔστι μία, εἰ δ' εἰσὶ καὶ πλείονες ὡσαύτως ποιεῖν, καὶ προστιθέναι τὰς ἐναπολιμπανομένας μοίρας μετὰ κύκλου ἢ κύκλων, ὡς ἔφημεν, περίοδον· καὶ ἀεὶ ἑξῆς προστιθέναι καὶ τοῦτο ποιεῖν δι' ὅλου, μέχρις ἂν τὰ ἐν μέσῳ πλήθη τῶν ἐτῶν δυνατὸν εἴη κατατέμνειν καὶ μερίζειν εἰς εἰκοσιπενταετηρίδας ἢ ὀκτωκαιδεκαετηρίδας. καὶ ὅταν ἐναπολιμπάνωνται ἔτη μὴ ἀπαρτίζοντα τὸν τῶν κε ἀριθμὸν ἢ τὸν τῶν ιη, οἷον δέκα ὄντα τυχὸν ἢ ις ἢ ἄλλα τινὰ ἐντὸς τοῦ τῶν κε ἢ τοῦ τῶν ιη ἀριθμοῦ, ταῦτα δὴ ἐπιλογίζεσθαι κατὰ ἀναλογίαν ὡς πρὸς τὴν τοῦ ἑνὸς ἔτους ἑκάστου τῶν ἀστέρων κίνησιν. διὰ τοῦτο καὶ πέντε κανόνας ἐπὶ τῶν ὁμαλῶν κινήσεων τοῦ τε ἡλίου καὶ τῆς σελήνης καὶ τῶν πέντε πλανωμένων ἐκτίθενται, ἐν οἷς δηλοποιοῦνται ὅ τε χρονικὸς ἀριθμὸς καὶ αἱ μοῖραι, ὅσας καθ' ἕκαστον αὐτὸν χρόνον οἱ ἀστέρες ἐπὶ τοῦ ζωδιακοῦ ὁμαλῶς παροδεύουσι.

126-27 Cf. Ptol. *Proch. Kan.* vol. 2 p. 66-68, 112-118 Halma; Ptol. *Alm.* 3.2 (vol. I.1, p. 210-211 Heiberg), 4.4 (vol. I.1, p. 282-285 Heiberg), 9.4 (vol. 1. 2, p. 220-221, 226-227, 232-233, 238-239 and 244-245 Heiberg) **139-41** Ptol. *Proch. Kan.* vol. 2 p. 66-77 Halma; Ptol. *Alm.* 3.2, 4.4 and 9.4

133 ὡς ἔφημεν post περίοδον transp. C

many revolutions intervened until the later time when we want to calculate and find the position on the ecliptic for the Sun, the Moon and the five planets. In addition, it requires a lot of work and effort to calculate the position of each star for each year. When many years intervened between the assumed initial time up to the time of the new calculation, it is good practice to combine the results of the computations and present values for several years. Theon in the *Handy Tables* tabulated the times and places where each star is located every 25 years, and Ptolemy tabulated then in the *Syntaxis* for 18 years. They computed precisely how many degrees of a circle or circles each star moves in 25 or 18 years together with the number of repeated revolutions, which may be one or four or however many, and accounted them [according to the convention] in the intervals of 25 years. Subsequently, they repeated this and added up the degrees for intervals divisible by 25 or 18. Any time the intervals did not comprise [a multiple] of 25 or 18, being for instance 10 or 16 or something else less than the number 25 or 18, they computed the rotation of each star in proportion to the one year intervals. For this reason they present five tables for the uniform rotation of the Sun, the Moon and the five planets in which appear the time-intervals and the degrees by which the stars rotate uniformly on the ecliptic.

Καὶ ἔστιν ὁ πρῶτος κανὼν ὁ τῶν εἰκοσιπενταετηρίδων, ὡς ἐν τῷ Προχείρῳ, ἢ ὁ τῶν ὀκτωκαιδεκαετηρίδων, ὡς ἐν τῇ Συντάξει καλούμενος, καθ᾽ ὃν ἐν τῷ πρώτῳ στίχῳ τίθεται ὅσας μοίρας ἐν τῷ πρώτῳ ἔτει τῆς ὑποτεθείσης καὶ δεδομένης ἀρχῆς, οἷον φέρε εἰπεῖν τῆς ἐν τοῖς Προχείροις Κανόσι, τῆς κατὰ τὸ πρῶτον ἔτος Φιλίππου τοῦ Ἀριδαίου μετὰ Ἀλέξανδρον, ἀπεῖχεν ἀπὸ τοῦ ἀπογείου τοῦ οἰκείου ἐκκέντρου, ὅτι δὴ καὶ ἀπὸ τοῦ ἀπογείου προσηκόντως οἱ ἀστρονόμοι ἔταξαν ψηφοφορεῖν τὰς κινήσεις τῶν ἀστέρων, ὁ ἥλιος ἢ ἡ σελήνη ἤ τις τῶν ε πλανωμένων. καὶ μετὰ τὸ ἓν καὶ πρῶτον ἔτος ἔπειτα τιθέασιν ἐν τῷ δευτέρῳ στίχῳ εἰκοσιπενταετηρίδα, ὡς γίνεσθαι ὁμοῦ σὺν τῷ πρώτῳ ἔτει ἀπὸ τῆς ὑποτιθεμένης ἀρχῆς κς ἔτη. καὶ παρατίθεται τοῖς τοιούτοις κς ἔτεσιν ἡ ποσότης τῶν μοιρῶν, ἃς κινούμενος ἐν αὐτοῖς ὁστισοῦν ἀστὴρ καὶ διεξελθὼν ἴσως τοσάσδε κυκλικὰς περιόδους, εἶτα ἐπιλαμβάνει πάλιν ταύτας ἀπὸ τοῦ ἀπογείου τοῦ οἰκείου ἐκκέντρου ἢ ι ἢ λ ἢ ν ἢ ὅσας δήποτε. καὶ οὕτω μὲν ὁ δεύτερος στίχος τὸν αὐτὸν δὲ τρόπον γίνεται καὶ ἑξῆς, οἷον τρίτος στίχος περιέχων τὸν τῶν να ἐτῶν ἀριθμὸν καὶ τέταρτος τὸν τῶν ος καὶ παρακειμένους τοὺς ἀνήκοντας τοῖς τοιούτοις ἀριθμοῖς τῶν ἐτῶν τῶν μοιρῶν ἀριθμούς, ἃς ἕκαστος τῶν ἀστέρων ἐπιλαμβάνει καὶ ἐπικινεῖται ἀπὸ τοῦ ἀπογείου τοῦ ἐκκέντρου αὐτοῦ μετὰ περιόδους κυκλικάς, ὅσας δήποτε φθάσας διεξῆλθε τοῖς καταγεγραμμένοις ἔτεσι. καὶ τοῦτο καθεξῆς μέχρις ἂν ἐγχωρῇ τιθέναι εἰκοσιπενταετηρίδων ἀριθμόν, ἕως τοῦ χρόνου, καθ᾽ ὃν ποιοῦμεν τὰς ψηφοφορίας. καὶ οὕτως μὲν ἐκτίθεται ὁ πρῶτος κανὼν τῶν εἰκοσιπενταετηρίδων.

Ἐφεξῆς δὲ τούτῳ δεύτερος κανὼν ὁ καλούμενος ἁπλῶν ἐτῶν, καθ᾽ ὃν ἐκτίθενται καθ᾽ εἱρμὸν κε ἔτη ἢ ὀκτωκαίδεκα, οἷον ἐν τῷ πρώτῳ στίχῳ ἓν ἔτος καὶ πόσαι μοῖραι ἀνήκουσιν αὐτῷ τῆς κινήσεως ἑκάστου τῶν ἀστέρων, ἐν τῷ δευτέρῳ στίχῳ δεύτερον ἔτος καὶ ἑξῆς τρίτον ἔτος καὶ ἑξῆς τέταρτον. καὶ τοῦτο κατὰ τάξιν μέχρι καὶ τοῦ κε ἢ ιη ἔτους, παρακειμένας ἔχοντα τὰς ἀνηκούσας μοίρας τῆς κινήσεως

148 Cf. Ptol., *Proch. Kan.* vol. 1, p. 1 Halma; Theon, *PC* p. 200 Tihon

157 ταύτας om. C **169** ἐκτίθεται C

In the *Handy Tables* the first table is for 25 years and in the *Syntaxis* for 18 years. In the first row is given the distance in degrees [that the star] travelled from the adopted origin, which in the *Handy Tables* is the first year of Philip Arrhidaios the successor of Alexander. The distance is from the apogee of its own eccentric because astronomers adopted the convention to count the rotation of the Sun, the Moon and of the five planets from the apogees of their eccentrics. After the first year they tabulate in the second row 25 years which together with the first year makes 26 years from the adopted origin of time. Next to the 26 years they give the position in degrees and how many revolutions the star rotated from the apogee: for instance, 10°, 30°, 40° or how many they may be. In this manner we complete the second row and the method is extended for the following rows. Thus, the third row corresponds to 51 years, the fourth to 76, and next to them are given the number of corresponding degrees, which each star covers in its rotation from the apogee of its eccentric, including the number of complete revolutions completed in these years. This is further repeated for the 25 year periods up to the year where the computation terminates. This way is completed the first table of 25 years.

Next appears the second table, the so called Table of Simple Years in which are tabulated in sequence 25 or 18 years and how many degrees belong to the rotation of each star. In the first row is the first year and how many degrees the star rotated. In the second row is the second year, in the third row the third year, then the fourth and up to either 25 or 18 years, respectively, and

ἑκάστου ἀστέρος, οὗ ἐστιν ὁ κανών, ὡς ἐν τῷ ἐκκειμένῳ ἐσχάτῳ ἔτει παρακείμενον εἶναι καὶ καταγεγραμμένον μοιρῶν ἀριθμόν, ὅσας ἐπικινεῖται καὶ ἐπιλαμβάνει ὁ ἀστὴρ μετὰ τὸ διεξελθεῖν τοσάσδε ἴσως κυκλικὰς περιόδους.

Ἐφεξῆς δὲ τούτοις τρίτος ἐστὶ κανὼν ὁ τῶν μηνῶν τῶν ιβ τριακονθημέρων καὶ τοῦ ἐλαχίστου καὶ καλουμένου ἐπαγομένου τοῦ πενθημέρου. ἐκτίθεται δὲ καὶ οὗτος ὁ κανὼν κατὰ τὴν αὐτὴν τάξιν. ἐν γὰρ τῷ πρώτῳ στίχῳ ἐκτίθεται ὁ πρῶτος μὴν καὶ ὅσος ἀνήκει αὐτῷ ἀριθμὸς μοιρῶν ἢ λεπτῶν ἢ καὶ ἀμφοτέρων κινήσεως οὑτινοσοῦν ἀστέρος. καὶ ἑξῆς ὁ δεύτερος μὴν καὶ ὁ τρίτος καὶ ὁ τέταρτος, μέχρι τῆς συμπληρώσεως τῶν ὅλων, παρακειμένους ἔχοντες ἀναλόγως τοὺς ἀνήκοντας ἀριθμοὺς τῶν μοιρῶν καὶ τῶν λεπτῶν τῆς κινήσεως ἑκάστου ἀστέρος, περὶ οὗ ὁ λόγος.

Τέταρτος ἐπὶ τούτοις κανὼν ὁ τῶν ἡμερῶν καλούμενος, περιέχων κατὰ τὴν αὐτὴν τάξιν ἀπὸ τῆς πρώτης ἡμέρας μέχρι καὶ τῆς τριακοστῆς τὰς ἀνηκούσας μοίρας καὶ λεπτὰ τῇ κινήσει οὑτινοσοῦν ἀστέρος.

Καὶ πέμπτος ἐστὶν ὁ τῶν ὡρῶν περιέχων κατὰ τὴν αὐτὴν τάξιν τῶν στίχων ἀπὸ πρώτης ὥρας ἕως κδ καὶ τοὺς ἀνήκοντας ὡσαύτως ἀριθμοὺς τῶν μοιρῶν καὶ τῶν λεπτῶν τῇ κινήσει τῇ κατ᾽ αὐτὰς τὰς ὥρας οὑτινοσοῦν ἀστέρος. ὑποτίθενται δὲ αἱ κδ αὗται ὧραι ὡς ἀκριβῶς ἰσημεριναί. ἔστι γὰρ διαφορὰ ὡρῶν ἀξιόλογος, ὡς ἐν τοῖς ἑξῆς διευκρινήσομεν.

Καὶ οὗτοι μὲν οἱ πέντε κανόνες ἐκτίθενται τόνδε τὸν τρόπον κατὰ λόγον ἑξῆς. ἀπαρτίζουσι γὰρ αἱ κδ ὧραι ἡμέραν μίαν, καὶ ἔστι τὸ τῶν κδ ὡρῶν κίνημα μιᾶς ἡμέρας κίνημα. καὶ ἀπαρτίζουσιν αἱ λ ἡμέραι μῆνα, καὶ τὸ τῶν λ ἡμερῶν κίνημα μηνός ἐστι κίνημα. καὶ ἀπαρτίζουσιν οἱ ιβ τριακονθήμεροι μῆνες καὶ ὁ ἐπαγόμενος ὁ πενθήμερος ἔτος αἰγυπτιακὸν ἕν, καὶ τὸ τῶν τοιούτων μηνῶν κίνημα ἔτους αἰγυπτιακοῦ ἐστι κίνημα. καὶ ἀπαρτίζουσι τὰ κε ἔτη ἢ τὰ ιη προδήλως εἰκοσιπενταετηρίδα μίαν καὶ κίνημα εἰκοσιπενταετηρίδος ἢ ὀκτωκαιδεκαετηρίδα καὶ κίνημα ὀκτωκαιδεκαετηρίδος. Αἰγυπτίων πάντως ἐτῶν τοῦτο γὰρ προληπτέον,

175 καὶ V[sl] : om. C **188-90** ἀπὸ τῆς πρώτης ἡμέρας - τὴν αὐτὴν τάξιν om. C[ac] (homoeotel.) **191** καὶ C[sl] : om. V

next to them is written the number of degrees that the corresponding star rotated. On the last year is tabulated a number of degrees covered by the star after the completion of full revolutions.

Next is the third table of the 12 months with 30 days each plus the minimal month, the one called *epagomenos* with five days, which is included in the same sequence. In the first row is tabulated the first month and the corresponding number of degrees or minutes or both for the appropriate month. Then appears the second month, the third and the fourth up to the end where the sequence is completed. Close to them is, of course, the number of degrees and minutes corresponding to the revolution of the star under consideration.

Fourth after them is the table of days, containing one after the other, the first day up to the thirtieth, and the degrees and minutes corresponding to the revolution of the appropriate star.

Fifth is the table of hours, containing in the sequence of the rows the first hour up to the 24th and the corresponding number of degrees and minutes for the hourly rotation of the star. We assume that the hours are equinoctial because there is a significant difference, as we shall explain later.

The five tables are presented in this manner and with the following logic. It is understood that 24 hours make one day, 30 days make a month and 12 months of 30 days each, plus the five days of the *epagomenos*, make one Egyptian year and the rotation of the stars in these months corresponds to one Egyptian year. Furthermore, 25 years make a 25-year-cycle, and 18 years make

ὅτι οἱ τοιοῦτοι κανόνες καὶ τὰς εἰκοσιπενταετηρίδας καὶ τὰ ἁπλᾶ ἔτη καὶ τοὺς μῆνας καὶ τὰς ἡμέρας τῶν μηνῶν κατὰ τὴν τῶν Αἰγυπτίων χρονικὴν ἀπαρίθμησιν ἐκτίθενται.

Καὶ φθάσαντες προείπαμεν τίς ἐστι διαφορὰ τῆς τῶν αἰγυπτιακῶν ἐτῶν ἀπαριθμήσεως πρὸς τὰ ἑλληνικὰ καὶ ῥωμαϊκά. ὅπως δὲ εὑρήσομεν τά τε αἰγυπτιακὰ ἔτη καὶ τοὺς μῆνας καὶ τὰς ἡμέρας ἀπὸ τῆς προϋποτεθειμένης καὶ δεδομένης χρονικῆς ἀρχῆς τῶν κινήσεων τῶν ἀστέρων μέχρι καὶ τοῦ καιροῦ, καθ᾽ ὃν ψηφοφοροῦμεν, ὅτι καὶ κατὰ τὰ τοιαῦτα ἔτη καὶ τοὺς μῆνας καὶ τὰς ἡμέρας οἱ εἰρημένοι κανόνες ἐκτίθενται, καὶ κατὰ ταῦτα τὰς ψηφοφορίας ποιούμεθα πολὺ τὸ διάφορον ἔχοντα ἐνίοτε πρὸς αὐτοὺς τοὺς χρόνους τοὺς ἑλληνικοὺς καὶ ῥωμαϊκούς, ἀναγκαῖον ἐρεῖν καὶ λέγομεν ἤδη. προέφημεν τοίνυν εἶναι τὴν διαφορὰν τοῦ αἰγυπτιακοῦ ἔτους πρὸς τὸ ἑλληνικὸν καὶ ῥωμαϊκόν, ὅτι οἱ μὲν Ἕλληνες καὶ Ῥωμαῖοι τὸ κατ᾽ αὐτοὺς ἔτος τοσούτου τίθενται χρόνου, δι᾽ ὅσου ὁ ἥλιος κυκλικὴν ποιεῖται περίοδον καὶ ἀπὸ τοῦ αὐτοῦ σημείου τοῦ ζῳδιακοῦ εἰς τὸ αὐτὸ ἀποκαθίσταται, ἤτοι διὰ τξε ἡμερῶν καὶ τετάρτου μέρους ἡμέρας, παρά τι βραχύτατον καὶ πολλοστὸν μόριον ἡμέρας καὶ σχεδὸν παντάπασιν ἀνεπιλόγιστον. οἱ δὲ Αἰγύπτιοι τξε ἡμερῶν τελείων μόνον καταλογίζονται τὸ οἰκεῖον ἔτος, τὸ τέταρτον μέρος τῆς ἡμέρας παρεῶντες, ὃ δὴ συντιθέντες κατὰ τέτταρα ἔτη ποιοῦσιν ἡμέραν καὶ καταλογίζονται ταύτην πρώτην καὶ ἀρχὴν τοῦ ἑξῆς ἔτους. καὶ διὰ ρκ ἐτῶν συντιθέντες αὖθις τὰς συναγομένας ἀπὸ τῶν τετραετηρίδων λ ἡμέρας, τίθενται ταύτας μῆνα ἕνα καὶ τίθενται τοῦτον πρῶτον μῆνα τοῦ ἑξῆς ἔτους, ὥστε προλαμβάνειν αὐτοῖς τηνικαῦτα τὸ ἔτος τοῦ ἑλληνικοῦ καὶ ῥωμαϊκοῦ ἔτους μῆνα ὁλόκληρον καὶ τὴν ἀρχὴν τοῦ δευτέρου μηνὸς παρ᾽ αὐτοῖς ἀρχὴν εἶναι παρ᾽ Ἕλλησι πρώτου μηνός· εἶτα διὰ ͵αυξ ἐτῶν ἀπὸ τῆς ἐπισυνθέσεως τῶν εἰρημένων ἡμερῶν τῶν ἀπὸ τῶν τετραετηρίδων ἀπαρτίζεσθαι καὶ συμπληροῦσθαι τέλειον αἰγυπτιακὸν ἔτος, ὃ καὶ προστίθεσθαι τοῖς τηνικαῦτα ἑλληνικοῖς καὶ

215 αὐτοὺς post τοὺς transp. C **221** μέρους post ἡμέρας transp. C **226-31** τὰ γὰρ ρκ ἔτη τριακοντάκις ἔχουσιν ἐν ἑαυτοῖς τὰς τετραετηρίδας ἢ τὰς ἀπὸ τῶν τεσσάρων τετάρτων συναγομένας ὁλοκλήρους ἡμέρας. αἱ δὲ τριάκοντα ἡμέραι μὴν εἰσὶν αἰγυπτιακός, ὥστε διὰ ρκ ἐτῶν, ὥσπερ εἴρηται, συνάγεται μὴν αἰγυπτιακὸς ἀπὸ τῶν τετραετηρίδων sch. in mg. C (Chort.)

an 18-year-cycle. We adopted the years to be Egyptian, that is the tables are for 25-year-cycles, for single years, for months and days presented in the Egyptian method for counting time.

We have arrived now at the point to state the difference between the Egyptian and Graeco-Roman years. How do we calculate the Egyptian years, months and days from a presupposed and given origin of counting time until to the time at which we make our calculations? Because the tables are tabulated for [Egyptian] years, months and days we make calculations according to them, which sometimes are very different from those in the Graeco-Roman calendar. This requires an explanation. We already mentioned the difference of the Egyptian from the Graeco-Roman year. The Greeks and the Romans define the length of the year as the period in which the Sun completes a revolution from one point of the ecliptic back to the same point. This contains 365 and a quarter day minus a very small fraction of a day, which is always almost negligible. The Egyptians, on the other hand, define their year as composed of 365 complete days, ignoring the one-quarter day, which when added up in four years make a whole day, and count this day as the first day of the next year. In 120 years the one-quarter days add up to one month of 30 days, which they consider to be first month of the next year. Thus their year is ahead of the Graeco-Roman year by one month. Then, in 1460 years from the time when one began counting the extra days, the sum of the four-year-cycles, constitute and complete an entire Egyptian year, which is added to the Graeco-Roman

ῥωμαϊκοῖς ἔτεσι, κἀντεῦθεν ἐπαύξεσθαι πάντως τοῖς Αἰγυπτίοις τὰ ἔτη καὶ πάλιν κοινὴν ἀρχὴν ποιεῖσθαι τοῦ ἔτους ὁμοῦ Αἰγυπτίους τὲ καὶ Ἕλληνας. τούτου δὲ οὕτως ἔχοντος γέγονεν ἀποκατάστασις ἔτους ὁλοκλήρου αἰγυπτιακοῦ διὰ τῶν τοιούτων ἡμερῶν τῶν ἀπὸ τῶν τετραετηρίδων, ἀπό τινος πρότερον ἀρχῆς ἐν τῷ πέμπτῳ ἔτει τῆς μοναρχίας Αὐγούστου. καὶ πάλιν ἐκεῖθεν ἤρξατο προστιθέναι ὁ ἐπιλογισμὸς ὁ αἰγύπτιος κατὰ τέτταρα ἔτη ἡμέραν μίαν, ἐκεῖθεν δὲ πολλῶν παραδραμόντων τῶν χρόνων μέχρι καὶ εἰς ἡμᾶς πολλαὶ καὶ ἀπὸ τῶν τετραετηρίδων ἡμέραι συντίθενται.

Ὅταν οὖν βουλώμεθα ψηφοφορίαν τινὰ ποιήσασθαι κινήσεως οὑτινοσοῦν ἀστέρος κατά τινα δεδομένον καὶ εἰρημένον χρόνον, καταλογιζόμεθα πόσα ἔτη εἰσὶ μέχρις αὐτοῦ τοῦ χρόνου ἀπὸ τοῦ πέμπτου ἔτους τῆς Αὐγούστου βασιλείας, ὅθεν ἔφημεν πάλιν εἶναι ἀρχὴν ἡμερῶν ἀπὸ τετραετηρίδων. καὶ εὑρόντες ταῦτα τὰ ἔτη λαμβάνομεν τὰ τέταρτα αὐτῶν καὶ ὅσος ἐστὶν ὁ ἀριθμὸς αὐτῶν ἀπογραφόμεθα καὶ τιθέαμεν ὡς ἡμέρας ἀπὸ τῶν τετραετηρίδων γινομένας κατὰ τὴν τῶν Αἰγυπτίων τάξιν. ἔπειτα καταλογιζόμεθα, ἵνα κατὰ τὴν παράδοσιν τῶν Προχείρων Κανόνων ἡ ψηφοφορία γένηται, πόσα ἔτη εἰσὶν ἀπὸ τῆς ἀρχῆς Φιλίππου μέχρι τοῦ ἀναδιδομένου καὶ εἰρημένου χρόνου. καὶ ἀνευρίσκοντες ταῦτα ἀπογραφόμεθα ἰδίᾳ ὅσα εἰσὶ καὶ τυχὸν μετὰ συμπλήρωσιν ὁλοκλήρων ἐτῶν εἰσι μέχρι τοῦ ἀναδοθέντος καὶ εἰρημένου χρόνου καὶ μῆνές τινες καὶ ἡμέραι μετὰ συμπλήρωσιν ὁλοκλήρων μηνῶν. οἷον καθ᾿ ὑπόθεσιν ἀναφαίνονται ͵ας ἔτη ὁλόκληρα ἀπὸ τῆς ἀρχῆς τοῦ Φιλίππου μέχρι τοῦ ἀναδιδομένου καὶ εἰρημένου χρόνου καὶ κατεπέκεινα τούτων μῆνες ὁλόκληροι τέτταρες καὶ ἡμέραι τοῦ πέμπτου μηνὸς δέκα μέχρι τῆς μεσημβρίας τῆς ἡμέρας τοῦ ὡρισμένου καιροῦ, εἰς ἣν ψηφοφοροῦμεν. ἴσως δὲ ἔστωσαν καὶ ὧραι τρεῖς ἢ τέτταρες μετὰ τὴν μεσημβρίαν, αἵτινες ὡς τῆς ια ἡμέρας τοῦ πέμπτου μηνὸς καταλογίζονται.

Ἰστέον γὰρ καὶ τοῦτο, ὅτι οἱ ἀστρονόμοι τὰς ἀρχὰς τῶν ἡμερῶν, μᾶλλον δὲ καὶ τῶν μηνῶν καὶ τῶν ἐτῶν, ἀπὸ μεσημβρίας ἐνόμισαν

241 post καὶ[2] add. αἱ C **243** ὑπόδειγμα ψηφοφορίας sch. in mg. C **244** εἰρημένον καὶ δεδομένον C **246** ὅθεν : ὅπερ C

years. This way the Egyptian years increase and produce again a common starting [date] for Egyptians and Greeks. This being so, the correction of the four-year-cycles produced an adjustment of an entire Egyptian year, added at the beginning of the fifth year of the reign of Emperor Augustus. From then on we began again adding to the Egyptian year one day every four years, and, since many years passed from that time up to now, many extra days have been accumulated.

When we wish to compute the movement of a star at a certain time, we count how many years passed from the fifth year of the reign Augustus — this date being the beginning for counting the one-quarter days — and, after we calculate this, we divide by four in order to obtain the number of extra days to be included in the Egyptian calendar. We write them down in the Egyptian counting as the additional days arising from the one-quarter days. Since we shall be performing the calculation following the *Handy Tables*, we count how many years passed from the reign of Philip up to the time of the calculation. Once we calculate them and add the supplement, we write them down as complete years and a few months and the remaining days. Let us consider 1006 full years passed from the beginning of the reign of Philip up to the date under consideration, and also four full months and 10 days from the fifth month up to the midday of the day that we are calculating. There may be also three or four hours after midday, which are counted in the 11th day of the fifth month.

It is worth mentioning that astronomers count the beginning of days, as well as of months and years from the middle of the day, which they find it

ποιεῖσθαι, διὰ πολλὰ χρήσιμα τῇ κατ᾽ αὐτοὺς προθέσει καὶ τῷ σκοπῷ, περὶ ὧν οὐκ ἔστιν ἀναγκαῖον νῦν ἡμᾶς διατρίβειν καὶ ἀκαίρως πλατύνεσθαι, πλήν γε ὅτι καὶ ὁ Πτολεμαῖος ἐν τῇ Συντάξει καὶ ὁ Θέων ἐν τοῖς Προχείροις Κανόσι τὴν χρονικὴν ἀρχὴν ὑποτίθενται ἐν τῷ πρώτῳ ἔτει, ὁ μέν, καθὼς προέφημεν, Ναβονασάρου, ὁ δὲ Φιλίππου νεομηνία τοῦ πρώτου μηνός, ὃν Αἰγύπτιοι Θὼθ καλοῦσιν, ἀπὸ μεσημβρίας. ἀλλ᾽, ὅπερ ἐλέγομεν, ἔχομεν ἀπὸ τῆς ἀρχῆς Φιλίππου μέχρι τοῦ ἀναδιδομένου καὶ ὡρισμένου χρόνου ἔτη ͵ας, τέτταρας μῆνας, κατὰ τὴν ἑλληνικὴν τάξιν, ἡμέρας δέκα καὶ ὥρας ἐπὶ ταῖς δέκα ἡμέραις τρεῖς ἢ τέτταρας. ἐπεὶ δὲ κατὰ τὴν τῶν Αἰγυπτίων τάξιν ποιούμεθα τὰς ψηφοφορίας, ὀφείλομεν προσθεῖναι τοῖς ἔτεσι τοῖς ͵ας ἕτερον ἔτος ἕν, ὅπερ πάντως ἀποκατέστη ἀπὸ τῶν τετραετηρίδων, ὡς ἔφημεν, κατὰ τὸ πέμπτον ἔτος τῆς Αὐγούστου μοναρχίας, ἐν μέσῳ τοῦ εἰρημένου χρόνου πάντως, ἀπὸ τῆς ἀρχῆς Φιλίππου, μέχρις οὗ τὴν ψηφοφορίαν ποιούμεθα, καὶ γίνονται τὰ ἔτη αζ κατ᾽ Αἰγυπτίους.

Καὶ πρός γε ἔτι ὀφείλομεν προσθεῖναι ταῖς ἡμέραις ἃς ἀπεγραψάμεθα, ὡς προέφημεν, τὰς γινομένας ἀπὸ τῶν τετραετηρίδων ἀπὸ τοῦ πέμπτου ἔτους τῆς μοναρχίας τοῦ Αὐγούστου, ὡς ἂν εὕρωμεν ὁπόσοι ἀναφαίνονται μῆνες αἰγύπτιοι καὶ ἡμέραι ἑξῆς. ἐπεὶ γοῦν ἀπὸ τῶν ͵αζ ἐτῶν, ὧν ἔφημεν, ἀπὸ τῆς ἀρχῆς τοῦ Φιλίππου συνεισαγομένου δηλονότι αὐτοῖς καὶ τοῦ ἀπὸ τῶν τετραετηρίδων εἰρημένου ἔτους, ὅπερ καὶ ἐμβόλιμον καλεῖται, μέχρι τοῦ πέμπτου ἔτους τῆς μοναρχίας Αὐγούστου τριακόσια ἔτη παρέδραμον (ἔστι γὰρ τοῦτο δῆλον ἐκ τοῦ κανόνος τοῦ ἐν τῷ Προχείρῳ), τῶν βασιλειῶν ὑπεξαιρουμένων τούτων καταλείπονται ψζ ἔτη, ὧν τὰ τέταρτα ρος. καὶ εἰσὶ λοιπὸν ἡμέραι ἐκ τῶν τετραετηρίδων αὗται αἱ ρος. ἀπολείπονται καὶ ἔτη γ, ἅπερ χρὴ καὶ παρεᾶν κατὰ τὴν ἀστρονομικὴν παράδοσιν. ἡ γὰρ παράδοσις οὕτως ἔχει καὶ τοιαύτη ἐστίν, ὡς εἰ μὲν ἀπαρτίζεται ὁ ἀριθμὸς εἰς τέτταρα ἔτη διαιρούμενος, χρὴ ἐπιλογίζεσθαι καὶ ποιεῖν κατὰ τὴν τῶν Αἰγυπτίων

267 Cf. Ptol. *Alm.* 3.7 (vol. 1.1, p. 254 Heiberg) **267-68** Cf. Ptol., *Proch. Kan.* vol. 1, p. 1 Halma; Theon, *PC* p. 200 Tihon

272 ἀναδιδομένου καὶ ὡρισμένου : ὡρισμένου καὶ ἀναδιδομένου C[ac] **276** πάντως post ἀποκατέστη transp. C

useful for their intentions and purposes. It is not necessary to further elaborate this point now since it will take us away from our example. We only mention that Ptolemy in the *Syntaxis* and Theon in the *Handy Tables* set the origin for counting time the first year of Nabonassar and of Philip, respectively, and at the midday of the first day of the first month, which the Egyptians call Thoth. As we were mentioning, from the beginning of Philip up to the time of calculation passed 1006 years, four months in the Greek calendar, 10 days and three or four hours after the 10th day. When we use the Egyptian calendar we must add one more year to 1006, for the adjustment of the four-year-cycles that took place at the fifth year of Augustus, because the date of Augustus falls between Philip and the date of our calculation. This makes them 1007 Egyptian years.

Furthermore, we must add the days obtained from the four-year intervals after the fifth year of Augustus in order to find out the additional Egyptian months and days. In the 1007 years from the beginning of Philip up to the fifth year of the reign of Augustus is also included one year from the four-year cycles, which we call intercalary (*emvolimon*). Three hudrend years intervened up to the fifth year of Augustus, as is evident from the Table of Kings in the *Handy Tables*. Subtracting them remain 707 years which contain 176 four-year cycles plus a remainder of three days which we neglect in accordance with astronomical practice. The traditional practice says, if the number of years is exactly divisible by four, we must count them and construct

τάξιν τὰς ἐκ τῶν τετραετηρίδων ἡμέρας. εἰ δὲ οὐκ ἀπαρτίζεται εἰς τέτταρα ἔτη, ἀλλ᾽ ὑπολείπονται ἓν ἔτος ἢ δύο ἢ τρία, χρὴ παρεᾶν ταῦτα. παρεῶμεν οὖν, ὡς εἴρηται, καὶ ἡμεῖς, ἐπὶ τοῦ ὑποδείγματος ἐνταῦθα τὰ ὑπολειφθέντα τρία ἔτη. ἐπεὶ γοῦν εἴχομεν δέκα ἡμέρας τοῦ πέμπτου μηνός, προστιθέαμεν αὐταῖς καὶ τὰς ἐκ τῶν τετραετηρίδων γινομένας ρος, καὶ γίνονται ρπς, αἵτινες ποιοῦσι μῆνας ς καὶ ἡμέρας ς. εἴχομεν οὖν τέτταρας μῆνας κατὰ τὴν τῶν Ἑλλήνων τάξιν καὶ ἡμέρας δέκα. καὶ ἀναφαίνονται κατὰ τὴν τῶν Αἰγυπτίων τάξιν, προστιθεμένων αὐτοῖς τῶν εἰρημένων ς μηνῶν τῶν ἀπὸ τῶν τετραετηρίδων, μῆνες δέκα καὶ ἡμέραι ς καὶ ὧραι τρεῖς ἢ τέτταρες, ὡς ἔφημεν.

Εἰσάγομεν οὖν τὸν τῶν ἐτῶν ἀριθμὸν εἰς τὸν ἀνωτέρω εἰρημένον κανόνα τῶν εἰκοσιπενταετηρίδων καὶ εὑρίσκομεν ἐν αὐτῷ τὰ ἐγγὺς ἐλάττονα τῶν ͵αζ καταγεγραμμένα ͵α ἓν μόνα, ὅτι μὴ εὑρίσκεται ὁ ἀριθμὸς τῶν ͵αζ εἰς τὸν κανόνα τῶν εἰκοσιπενταετηρίδων. καὶ ἀπογραφόμεθα τὰς προκειμένας μοίρας καὶ λεπτὰ ἴσως τῷ ἀριθμῷ τῶν ͵α ἑνὸς ἐν τῷ κανόνι τῶν εἰκοσιπενταετηρίδων. εἶτα εἰσάγομεν τὰ λοιπὰ ς ἔτη εἰς τὸν κανόνα τῶν ἁπλῶν ἐτῶν καὶ τοὺς δέκα μῆνας εἰς τὸν κανόνα τῶν μηνῶν καὶ τὰς ς ἡμέρας εἰς τὸν κανόνα τῶν ἡμερῶν καὶ τὰς τρεῖς ἢ τέτταρας ὥρας εἰς τὸν κανόνα τῶν ὡρῶν. καὶ ἀνευρίσκομεν καὶ ἀπογραφόμεθα τὰς ἑκάστοις παρακειμένας μοίρας ἢ καὶ λεπτὰ καὶ συντιθέαμεν τὰς ὅλας μοίρας. καὶ εἴπερ εἰσὶν ἐλάττονες τξ, εἰς ὅσας ἕκαστος δηλαδὴ κύκλος διαιρεῖται, ἀποφαινόμεθα τοσαύτας κεκινῆσθαι μοίρας τὸν ἀστέρα ἀπὸ τοῦ ἀπογείου τοῦ οἰκείου ἐκκέντρου κύκλου καὶ ταύτην εἶναι τὴν ὁμαλὴν αὐτοῦ κίνησιν. εἰ δὲ πλείους εἰσὶν αἱ συναγόμεναι μοῖραι τῶν τξ, ἀφαιροῦμεν κυκλικὴν περίοδον, ἤτοι τξ μοίρας, ἢ καὶ δευτέραν ἴσως ἢ καὶ τρίτην καὶ τὰς ἐναπολειφθείσας μοίρας ταύτας ἐπιλογιζόμεθα κεκινῆσθαι καὶ προκεχωρηκέναι τὸν ἀστέρα ἀπὸ τοῦ ἀπογείου τοῦ ἐκκέντρου κύκλου. καὶ τοῦτο κατὰ λόγον πρόδηλον. ἐπεὶ γὰρ ὁ κύκλος αὐτοῦ τξ μοιρῶν ἐστιν, εἴπερ κατὰ τὸ εἰρημένον χρονικὸν διάστημα εὑρίσκεται ὁ ἀστὴρ κεκινημένος πλείους τῶν τξ,

294 ἀπαρτίζεται : ἀπαρτίζονται C **302** τῶν εἰρημένων ς μηνῶν V : τῶν ς μηνῶν τῶν εἰρημένων C **311** ε V C : correxi **315** ἕκαστος post κύκλος transp. C **316** κεκινῆσθαι post μοῖρας transp. C

the [correction] to the Egyptian calendar from the one-quarter days. But whenever there is a remainder of one, two or three years, we ignore the remainder, as is the case in the present example. Since we had 10 days in the fifth month, we add to it 176 days from the one-quarter days to become 186, i.e. six months and six days. In the Greek calendar we had four months and ten days, which in the Egyptian counting are ten months, six days and three or four hours.

We introduce the number of the years in the Table for the 25-year cycle, and search there for a number smaller than 1007 but closest to it. The tabulated number is 1001 years since we do not find the number 1007 in the table of 25-years. We record the degrees and minutes corresponding to it being equal to those for one year. Then, we introduce the remaining six years in the Table of Simple Years, ten months in the Table of Months, the six days in the Table of Days, the three or four hours in the Table of Hours and we find the tabulated degrees and/or minutes next to them and add them up. If the degrees are less than 360°, into which each circle is subdivided, we conclude that the star travelled in its uniform rotation so many degrees from the apogee in the eccentric. If the entry in the table is more than 360 degrees, we subtract a complete revolution that is 360° or a second or a third, and the remaining degrees we count as the degrees that the star advanced from the apogee of the eccentric. This is obvious because each circle has 360° and when the star

εὔδηλον ὡς περιώδευσε τὸν οἰκεῖον κύκλον. καὶ πάλιν ἐκ δευτέρας ἀρχῆς ἤρξατο παροδεύειν αὐτόν, καὶ τὰς ἐναπολειφθείσας λοιπὸν μοίρας ἔστιν ἀποφαίνεσθαι προκεχωρηκέναι αὐτὸν ἀπὸ τοῦ ἀπογείου αὐτοῦ. καθὼς δὲ εἴρηται, ἐνίοτε τοσοῦτός ἐστιν ὁ συναγόμενος ἀριθμός, ὡς οὐ μίαν μόνον ἀφαιρεῖν χρῆναι κυκλικὴν περίοδον, ἀλλὰ καὶ πρός γε ἔτι δευτέραν ἢ καὶ τρίτην, καὶ τὰς μετὰ τὴν ἀφαίρεσιν τῶν αὐτῶν κυκλικῶν περιόδων ἐναπολειφθείσας μοίρας, ταύτας ἀποφαίνεσθαι κεκινῆσθαι τὸν ἀστέρα ἀπὸ τοῦ ἀπογείου αὐτοῦ.

Καὶ ἡ μὲν ὑποτύπωσις καὶ ἡ χρῆσις τῶν ε χρονικῶν κεφαλαίων, ὧν καὶ οἱ κανόνες, ὡς ἔφημεν, εἰσίν, ἤτοι τῶν εἰκοσιπενταετηρίδων ἢ τῶν ὀκτωκαιδεκαετηρίδων τῶν ἁπλῶν ἐτῶν, τῶν μηνῶν, τῶν ἡμερῶν, τῶν ὡρῶν, τόνδε τὸν τρόπον μεθοδεύεται ἐν ταῖς ψηφοφορίαις. συμβαίνει μέντοι ἐνίοτε μὴ καὶ τῶν πέντε κεφαλαίων χρῆσιν εἶναι ἐν ταῖς ψηφοφορίαις τῶν κινήσεων τῶν ἀστέρων, ἀλλ᾽ ἐκλείπειν τι ἤ τινα ἐξ αὐτῶν, οἷον ὅταν αἱ εἰκοσιπενταετηρίδες ἢ ὀκτωκαιδεκαετηρίδες ἀπαρτίζωνται συμπληροῦσαι τὰ δοθέντα ὅλα ἔτη καὶ οὐχ ὑπολείπηται ἔτος οὐδέν—τηνικαῦτα γὰρ τὸ τῶν ἁπλῶν ἐτῶν ἐκλείπει κεφάλαιον—ἢ ὅταν μετὰ τὴν ἔκθεσιν τῶν εἰκοσιπενταετηρίδων ἢ τῶν ὀκτωκαιδεκαετηρίδων τὰ ἐπιτιθέμενα ἁπλᾶ ἔτη, ὅσα εὑρίσκονται μετὰ τῆς συνθέσεως δηλονότι τῶν ἀπὸ τῶν τετραετηρίδων ἡμερῶν, οὐδὲν καταλείπωσι μηνῶν τινων ἢ μηνὸς ἡμέρας, ὡς ἐντεῦθεν ἐκλείπειν πάλιν τὸ τῶν μηνῶν κεφάλαιον· ἢ ὅταν ἀναφαίνωνται μὲν μετὰ τὰς εἰκοσιπενταετηρίδας ἢ ὀκτωκαιδεκαετηρίδας καὶ ἁπλῶν ἐτῶν ἀριθμὸς καὶ μηνῶν, οἱ μῆνες δὲ συμπληρῶνται μὴ καταλείποντες ἡμέρας τινὰς ἢ ἡμέραν· ἢ ὅταν αἱ ἀναφαινόμεναι ἡμέραι ἢ ἡμέρα καταντᾷ εἰς τὴν μεσημβρίαν αὐτήν, ὡς μηδεμίαν ὑπολείπεσθαι ὥραν ἐν ταῖς ψηφοφορίαις. ὅταν γὰρ ἐπισυμβαίνῃ τι τούτων ἤ τινα, τηνικαῦτα πάντως, ὡς ἔστιν αὐτόθεν δῆλον, ἐκλείπει κεφάλαιον ἓν ἢ καὶ πλείω ἴσως ἐνίοτε ἀπὸ τῶν ε χρονικῶν κεφαλαίων, ὧν εἰσιν οἱ κανόνες. καὶ οὕτω μὲν ἐν ταῖς ψηφοφορίαις κατὰ τοὺς ἀναδιδομένους ὡρισμένους χρόνους ἑκάστοτε χρώμεθα ταῖς ἐπισυναγωγαῖς τῶν μοιρῶν καὶ τοῖς ἀριθμοῖς τῶν κινήσεων τῶν ἀστέρων, κατὰ τοὺς κανόνας τῶν

348 ἀναφαινόμεναι : φαινόμεναι C **350** ἐπισυμβαίνει C^{ac}

rotated more than 360° it means it revolved around its own circle and then it started rotating for a second time around it with the remaining degrees indicating the distance from the apogee. As we mentioned, sometimes this number is such that we need to subtract not only one period but a second and a third, considering the remaining degrees as the distance the star rotated from the apogee.

This is the presentation and use of the five rules and their tables which are for 25 or 18 years, for simple years, months, days, hours, and this is the method used in the calculations. In the calculations, it happens sometimes that we do not need to use all five tables. This happens whenever the 25 and 18 year intervals complete full years without a remainder, in which case the table of simple years is omitted, or when after the counting of 25 or 18 years the additional simple years which are obtained from the supplement of the one-quarter days do not leave a reminder of months or days. Then the table of months is omitted. However, when after the counting of the 25 or 18 years the number of simple years and months are complete without a remainder or when the counting of days ends up at midday without a remaining hour; when one or more of these takes place then, as is obvious, one or more of the five rules that are associated with the tables is omitted. This way we carry the calculations for given specific times using the collected number of degrees and

εἰρημένων ε χρονικῶν κεφαλαίων. ἀκόλουθον δὲ ἤδη λοιπὸν καὶ αὐτοὺς ἐκθέσθαι τοὺς πέντε κανόνας, δι᾽ ὧν ἔσται ψηφοφορεῖν ἑκάστοτε καὶ ἀνευρίσκειν τὰς τῆς τῶν ἀστέρων κινήσεως ἐποχὰς καὶ τοὺς τόπους καὶ προηγουμένως αὐτοὺς μάλιστα, ὡς ἔφημεν, τοὺς περὶ τῶν τοῦ ἡλίου κινήσεων.

Ἐπεὶ τοίνυν ὁ μὲν Πτολεμαῖος, ὡς εἴρηται, κατὰ ὀκτωκαιδεκαετηρίδας ἐν τῇ Συντάξει ποιεῖται τὴν κατατομὴν καὶ τὸν μερισμὸν τῶν πολλῶν ἐτῶν, προστίθησι δὲ εἰς τοὺς κανόνας αὐτοῦ τῷ τῶν μοιρῶν ἀριθμῷ, οὐ μόνον καὶ ὅσα ἐπιβάλλει ἑκάστοτε ἑξηκοστὰ πρῶτα, ἀλλὰ καὶ δεύτερα καὶ τρίτα καὶ μέχρι τῶν ἕκτων· ὁ δὲ Θέων ἐν τῷ Προχείρῳ τὴν τοιαύτην διαίρεσιν τῶν πολλῶν ἐτῶν ποιεῖται κατὰ εἰκοσιπενταετηρίδας, ὡς ἔφημεν, ἐν δὲ τῇ ἐπιβαλλούσῃ παραθέσει τῶν μοιρῶν διὰ πάντων τῶν κανόνων αὐτοῦ μόνα πρῶτα ἑξηκοστὰ ταῖς μοίραις προσπαρατίθησι καὶ οὐ πρὸς τούτοις καὶ ἕτερα, κἀντεῦθεν πάντως ἀσφαλεστέρα καὶ ἀκριβεστέρα ἐστὶν ἡ κανονοποιία τοῦ Πτολεμαίου· δοκεῖ μοι κρεῖττον εἶναι νῦν ἐν τῇ παρούσῃ ἡμῶν πραγματείᾳ τοὺς περὶ ἡλίου κανόνας ἐκθέσθαι ὡς ὁ Πτολεμαῖος ἐν τῇ Συντάξει, καὶ τὸν πρῶτον αὐτὸν μάλιστα κανόνα ὡς ἐκεῖνος κατὰ ὀκτωκαιδεκαετηρίδας καταγράψασθαι. καὶ εἰσὶν οὗτοι:

362 Cf. Ptol. *Alm.* 3.2

356 εἰρημένων : προειρημένων C

numbers for the revolutions of the stars by following the five tables. As a consequence it is necessary to present the tables that we use in order to calculate and find the longitudes and positions of the stars and first of all the rotations of the Sun.

Ptolemy in his *Syntaxis* chose to classify and divide the multiple years into groups of 18, and adds to the tables of degrees not only degrees and minutes but also seconds and thirds up to sixths, while Theon in the *Handy Tables* makes the division of multiple years into groups of 25, and for the required angles in all his tables includes only degrees and minutes and nothing more. It follows that the tabulation of Ptolemy is more certain and accurate. I think it is better in this Treatise to present for the tables of the Sun those in the *Syntaxis* of Ptolemy. The first table is for intervals of the 18 years. The tables now follow:

21 κανόνες τῶν ὁμαλῶν τοῦ ἡλίου κινήσεων

κανὼν ὀκτωκαιδεκαετηρίδων

ὀκτωκαίδεκα ἔτη	μοῖραι	λεπτά α′	δεύτερα	τρίτα	τέταρτα	πέμπτα	ἕκτα
ιη	τνε	λζ	κε	λς	κ	λδ	λ
λς	τνα	ιδ	να	ιβ	μα	θ	ō
νδ	τμς	νβ	ις	μθ	α	μγ	λ
οβ	τμβ	κθ	μβ	κε	κβ	ιη	ō
ϟ	τλη	ζ	η	α	μβ	νβ	λ
ρη	τλγ	μδ	λγ	λη	γ	κζ	ō
ρκς	τκθ	κα	νθ	ιδ	κδ	α	λ
ρμδ	τκδ	νθ	κδ	ν	μδ	λς	ō
ρξβ	τκ	λς	ν	κζ	ε	ι	λ
ρπ	τις	ιδ	ις	γ	κε	με	ō
ρϟη	τια	να	μα	λθ	μς	ιθ	λ
σις	τζ	κθ	ζ	ις	ς	νδ	ō
σλδ	τγ	ς	λβ	νβ	κζ	κη	λ
σνβ	σϟη	μγ	νη	κη	μη	γ	ō
σο	σϟδ	κα	κδ	ε	η	λζ	λ
σπη	σπθ	νη	μθ	μα	κθ	ιβ	ō
τς	σπε	λς	ιε	ιζ	μθ	μς	λ
τκδ	σπα	ιγ	μ	νδ	ι	κα	ō
τμβ	σος	να	ς	λ	να	νε	λ
τξ	σοβ	κη	λβ	ς	ιβ	λ	ō

21 Tables for the solar mean motion

Table of 18-year periods

18 years	degrees	minutes	seconds	thirds	fourths	fifths	sixths
18	355	37	25	36	20	34	30
36	351	14	51	12	41	9	0
54	346	52	16	49	1	43	30
72	342	29	42	25	22	52	0
90	338	7	8	1	42	52	30
108	333	44	33	38	3	27	0
126	329	21	59	14	24	1	30
144	324	59	24	50	44	36	0
162	320	36	50	27	5	10	30
180	316	14	16	3	25	45	0
198	311	51	41	39	46	19	30
216	307	29	7	16	6	54	0
234	303	6	32	52	27	28	30
252	298	43	58	28	48	3	0
270	294	21	24	5	8	37	30
288	289	58	49	41	29	12	0
306	285	36	15	17	49	46	30
324	281	13	40	54	10	21	0
342	276	51	6	30	30	55	30
360	272	28	32	6	51	30	0

κανὼν ἁπλῶν ἐτῶν

ἔτη ἁπλᾶ	μοῖραι	λεπτά α	δεύτερα	τρίτα	τέταρτα	πέμπτα	ἕκτα
α	τνθ	με	κδ	με	κα	η	λε
β	τνθ	λ	μθ	λ	μβ	ιζ	ι
γ	τνθ	ις	ιδ	ις	γ	κε	με
δ	τνθ	α	λθ	α	κδ	λδ	κ
ε	τνη	μζ	γ	μς	με	μβ	νε
ς	τνη	λβ	κη	λβ	ς	να	λ
ζ	τνη	ιζ	νγ	ιζ	κη	ō	ε
η	τνη	γ	ιη	β	μθ	η	μ
θ	τνζ	μη	μβ	μη	ι	ιζ	ιε
ι	τνζ	λδ	ζ	λγ	λα	κε	ν
ια	τνζ	ιθ	λβ	ιη	νβ	λδ	κε
ιβ	τνζ	δ	νζ	δ	ιγ	μγ	ο
ιγ	τνς	ν	κα	μθ	λδ	να	λε
ιδ	τνς	λε	μς	λδ	νς	ο	ι
ιε	τνς	κα	ια	κ	ιζ	η	με
ις	τνς	ς	λς	ε	λη	ιζ	κ
ιζ	τνε	νβ	ō	ν	νθ	κε	νε
ιη	τνε	λζ	κε	λς	κ	λδ	λ

Table for simple years

simple years	degrees	minutes	seconds	thirds	fourths	fifths	sixths
1	359	45	24	45	21	8	35
2	359	30	49	30	42	17	10
3	359	16	14	16	3	25	45
4	359	1	39	1	24	34	20
5	358	47	3	46	45	42	45
6	358	32	28	32	6	51	30
7	358	17	53	17	28	0	5
8	358	3	18	2	49	8	40
9	357	48	42	48	10	17	15
10	357	34	7	33	31	25	50
11	357	19	32	18	52	34	25
12	357	4	57	4	13	43	0
13	356	50	21	49	34	51	35
14	356	35	46	34	56	0	10
15	356	21	11	20	17	8	45
16	356	6	36	5	38	17	20
17	355	52	0	50	59	25	55
18	355	37	25	36	20	34	30

κανὼν μηνῶν

μῆνες αἰγυπτιακοί	μοῖραι	λεπτά α	δεύτερα	τρίτα	τέταρτα	πέμπτα	ἕκτα
λ ξ	κθ νθ	λδ η	η ιζ	λς ιγ	λς ιβ	ιε λα	λ ō
ϛ ρκ	πη ριη	μβ ις	κε λδ	μθ κς	μη κε	μς β	λ ō
ρν ρπ	ρμζ ροζ	ν κδ	μγ να	γ λθ	α λζ	ιζ λγ	λ ō
σι σμ	σς σλς	νθ λγ	ο η	ις νβ	ιγ ν	μη δ	λ ō
σο τ	σξς σϛε	ζ μα	ιζ κς	κθ ς	κς β	ιθ λε	λ ō
τλ τξ	τκε τνδ	ιε μθ	λδ μγ	μβ ιθ	λη ιε	ν ς	λ ō

Table of months

Egyptian months expressed in days	degrees	minutes	seconds	thirds	fourths	fifths	sixths
30	29	34	8	36	36	15	30
60	59	8	17	13	12	31	0
90	88	42	25	49	48	46	30
120	118	16	34	26	25	2	0
150	147	50	43	3	1	17	30
180	177	24	51	39	37	33	0
210	206	59	0	16	13	48	30
240	236	33	8	52	50	4	0
270	266	7	17	29	26	19	30
300	295	41	26	6	2	35	0
330	325	15	24	42	38	50	30
360	354	49	43	19	15	6	0

κανὼν ἡμερῶν

ἡμέραι	μοῖραι	λεπτά α	δεύτερα	τρίτα	τέταρτα	πέμπτα	ἕκτα
α	ō	νθ	η	ιζ	ιγ	ιβ	λα
β	α	νη	ις	λδ	κς	κε	β
γ	β	νζ	κδ	να	λθ	λζ	λγ
δ	γ	νς	λγ	η	νβ	ν	δ
ε	δ	νε	μα	κς	ς	β	λε
ς	ε	νδ	μθ	μγ	ιθ	ιε	ς
ζ	ς	νγ	νη	ō	λβ	κζ	λζ
η	ζ	νγ	ς	ιζ	με	μ	η
θ	η	νβ	ιδ	λδ	νη	νβ	λθ
ι	θ	να	κβ	νβ	ιβ	ε	ι
ια	ι	ν	λα	θ	κε	ιζ	μα
ιβ	ια	μθ	λθ	κς	λη	λ	ιβ
ιγ	ιβ	μη	μζ	μγ	να	μβ	μγ
ιδ	ιγ	μζ	νς	α	δ	νε	ιδ
ιε	ιδ	μζ	δ	ιη	ιη	ζ	με
ις	ιε	μς	ιβ	λε	λα	κ	ις
ιζ	ις	με	κ	νβ	μδ	λβ	μζ
ιη	ιζ	μδ	κθ	θ	νζ	με	ιη
ιθ	ιη	μγ	λζ	κζ	ι	νζ	μθ
κ	ιθ	μβ	με	μδ	κδ	ι	κ
κα	κ	μα	νδ	α	λζ	κβ	να
κβ	κα	μα	β	ιη	ν	λε	κβ
κγ	κβ	μ	ι	λε	γ	μζ	νγ
κδ	κγ	λθ	ιη	νγ	ιζ	ō	κδ
κε	κδ	λη	κζ	ι	λ	ιβ	νε
κς	κε	λζ	λε	κζ	μγ	κε	κς
κζ	κς	λς	μγ	μδ	νς	λζ	νζ
κη	κζ	λε	νβ	β	θ	ν	κη
κθ	κη	λε	ō	ιθ	κγ	β	νθ
λ	κθ	λδ	η	λς	λς	ιε	λ

Table of days

days	degrees	minutes	seconds	thirds	fourths	fifths	sixths
1	0	59	8	17	13	12	31
2	1	58	16	34	26	25	2
3	2	57	24	51	39	37	33
4	3	56	33	8	52	50	4
5	4	55	41	26	6	2	35
6	5	54	49	43	19	15	6
7	6	53	58	0	32	27	37
8	7	53	6	17	45	40	8
9	8	52	14	34	58	52	39
10	9	51	22	52	12	5	10
11	10	50	31	9	25	17	41
12	11	49	39	26	38	30	12
13	12	48	47	43	51	42	43
14	13	47	56	1	4	55	14
15	14	47	4	18	18	7	45
16	13	46	12	35	31	20	16
17	16	45	20	52	44	32	47
18	17	44	29	9	57	45	18
19	18	43	37	27	10	57	49
20	19	42	45	44	24	10	20
21	20	41	54	1	37	22	51
22	21	41	2	18	50	35	22
23	22	40	10	35	3	47	53
24	23	39	18	53	17	0	24
25	24	38	27	10	30	12	55
26	25	37	35	27	43	25	26
27	26	36	43	44	56	37	57
28	27	35	52	2	9	50	28
29	28	35	0	19	23	52	59
30	29	34	8	36	36	15	30

κανὼν ὡρῶν

ὧραι	μοῖραι ἡλίου	λεπτά α	δεύτερα	τρίτα	τέταρτα	πέμπτα	ἕκτα
α	ō	β	κζ	ν	μγ	γ	α
β	ō	δ	νε	μα	κς	ς	β
γ	ō	ζ	κγ	λβ	θ	θ	γ
δ	ō	θ	να	κβ	νβ	ιβ	ε
ε	ō	ιβ	ιθ	ιγ	λε	ιε	ς
ς	ō	ιδ	μζ	δ	ιη	ιη	ζ
ζ	ō	ιζ	ιδ	νε	α	κα	θ
η	ō	ιθ	μβ	με	μδ	κδ	ι
θ	ō	κβ	ι	λς	κζ	κζ	ια
ι	ō	κδ	λη	κζ	ι	λ	ιβ
ια	ō	κζ	ς	ιζ	νγ	λγ	ιδ
ιβ	ō	κθ	λδ	η	λς	λς	ιε
ιγ	ō	λβ	α	νθ	ιθ	λθ	ις
ιδ	ō	λδ	κθ	ν	β	μβ	ιη
ιε	ō	λς	νζ	μ	με	με	ιθ
ις	ō	λθ	κε	λα	κη	μη	κ
ιζ	ō	μα	νγ	κβ	ια	να	κα
ιη	ō	μδ	κα	ιβ	νδ	νδ	κγ
ιθ	ō	μς	μθ	γ	λζ	νζ	κδ
κ	ō	μθ	ις	νδ	κα	ō	κε
κα	ō	να	μδ	με	δ	γ	κζ
κβ	ō	νδ	ιβ	λε	μζ	ς	κη
κγ	ō	νς	μ	κς	λ	θ	κθ
κδ	ō	νθ	η	ιζ	ιγ	ιβ	λα

Table of hours

hours	degrees of the sun	minutes	seconds	thirds	fourths	fifths	sixths
1	0	2	27	50	43	3	1
2	0	4	55	41	26	6	2
3	0	7	23	32	9	9	3
4	0	9	51	22	52	12	5
5	0	12	19	13	35	15	6
6	0	14	47	4	18	18	7
7	0	17	14	55	1	21	9
8	0	19	42	45	44	24	10
9	0	22	10	36	27	27	11
10	0	24	38	27	10	30	12
11	0	27	6	17	53	33	14
12	0	29	34	8	36	36	15
13	0	32	1	59	19	39	17
14	0	34	29	50	2	42	18
15	0	36	57	40	45	45	19
16	0	39	25	31	28	48	20
17	0	41	53	22	11	51	21
18	0	44	21	12	54	54	23
19	0	46	49	3	37	57	24
20	0	49	16	54	21	0	25
21	0	51	44	42	4	3	27
22	0	54	12	35	47	6	28
23	0	56	40	26	30	9	29
24	0	59	8	17	13	12	31

22 διασάφησις τῆς ἐκθέσεως τῶν αὐτῶν κανόνων

Ἐν μὲν οὖν τῷ πρώτῳ κανόνι, τῷ τῶν ὀκτωκαιδεκαετηρίδων ἐπιγραφομένῳ, κατὰ τὸν πρῶτον στίχον ἐκτίθεται εὐθὺς ἐν τῷ πρώτῳ σελιδίῳ ὁ τῶν ὀκτωκαίδεκα ἐτῶν ἀριθμός, ἐν δὲ τοῖς μετὰ ταῦτα σελιδίοις αἱ ἐπιβάλλουσαι τῆς ἡλιακῆς κινήσεως μοῖραι τοῖς τοσούτοις ἔτεσιν, ἀπό τινος δηλονότι τοῦ κύκλου ἀρχῆς, μετὰ ἀφαίρεσιν πάντως κυκλικῶν οὐκ ὀλίγων περιόδων, ἔτι δὲ ἑξηκοστὰ πρῶτα, δεύτερα, τρίτα καὶ μέχρι τῶν ἕκτων· κατὰ δὲ τὸν δεύτερον στίχον, αὖθις ἐν τῷ πρώτῳ σελιδίῳ, ὁ ἀριθμὸς τῶν ἐτῶν συντεθεὶς ἐκ δὶς ὀκτωκαίδεκα ἐτῶν, ἤτοι ὁ τῶν λς, καὶ ἐν τοῖς ἑξῆς σελιδίοις τὸν αὐτὸν τρόπον αἱ ἐπιβάλλουσαι μοῖραι καὶ τὰ λεπτά· κατὰ δὲ τὸν τρίτον στίχον ὡσαύτως, ἐν μὲν τῷ πρώτῳ σελιδίῳ ὁ ἀριθμὸς αὖθις τῶν ἐτῶν συντεθεὶς ἐκ τρὶς ὀκτωκαίδεκα ἐτῶν, ἤτοι ὁ τῶν νδ, καὶ ἐν τοῖς ἑξῆς σελιδίοις αἱ ἐπιβάλλουσαι μοῖραι καὶ τὰ λεπτά. καὶ τὸν αὐτὸν τρόπον ἐκτίθεται τὰ ἔτη ἐν τῷ πρώτῳ σελιδίῳ μέχρις ἂν βούληταί τις ἀεὶ συντιθέμενα κατὰ ὀκτωκαίδεκα. καὶ ἔστιν ἡ προχώρησις αὕτη τῶν ἀριθμῶν καὶ ἡ σύνθεσις οὕτως ἔχουσα, καὶ αἱ ἐπιβάλλουσαι ἑκάστῳ συναγομένῳ πάντως ἀριθμῷ ἐν τοῖς ἑξῆς καθ᾽ ἕκαστον στίχον σελιδίοις μοῖραι καὶ ἑξηκοστά.

Εἰσὶ δὲ καὶ αἱ ἀνήκουσαι παραθέσεις αὗται τῶν μοιρῶν καὶ τῶν ἑξηκοστῶν ῥᾳδίως ἐπιλογιζόμεναι τὸν τρόπον τοῦτον. ὁ γὰρ παρακείμενος ἀριθμὸς τῶν μοιρῶν καὶ τῶν λεπτῶν τῇ πρώτῃ ὀκτωκαιδεκαετηρίδι αὐτὸς συντιθέμενος ἐφ᾽ ἑαυτὸν ποιεῖ τὸν ἀριθμὸν τῶν μοιρῶν καὶ τῶν λεπτῶν, ὅσαι ἀνήκουσι τῇ κινήσει τοῦ ἡλίου ἐν τῇ δευτέρᾳ ὀκτωκαιδεκαετηρίδι. καὶ ἐὰν ὁ συντιθέμενος οὗτος ἀριθμὸς ἐλάττων ᾖ μοιρῶν τξ, αὐτὸς ὀφείλει τίθεσθαι ἐν τῷ δευτέρῳ στίχῳ, ἤτοι ἐν τῇ δευτέρᾳ ὀκτωκαιδεκαετηρίδι. ἐὰν δὲ ὑπερβαίνῃ τὰς τξ μοίρας, ὀφείλει ἀφαιρεῖσθαι κυκλικὴ πάροδος, ἤτοι αἱ τξ μοῖραι, καὶ τὰ καταλειπόμενα τίθεσθαι ὡς ἐπιβάλλοντα τῇ δευτέρᾳ ὀκτωκαιδεκαετηρίδι. εἶτα πάλιν τούτοις προσήκει συντιθέναι τὰ τῇ πρώτῃ ὀκτωκαιδεκαετηρίδι παρακείμενα, ἤτοι μοίρας καὶ λεπτά, καὶ

26 περίοδος C

22 Explanation for the Contents of the Tables

In the first table, which has the title "Periods of 18 Years", in the first row of the first column is presented the number of eighteen years, and in the other columns the position of the Sun in degrees corresponding to that many years from the beginning of the cycle, with the proviso that we subtract complete periods. The positions are given in minutes, seconds, thirds up to sixths. In the second row and on first column is presented the number of twice 18 years, that is 36 years, and in the adjacent columns and in the same manner the corresponding degrees and minutes. Similarly, in the third row and in the first column is the number of the subsequent years made up of thrice 18 years, i.e. 54, and in the following columns the degrees and minutes. In this manner are enumerated in the first column the years up to the desired 18 years we wish to calculate. The development of the numbers and their synthesis proceeds in this manner, tabulating in each row of the columns the corresponding number of degrees and minutes.

The entries of degrees and minutes are computed appropriately and easily in the following manner. The reported number of degrees and minutes for the first 18 years added to itself produces the degrees and minutes which correspond to the movement of the Sun for the second 18 years. If the resulting number is smaller or equal to 360° it must be included in the second row, that is for the second 18 years. If it exceeds 360°, we subtract a full period, and the remainder is recorded for the second cycle of 18 years. Then, adding again to this number, the degrees and minutes of the first row

ὁ ἐκ τῆς συνθέσεως αὐτῶν γινόμενος ἀριθμὸς ὀφείλει τίθεσθαι ἐν τῷ τρίτῳ στίχῳ, ὡς ἀνήκων τῇ τρίτῃ ὀκτωκαιδεκαετηρίδι τὸν αὐτὸν τρόπον. εἰ μὲν ἔστιν ἐντὸς τῶν τξ αὐτὸς παρατιθέμενος, εἰ δὲ ὑπερβαίνει τὸν τῶν τξ ἀριθμόν, ὀφείλει ὡσαύτως μετὰ τὴν ἀφαίρεσιν τῶν τξ μοιρῶν τίθεσθαι ὁ καταλιμπανομένος ἀριθμός. καὶ πάλιν τούτῳ συντιθέμενος ὁ ἀριθμὸς τῶν μοιρῶν καὶ τῶν λεπτῶν τῶν παρακειμένων τῇ πρώτῃ ὀκτωκαιδεκαετηρίδι γεννήσει τὸν ἑξῆς τέταρτον στίχον. καὶ οὕτω δὲ παραπλησίως ποιῶν τις εἰς ὅσον ἂν βούληται πλῆθος χρόνων ἐκτενεῖ τὸν κανόνα τῶν τοιούτων ὀκτωκαιδεκαετηρίδων. καὶ ὁ μὲν πρῶτος κανὼν τῶν χρονικῶν κεφαλαίων, ὁ τῶν ὀκτωκαιδεκαετηρίδων, ἔχει οὕτως.

Ἑξῆς δὲ ὁ δεύτερος κανών, ὁ καλούμενος τῶν ἁπλῶν ἐτῶν τῆς ἡλίου κινήσεως, ἔχει ἐν τῷ πρώτῳ στίχῳ κατὰ τὸ πρῶτον σελίδιον ἔτος ἓν καὶ ὅσα αὐτῷ ἐπιβάλλει ἐν τοῖς ἑξῆς σελιδίοις παρακείμενα, μοίρας δηλαδὴ καὶ λεπτά, ὡσαύτως ἐκτιθέμενα μέχρι καὶ τῶν ἕκτων. καὶ ἑξῆς δεύτερον ἔτος ἐν τῷ δευτέρῳ στίχῳ καὶ τρίτον ἐν τῷ τρίτῳ καὶ τέταρτον ἑξῆς καὶ οὕτω δὴ συνεχῶς μέχρι τῶν ὀκτωκαίδεκα. καὶ τοὺς μὲν ἀριθμοὺς τῶν ἐτῶν ἔχει καθ᾽ εἱρμὸν οὕτω τὸ παρὸν σελίδιον, τὰ δὲ ἑξῆς σελίδια τὰς ἐπιβαλλούσας ἑκάστῳ ἔτει τῆς κινήσεως τοῦ ἡλίου μοίρας καὶ λεπτά. καὶ ἔστιν ἡ σύνθεσις αὐτῶν καὶ ἡ καταγραφὴ κατὰ τὸν ἀνωτέρω τρόπον, ὃν ἔφημεν ἐν τῷ κανόνι τῶν ὀκτωκαιδεκαετηρίδων. καὶ ὁ παρακείμενος ἀριθμὸς τῶν μοιρῶν καὶ τῶν λεπτῶν ἐν τῷ ἐσχάτῳ στίχῳ, τῶν ὀκτωκαίδεκα ἐτῶν δηλονότι, ὁ αὐτὸς ἔσται πάντως τῷ παρακειμένῳ ἀριθμῷ ἐν τῷ κανόνι τῶν ὀκτωκαιδεκαετηρίδων τῇ πρώτῃ ὀκτωκαιδεκαετηρίδι. οὕτω δὴ καὶ ὁ δεύτερος κανὼν ἔχει τῶν ἁπλῶν ἐτῶν.

Ἑξῆς δὲ τούτων ὁ τρίτος κανών, ὁ ἐπιγραφόμενος τῶν μηνῶν, ὡσαύτως καὶ οὗτος ἔχων. εἰσὶ μὲν γὰρ τριακονθήμεροι δώδεκα καὶ τίθενται ἐν τῷ κανόνι κατὰ τὸ πρῶτον σελίδιον καθ᾽ εἱρμὸν καὶ οὗτοι ἀπὸ τοῦ πρώτου στίχου ἑξῆς, ὁ πρῶτος, ὁ δεύτερος, ὁ τρίτος μέχρι καὶ τῶν ιβ. καὶ ἐπεὶ οἱ τοιοῦτοι δώδεκα οὐδὲν ἀπαρτίζουσιν ἔτος αἰγυπτιακόν, ἀλλὰ χρήζουσι καὶ ἡμερῶν πέντε, ὅτι τὸ αἰγυπτιακὸν ἔτος τξε ἡμέρας ἀείποτε μόνας ἔχει, ὡς πολλάκις ἔφημεν, προσλαμβάνει ἀπὸ τοῦ μετ᾽ αὐτὸν κανόνος ἐπιγραφομένου τῶν ἡμερῶν πέντε ἡμερῶν

the resulting number must be introduced to the third row, thus arriving in this manner to the third 18 year period. If the number is smaller than 360 it is recorded. If it exceeds 360°, it is introduced after we subtract a full period. Then again, this number combined with the degrees and minutes of the first row of 18 years produces the fourth row. By following this method faithfully for as many years as one wishes, one produces the table of 18 years. This is the table of 18 years.

Next is the second table, called table for the motion of the Sun in simple years, and it has in the first column of the first row the first year and whatever [position] corresponds to it in the adjacent columns, that is the degrees and minutes up to sixths. Next comes the second year in the second row, the third year in the third row, fourth and so on up to the 18th year. The numbers of years are included [in sequence] one after the other in the first column, and in the subsequent columns are the degrees and minutes corresponding to the rotation of the Sun. The composition and presentation is obtained by the same method we described for the table of eighteen years. The values for degrees and minutes in the last row, i.e. that of 18 years, are always the same values as those in the first row of the table for 18 years. This is the second table of the simple years.

Next the third table, that of the months, is constructed in a similar way. There are twelve months of 30 days, and they are recorded one after the other in the first column of the table. Starting with the first row they are enumerated as first, second, third up to the twelfth. And since these months do not complete an Egyptian year but are missing five days (the Egyptian year always has 365 days as we mentioned many times), it borrows five days from the table of days.

ἐπιβάλλοντα παρακείμενα μοιρῶν καὶ λεπτῶν. καὶ συντίθησι ταῦτα ταῖς παρακειμέναις μοίραις καὶ λεπτοῖς τῷ ιβ μηνὶ καὶ ὁ συντιθέμενος ἐκ τούτων ἀριθμὸς τῶν μοιρῶν καὶ τῶν λεπτῶν, αὐτός ἐστι πάντως ὁ καὶ τῷ ἑνὶ ἔτει παρακείμενος.

Τέταρτος κανὼν ὁ ἐπιγραφόμενος τῶν ἡμερῶν, καὶ αὐτὸς ὡσαύτως ἔχων. ἐκτίθενται γὰρ καθ᾽ εἱρμὸν ἐν αὐτῷ κατὰ τὸ πρῶτον σελίδιον, ἀπὸ τοῦ πρώτου στίχου καὶ ἑξῆς, πρώτη ἡμέρα, δευτέρα, τρίτη καὶ μέχρι τῆς τριακοστῆς. καὶ παράκεινται ἑκάστῃ ἐν τοῖς ἑξῆς σελιδίοις αἱ ἐπιβάλλουσαι μοῖραι καὶ τὰ λεπτὰ τῆς ἡλιακῆς κινήσεως. καὶ τὰ παρακείμενα τῇ τριακοστῇ ἡμέρᾳ, ἤτοι μοῖραι καὶ λεπτά, τὰ αὐτά εἰσι πάντως, ἃ καὶ τῷ ἑνὶ μηνὶ παράκεινται.

Πέμπτος ἐπὶ πᾶσι κανὼν ὁ ἐπιγραφόμενος τῶν ὡρῶν, καὶ αὐτὸς ὡσαύτως ἔχων κατὰ τὸ πρῶτον σελίδιον τὸν ἀριθμὸν αὐτὸν τῶν ὡρῶν καθ᾽ εἱρμὸν ἑξῆς, πρώτη, δευτέρα, τρίτη, μέχρι καὶ τῆς κδ, ὅσαι δὴ καὶ ἀπαρτίζουσι τὸν τοῦ ἡμερονυκτίου χρόνον, ὡς ἰσημεριναὶ λογιζόμεναι. πρὸς γὰρ ὥρας ἰσημερινάς, ὡς προέφημεν, ὁ κανὼν οὗτος τοῖς ἀστρονόμοις ἐκτίθεται, περὶ δὲ τῆς ὡρῶν διαφορᾶς μετ᾽ ὀλίγον ἐροῦμεν. ἐν δὲ τοῖς ἑξῆς μετὰ τὸ πρῶτον ἐν τῷ τοιούτῳ κανόνι σελιδίοις παράκεινται τὸν αὐτὸν τρόπον, καθὼς καὶ ἐν τοῖς ἄλλοις κανόσι, κατὰ τάξιν τὰ ἐπιβάλλοντα ἑκάστῃ ὥρᾳ λεπτά. κἀνταῦθα πάλιν τὰ ἐπιβάλλοντα τῇ κδ ὥρᾳ λεπτὰ τὰ αὐτά εἰσι τοῖς παρακειμένοις τῇ μιᾷ ἡμέρᾳ. καὶ ἡ μὲν μέθοδος τῶν πέντε κανόνων οὕτως ἔχει.

Five days with the associated degrees and minutes are added to the twelfth month and the resulting number is the value for one year.

The fourth table has the title "table of the days" and is very similar. In the table are presented in the first column [starting with the first row] the first day, second, third up to the thirtieth and in the adjacent columns the degrees and minutes corresponding to the movement of the Sun. The entry for the thirtieth day (in degrees and minutes) is always the same as that recorded for the first month.

Fifth among them is the table with the title for hours and this, in the same manner, has in the first column the sequence of hours: first, second, third up to the 24th, which comprise a day and a night and are equinoctial hours. As we mentioned, the table is presented for the benefit of astronomers in equinoctial hours. The differences among the definitions of hours will be discussed a little later. Adjacent to the first column are presented, in the same way as in other tables, the minutes corresponding to each hour. Again, the minutes corresponding to the 24th hour are the same as those for one day. This is the method for constructing the five tables.

23 περὶ τῆς ἀνωμάλου τοῦ ἡλίου κινήσεως καὶ ἔκθεσις κανόνος τῆς τοιαύτης ἀνωμάλου αὐτοῦ κινήσεως

Ἐπεὶ δὲ οὕτως αἱ καθ' ὁντιναοῦν χρόνον ὁμαλαὶ τοῦ ἡλίου κινήσεις καὶ ἐποχαὶ διὰ τῶν εἰρημένων πέντε κανόνων ῥᾳδίως δύνανται ἐπιλογίζεσθαι (προέφημεν δὲ καὶ ἀνώμαλον θεωρεῖσθαι τὴν κίνησιν μὴ μόνον τῶν ε πλανωμένων καὶ τῆς σελήνης, ἀλλὰ καὶ αὐτοῦ τοῦ ἡλίου—ἁπλουστέραν μέντοι ἐπὶ τοῦ ἡλίου θεωρεῖσθαι τὴν ἀνώμαλον κίνησιν καὶ ὀλιγομερεστέραν—καὶ τὸν τρόπον δὲ δι' ὃν (καὶ τὴν αἰτίαν) αἱ ἀνώμαλοι κινήσεις γίνονται καθόλου ἐπὶ πάντων καὶ ὅπως ἁπλούστερον ἐπὶ τοῦ ἡλίου γίνεται προέφημεν), ἀναγκαῖον διευκρινῆσαι πῶς οἷον τέ ἐστι καὶ τὴν τῆς ἀνωμάλου κινήσεως ἑκάστοτε διαφορὰν ἐπὶ τῆς ἡλιακῆς κινήσεως πρὸς τὴν ὁμαλὴν αὐτοῦ κίνησιν ἐπιλογίσασθαι, καὶ ταύτην τὴν διαφορὰν ἢ ἀφελεῖν πάντως ἢ προσθεῖναι τῇ ἐπιλογιζόμενῃ διὰ τῶν ἀνωτέρω εἰρημένων κανόνων ὁμαλῇ αὐτοῦ κινήσει.

Ἐπειδὴ καὶ ἐξανάγκης ἡ τοιαύτη διαφορὰ οὐκ ἀεὶ ἢ προστίθεται ἢ ἀφαιρεῖται τῆς ὁμαλῆς κινήσεως αὐτοῦ, ἀλλ' ἐπ' ἐνίων μὲν τμημάτων τῆς ὁμαλῆς αὐτοῦ κινήσεως προστίθεται καὶ ἔστιν ἐπαυξητικὴ τηνικαῦτα τῆς ἐπιλογιζόμενης διὰ τῶν εἰρημένων κανόνων ὁμαλῆς αὐτοῦ κινήσεως, ἐπ' ἐνίων δὲ αὖθις τμημάτων ἀφαιρεῖται ἡ τοιαύτη διαφορὰ τῆς ἀνωμάλου κινήσεως, ὡς τηνικαῦτα θεωρεῖσθαι ἐλάττονα τὴν ἀνώμαλον καὶ ἀληθῆ καὶ ἀκριβῆ κίνησιν τοῦ ἡλίου τῆς ὁμαλῆς αὐτοῦ κινήσεως, ᾗ δὴ καὶ χρώμεθα κατὰ τὴν προκειμένην ἑκάστοτε χρείαν, ἐπειδὴ αἱ ἀνώμαλοι κινήσεις τῶν ἀστέρων αὗταί εἰσιν αἱ ἀληθεῖς καὶ ἀκριβεῖς τοῖς καταλογιζομένοις ταύτας ἡμῖν, ὡς ἀπὸ τῆς γῆς καὶ τοῦ κέντρου τοῦ κόσμου, ὡς προέφημεν, θεωροῦσιν αὐτὰς ἐπὶ τοῦ ζωδιακοῦ μεγίστου κύκλου.

Καταλαμβάνεται τοίνυν τῷ Πτολεμαίῳ διὰ γραμμικῶν ἀποδείξεων ἀναντίρρητα καὶ ὡς ἔστι μάλιστα ἀσφαλέστατα, ὅτι ἐν τοῖς ἀπογείοις τοῦ οἰκείου ἐκκέντρου φερόμενος ὁ ἥλιος διαφορὰν ἔχει ἀνωμάλου κινήσεως ἀφαιρετικήν, οἷον ὅτι πλείων ἐστὶν ἡ ὁμαλὴ

25 Cf. Ptol. *Alm.* 3.5 (vol. 1.1, p. 241-243, 246-248 Heiberg)

23 On the Anomalous Motion of the Sun and the corresponding table

Using the five tables it is easy to calculate at any time the uniform movement and longitude of the Sun. As mentioned already we regard as anomalous not only the rotation of the five planets and the Moon, but also that of the Sun. In contrast, the anomalous rotation of the Sun is simpler and has fewer components; the cause and the method that in general create the anomaly is simpler for the Sun. Furthermore, it is necessary to explain the manner [in which they are different], and to calculate the amount by which the anomalous motion differs from the uniform so that we subtract or add this amount to the uniform motion, computed from the above tables.

Since the difference is not always added or subtracted to the uniform motion, but in some regions is added to the uniform, thus making the final movement larger than the regular, and sometimes is subtracted making at this particular time the real, anomalous and precise motion smaller than the regular, we practically need the correction for each case in order to obtain the anomalous motions. These are the true and precise movements of the stars on the ecliptic, as we observe them from the Earth the centre of the world.

Using geometric methods, Ptolemy understood reliably and with large certainty that, for the Sun at its apogee of the eccentric, the difference caused by the anomaly must be subtracted, which means the regular motion on the

κίνησις αὐτοῦ, ἤτοι ἡ ἐπὶ τοῦ οἰκείου κύκλου τοῦ ἐκκέντρου, παρὸ καταλαμβάνεται ἡ ἐπὶ τοῦ ζῳδιακοῦ θεωρουμένη· ἐν δὲ τοῖς περιγείοις μέρεσιν αὐτοῦ φερομένου τὸ ἐναντίον, ἡ διαφορὰ τῆς ἀνωμάλου κινήσεως αὐτοῦ ἐστι προσθετική, οἷον ὅτι πλεῖον θεωρεῖται κινούμενος ὁ ἥλιος ἐπὶ τοῦ ζῳδιακοῦ κύκλου ἤπερ ἐπὶ τοῦ οἰκείου ἐκκέντρου ὁμαλῶς καταλαμβάνεται φερόμενος. καὶ τοῦτο δείκνυται μὲν τῷ Πτολεμαίῳ καθολικῷ τῷ λόγῳ ἐπὶ τῶν ἀπογείων καὶ περιγείων, καὶ κατὰ μέρος δὲ διὰ τῶν αὐτῶν ἀποδείξεων γραμμικῶν ἀνευρὼν ὅση διαφορὰ ἀφαιρετική, ὡς ἔφημεν, ἀνήκει ταῖς τοσαῖσδε μοίραις τῆς ὁμαλῆς κινήσεως τοῦ ἡλίου, τρισὶν ἢ τέτταρσιν ἢ ἄλλαις ὅσαις δή τισιν ἀπὸ τοῦ ἀπογείου φερομένου πρὸς τὴν ἀνώμαλον αὐτοῦ κίνησιν, τὴν ἐπὶ τοῦ ζῳδιακοῦ δηλονότι θεωρουμένην καὶ ὅση αὖθις διαφορὰ ἀνήκει ταῖς τοσαῖσδε μοίραις τῆς ὁμαλῆς κινήσεως τοῦ ἡλίου, τρισὶν ἢ τέτταρσιν ἢ ἄλλαις ὅσαις δή τισιν ἀπὸ τοῦ περιγείου φερομένου πρὸς τὴν ἀνώμαλον αὐτοῦ κίνησιν, τὴν ἐπὶ τοῦ ζῳδιακοῦ δηλονότι θεωρουμένην.

Ταῦτα ἐπιλογισάμενος καὶ ἀνευρών, ὡς ἔφην, δείξεσι γραμμικαῖς πολὺ τὸ πιστὸν ἐχούσαις, ἔπειτα ἐκτίθεται κανόνα ἴδιον ἕτερον, ὃν ἐπιγράφει κανόνα ἀνωμαλίας τῆς τοῦ ἡλίου κινήσεως, ἐν ᾧ ἐκτιθέμενος τὴν ἀπὸ τοῦ ἀπογείου κίνησιν τοῦ ἡλίου, οὐ κατὰ μοῖραν ἑκάστην τῆς κινήσεως παρατίθησι τὴν ἀνήκουσαν διαφορὰν εἴτε ἀφαιρετικὴν εἴτε προσθετικήν, ἀλλὰ κατὰ πλείους. ὅ γε μὴν Θέων ἐν τοῖς Προχείροις τὸν αὐτὸν κανόνα τῆς τοῦ ἡλίου ἀνωμαλίας καταγράφων κατὰ μίαν ἑκάστην μοῖραν τῆς ἀπὸ τοῦ ἀπογείου τοῦ ἡλίου κινήσεως παρατίθησι τὴν ἀνήκουσαν διαφορὰν εἴτε ἀφαιρετικὴν εἴτε προσθετικήν. διὰ ταῦτα δὴ καὶ ἡμεῖς ἐν τῇ παρούσῃ συντάξει τὸν τοῦ Θέωνος κανόνα προειλόμεθα νῦν ἐκθέσθαι, ὡς ἄν, ὡς βούλοιτό τις, ῥᾳδίως ἐντεῦθεν καὶ παντάπασιν ἀπόνως ἐξείη πορίζεσθαι καὶ ἀνευρίσκειν τὰς ἐν πᾶσι τμήμασι καὶ τόποις τῆς ἀπὸ τοῦ ἀπογείου ὁμαλῆς τοῦ ἡλίου κινήσεως διαφορᾶς πρὸς τὴν ἀνώμαλον. καὶ ἔστιν ὁ κανὼν οὗτος:

35 Cf. Ptol. *Alm.* 3.5 (vol. 1.1, p. 241-243, 246-248 Heiberg) **45-50** Cf. Ptol. *Alm.* 3.6 (vol. 1.1, p. 253 Heiberg) **50** Cf. Ptol. *Proch. Kan.* (vol. 2, p. 78-89 Halma)

38 τέταρσιν C **40-44** καὶ ὅση ... θεωρουμένην om. G[ac] (homoeotel.), corr. in mg. Greg.

eccentric is larger than the one observed on the ecliptic. The opposite happens in regions close to perigee where the difference of the anomalous motion must be added, which means that the Sun appears to move faster in the ecliptic than on its eccentric. This is proven by Ptolemy with complete arguments for the apogee, perigee and for other places where he found geometrically how large a difference must be subtracted from the uniform motion as the Sun moves from its apogee, three, four or however many degrees, in order to obtain the anomalous motion we observe on the ecliptic. Then, again [he found] the correction corresponding to three, four or however many degrees from the perigee in order to obtain the anomalous rotation observed on the ecliptic.

As we mentioned, once Ptolemy reckoned and discovered all these, he presented convincing geometrical proofs, and then produced another special table, which he calls "table of the anomalous motions of the Sun". In this [table] he presents the appropriate difference (subtractive or additive) as a function of the distance of the Sun from the apogee, not in single but in more degrees. Theon, on the other hand, in his *Handy Tables*, records the anomaly as a function of the Sun from the apogee in single degrees, being subtractive or additive. For this reason we chose to present in this *Syntaxis* Theon's Table, because, if one wishes he can easily retrieve and find out from [the Table], for every time and at any location from the apogee, the difference of the anomalous from the regular motion. Next follows the Table:

Κανὼν ἀνωμάλου κινήσεως ἡλίου

Μ = Μοῖραι, Λ = Λεπτά

Ἀριθμοί κοινοί		Μῆκος ἡλίου		Ἀριθμοί κοινοί		Μῆκος ἡλίου		Ἀριθμοί κοινοί		Μῆκος ἡλίου		Ἀριθμοί κοινοί		Μῆκος ἡλίου		Ἀριθμοί κοινοί		Μῆκος ἡλίου	
		Μ	Λ			Μ	Λ			Μ	Λ			Μ	Λ			Μ	Λ
α	τνθ	ō	γ	λζ	τκγ	α	κγ	ογ	σπζ	β	ιε	ρθ	σνα	β	ιζ	ρμε	σιε	α	κε
β	τνη	ō	ε	λη	τκβ	α	κε	οδ	σπς	β	ις	ρι	σν	β	ις	ρμς	σιδ	α	κγ
γ	τνζ	ō	ζ	λθ	τκα	α	κζ	οε	σπε	β	ιζ	ρια	σμθ	β	ις	ρμζ	σιγ	α	κα
δ	τνς	ō	ι	μ	τκ	α	κθ	ος	σπδ	β	ιη	ριβ	σμη	β	ιε	ρμη	σιβ	α	ιθ
ε	τνε	ō	ιβ	μα	τιθ	α	λα	οζ	σπγ	β	ιθ	ριγ	σμζ	β	ιδ	ρμθ	σια	α	ιζ
ς	τνδ	ō	ιδ	μβ	τιη	α	λγ	οη	σπβ	β	ιθ	ριδ	σμς	β	ιγ	ρν	σι	α	ιδ
ζ	τνγ	ō	ιζ	μγ	τιζ	α	λε	οθ	σπα	β	κ	ριε	σμε	β	ιβ	ρνα	σθ	α	ιβ
η	τνβ	ō	ιθ	μδ	τις	α	λς	π	σπ	β	κα	ρις	σμδ	β	ια	ρνβ	ση	α	ι
θ	τνα	ō	κα	με	τιε	α	λη	πα	σοθ	β	κα	ριζ	σμγ	β	ι	ρνγ	σζ	α	ζ
ι	τν	ō	κδ	μς	τιδ	α	μ	πβ	σοη	β	κβ	ριη	σμβ	β	θ	ρνδ	σς	α	ς
ια	τμθ	ō	κς	μζ	τιγ	α	μβ	πγ	σοζ	β	κβ	ριθ	σμα	β	η	ρνε	σε	α	γ
ιβ	τμη	ō	κη	μη	τιβ	α	μγ	πδ	σος	β	κβ	ρκ	σμ	β	ζ	ρνς	σδ	α	ō
ιγ	τμζ	ō	λα	μθ	τια	α	με	πε	σοε	β	κβ	ρκα	σλθ	β	ς	ρνζ	σγ	ō	νη
ιδ	τμς	ō	λγ	ν	τι	α	μζ	πς	σοδ	β	κγ	ρκβ	σλη	β	ε	ρνη	σβ	ō	νς
ιε	τμε	ō	λε	να	τθ	α	μη	πζ	σογ	β	κγ	ρκγ	σλζ	β	γ	ρνθ	σα	ō	νδ
ις	τμδ	ō	λη	νβ	τη	α	ν	πη	σοβ	β	κγ	ρκδ	σλς	β	β	ρξ	σ	ō	να
ιζ	τμγ	ō	μ	νγ	τζ	α	νβ	πθ	σοα	β	κγ	ρκε	σλε	β	α	ρξα	ρϟθ	ō	μθ
ιη	τμβ	ō	μγ	νδ	τς	α	νγ	ϟ	σο	β	κγ	ρκς	σλδ	α	νθ	ρξβ	ρϟη	ō	μς
ιθ	τμα	ō	με	νε	τε	α	νε	ϟα	σξθ	β	κγ	ρκζ	σλγ	α	νη	ρξγ	ρϟζ	ō	μδ
κ	τμ	ō	μζ	νς	τδ	α	νς	ϟβ	σξη	β	κγ	ρκη	σλβ	α	νς	ρξδ	ρϟς	ō	μβ
κα	τλθ	ō	μθ	νζ	τγ	α	νζ	ϟγ	σξζ	β	κγ	ρκθ	σλα	α	νδ	ρξε	ρϟε	ō	λθ
κβ	τλη	ō	να	νη	τβ	α	νθ	ϟδ	σξς	β	κγ	ρλ	σλ	α	νγ	ρξς	ρϟδ	ō	λζ
κγ	τλζ	ō	νδ	νθ	τα	β	ō	ϟε	σξε	β	κγ	ρλα	σκθ	α	να	ρξζ	ρϟγ	ō	λδ
κδ	τλς	ō	νς	ξ	τ	β	α	ϟς	σξδ	β	κγ	ρλβ	σκη	α	μθ	ρξη	ρϟβ	ō	λα
κε	τλε	ō	νη	ξα	σϟθ	β	γ	ϟζ	σξγ	β	κγ	ρλγ	σκζ	α	μη	ρξθ	ρϟα	ō	κθ
κς	τλδ	α	α	ξβ	σϟη	β	δ	ϟη	σξβ	β	κβ	ρλδ	σκς	α	μς	ρο	ρϟ	ō	κς
κζ	τλγ	α	γ	ξγ	σϟζ	β	ε	ϟθ	σξα	β	κβ	ρλε	σκε	α	μδ	ροβ	ρπθ	ō	κγ
κη	τλβ	α	ε	ξδ	σϟς	β	ς	ρ	σξ	β	κβ	ρλς	σκδ	α	μγ	ροβ	ρπη	ō	κα
κθ	τλα	α	ζ	ξε	σϟε	β	ζ	ρα	σνθ	β	κβ	ρλζ	σκγ	α	μα	ρογ	ρπζ	ō	ιθ
λ	τλ	α	θ	ξς	σϟδ	β	η	ρβ	σνη	β	κα	ρλη	σκβ	α	λθ	ροδ	ρπς	ō	ις
λα	τκθ	α	ια	ξζ	σϟγ	β	θ	ργ	σνζ	β	κα	ρλθ	σκα	α	λζ	ροε	ρπε	ō	ιδ
λβ	τκη	α	ιγ	ξη	σϟβ	β	ι	ρδ	σνς	β	κ	ρμ	σκ	α	λε	ρος	ρπδ	ō	ια
λγ	τκζ	α	ιε	ξθ	σϟα	β	ια	ρε	σνε	β	ιθ	ρμα	σιθ	α	λγ	ροζ	ρπγ	ō	η
λδ	τκς	α	ιζ	ο	σϟ	β	ιβ	ρς	σνδ	β	ιθ	ρμβ	σιη	α	λα	ροη	ρπβ	ō	ς
λε	τκε	α	ιθ	οα	σπθ	β	ιγ	ρζ	σνγ	β	ιη	ρμγ	σιζ	α	κθ	ροθ	ρπα	ō	γ
λς	τκδ	α	κα	οβ	σπη	β	ιδ	ρη	σνβ	β	ιη	ρμδ	σις	α	κζ	ρπ	ρπ	ō	ō

Table of the anomalous motion of the Sun

D = Degress, M = Minutes

Common Numbers		Equation D	Equation M	Common Numbers		Equation D	Equation M	Common Numbers		Equation D	Equation M	Common Numbers		Equation D	Equation M	Common Numbers		Equation D	Equation M
1	359	0	3	37	323	1	23	73	287	2	15	109	251	2	17	145	215	1	25
2	358	0	5	38	322	1	25	74	286	2	16	110	250	2	16	146	214	1	23
3	357	0	7	39	321	1	27	75	285	2	17	111	249	2	16	147	213	1	21
4	356	0	10	40	320	1	29	76	284	2	18	112	248	2	15	148	212	1	19
5	355	0	12	41	319	1	31	77	283	2	19	113	247	2	14	149	211	1	17
6	354	0	14	42	318	1	33	78	282	2	19	114	246	2	13	150	210	1	14
7	353	0	17	43	317	1	35	79	281	2	20	115	245	2	12	151	209	1	12
8	352	0	19	44	316	1	36	80	280	2	21	116	244	2	11	152	208	1	10
9	351	0	21	45	315	1	38	81	279	2	21	117	243	2	10	153	207	1	7
10	350	0	24	46	314	1	40	82	278	2	22	118	242	2	9	154	206	1	6
11	349	0	26	47	313	1	42	83	277	2	22	119	241	2	8	155	205	1	3
12	348	0	28	48	312	1	43	84	276	2	22	120	240	2	7	156	204	1	0
13	347	0	31	49	311	1	45	85	275	2	22	121	239	2	6	157	203	0	58
14	346	0	33	50	310	1	47	86	274	2	23	122	238	2	5	158	202	0	56
15	345	0	35	51	309	1	49	87	273	2	23	123	237	2	3	159	201	0	54
16	344	0	38	52	308	1	50	88	272	2	23	124	236	2	2	160	200	0	51
17	343	0	40	53	307	1	52	89	271	2	23	125	235	2	1	161	199	0	49
18	342	0	43	54	306	1	53	90	270	2	23	126	234	1	59	162	198	0	46
19	341	0	45	55	305	1	55	91	269	2	23	127	233	1	58	163	197	0	44
20	340	0	47	56	304	1	56	92	268	2	23	128	232	1	56	164	196	0	42
21	339	0	49	57	303	1	57	93	267	2	23	129	231	1	54	165	195	0	39
22	338	0	51	58	302	1	59	94	266	2	23	130	230	1	53	166	194	0	37
23	337	0	54	59	301	2	0	95	265	2	23	131	229	1	51	167	193	0	34
24	336	0	56	60	300	2	1	96	264	2	23	132	228	1	49	168	192	0	31
25	335	0	58	61	299	2	3	97	263	2	23	133	227	1	48	169	191	0	29
26	334	1	1	62	298	2	4	98	262	2	22	134	226	1	46	170	190	0	26
27	333	1	3	63	297	2	5	99	261	2	22	135	225	1	44	171	189	0	23
28	332	1	5	64	296	2	6	100	260	2	22	136	224	1	43	172	188	0	21
29	331	1	7	65	295	2	7	101	259	2	22	137	223	1	41	173	187	0	19
30	330	1	9	66	294	2	8	102	258	2	21	138	222	1	39	174	186	0	16
31	329	1	11	67	293	2	9	103	257	2	21	139	221	1	37	175	185	0	14
32	328	1	13	68	292	2	10	104	256	2	20	140	220	1	35	176	184	0	11
33	327	1	15	69	291	2	11	105	255	2	19	141	219	1	33	177	183	0	8
34	326	1	17	70	290	2	12	106	254	2	19	142	218	1	31	178	182	0	6
35	325	1	19	71	289	2	13	107	253	2	18	143	217	1	29	179	181	0	3
36	324	1	21	72	288	2	14	108	252	2	18	144	216	1	27	180	180	0	0

24 διασάφησις τοῦ αὐτοῦ κανόνος τῆς ἀνωμάλου τοῦ ἡλίου κινήσεως

Τούτου δὴ τοῦ κανόνος ὡς ὁρᾶν ἔξεστι τὰ μὲν δύο πρῶτα σελίδια, εἴτουν τμήματα, ἐκτίθεται καθ' εἱρμὸν τοὺς ἀριθμοὺς τῶν ἀπὸ τοῦ ἀπογείου μοιρῶν, ἐν μὲν τῷ ἑνὶ ἀπὸ τοῦ πρώτου στίχου καὶ καθεξῆς τὰς ἀπὸ τοῦ ἀπογείου μοίρας εἰς τὰ ἑπόμενα, μίαν, δευτέραν, τρίτην, μέχρι καὶ τῶν ρπ, ἐν δὲ τῷ ἑτέρῳ ἀπὸ τοῦ πρώτου στίχου ἑξῆς μέχρι τοῦ ἐσχάτου τὰς ἀπὸ τοῦ αὐτοῦ ἀπογείου εἰς τὰ ἡγούμενα μοίρας ἀνάπαλιν, ἤτοι τξ, τνθ, τνη καὶ οὕτω καθ' εἱρμὸν καὶ ἐν τῷ ἐσχάτῳ αὖθις στίχῳ τὰς ρπ. τὸ δὲ τρίτον τμῆμα, εἴτουν σελίδιον, περιέχει τὰς ἐπιβαλλούσας διαφορὰς ἑκάστῃ ἀπὸ τοῦ ἀπογείου μοίρᾳ, κοινὰς οὔσας τοῖς τε ἐν τῷ πρώτῳ σελιδίῳ κατὰ τοὺς αὐτοὺς στίχους ἐκκειμένοις ἀριθμοῖς τῶν μοιρῶν καὶ τοῖς ἐν τῷ δευτέρῳ, ἐπὶ μὲν τοῦ πρώτου σελιδίου ἀφαιρετικῶς καταλαμβανομένας, ἐπὶ δὲ τοῦ δευτέρου προσθετικῶς.

Ἀπὸ γὰρ τῆς μιᾶς μέχρι τῶν ρπ μοιρῶν αἱ διαφοραὶ τῆς ἀνωμαλίας ἀφαιροῦνται, ἀπὸ δὲ τῶν ρπα αἱ αὐταὶ διαφοραὶ προστίθενται, κατ' ἰσότητα ἀφαιρούμεναι καὶ προστιθέμεναι τοῖς ἐπὶ τοῦ αὐτοῦ στίχου ἀριθμοῖς, ἤγουν τῶν μὲν ἐν τῷ πρώτῳ σελιδίῳ, ὡς εἴρηται, ἀφαιρούμεναι, τοῖς δὲ ἐν τῷ δευτέρῳ σελιδίῳ προστιθέμεναι. διὰ τοῦτο καὶ τὰ μὲν δύο πρῶτα σελίδια ἐν τῷ κανόνι ἐπιγράφονται κοινοὶ ἀριθμοί, τὸ δὲ τρίτον προσθαφαιρέσεις. ἐπεὶ γὰρ εἰς τέτταρα τμήματα τέμνεται ὁ ὅλος κύκλος, ἐφ' οὗ φέρεται ὁ ἥλιος, δύο ἑκατέρωθεν τοῦ ἀπογείου καὶ δύο ἑκατέρωθεν τοῦ περιγείου κατὰ ϟ μοίρας, τὸ μὲν ἀπὸ τῆς μιᾶς μοίρας τεταρτημόριον μέχρι τῶν ϟ διέρχεται ὁ ἥλιος ἀπὸ τοῦ ἀπογείου, καὶ ἔστιν ἐλάττων ἡ καταλογιζομένη κίνησις αὐτοῦ ἐπὶ τοῦ ζῳδιακοῦ, ὡς συνάγεσθαι τὸ πλεῖστον ἔλλειμμα τῆς τοῦ ἡλίου ἀνωμάλου κινήσεως, ἤτοι τῆς θεωρουμένης ἐπὶ τοῦ ζῳδιακοῦ πρὸς τὴν τῶν ϟ μοιρῶν, ὡς εἴρηται, ὁμαλὴν κίνησιν, μοίρας δύο καὶ λεπτὰ κγʹ. ἐντεῦθεν δὲ μετὰ ἐνενηκοστὴν μοῖραν προχωρῶν εἰς τὰ περίγεια αὐτοῦ πλείω καταλογίζεται κινεῖσθαι ἐπὶ τῆς ἀνωμάλου κινήσεως, ἤτοι ἐπὶ

1 δύο post πρῶτα transp. C **6** μοίρας ante εἰς τὰ ἡγούμενα transp. C **6** τξ deest C (rasura)

24 Explanation of the table for the anomalous movement of the Sun

It is possible to see in this table that in the first two columns are displayed in sequence the distances from the apogee measured in degrees. In the first column of the first row are displayed the distances in degrees from the apogee toward the subsequent [stars]: one, two and three up to 180°. In the other [entry] of the first row are tabulated the distances of the Sun from the apogee toward the preceding [stars] in degrees: 360°, 359° and 358° up to the last row of 180°. The third column contains the appropriate differences for each degree measured from the apogee, and they are the same for the first and second column, and with the understanding that for the first column are subtractive and for the second column additive.

Thus from 1° to 180° the difference introduced by the anomaly is subtracted, and from 181° on the difference is added, with the subtracted or added number having the same value occurring in that row. For the first column they are subtracted and for the second added. For these reasons, the first two columns of the table are denoted as common numbers and the third one as *prosthaphaeresis*. The orbit of the Sun is divided into four quadrants, two on each side of the apogee and two on the side of the perigee, each of them being 90°. As the Sun rotates on the [first] quadrant from 1 up to 90 degrees, the computed rotation on the ecliptic is slower than the mean and its anomalous rotation on the ecliptic produces a maximum deficit of 2° and 23′ in the neighbourhood of 90°. From then on after the 90°, as the Sun advances toward the perigee the computed rotation on the ecliptic is faster

τοῦ ζωδιακοῦ ἤπερ ἐπὶ τῆς ὁμαλῆς αὐτοῦ περιόδου, ὡς ἀποκαθίστασθαι πάλιν μέχρι τῶν ρπ μοιρῶν, τουτέστι μέχρις αὐτοῦ τοῦ περιγείου καὶ τὴν ἐπὶ τοῦ ζωδιακοῦ καταλογιζομένην κίνησιν, καὶ ἀναπληροῦσθαι τὸ πλεῖστον ἔλλειμμα πάλιν, ἤτοι τὰς εἰρημένας μοίρας δύο καὶ λεπτὰ κγ΄, καὶ συναποκαθίστασθαι κατ᾽ ἰσότητα ρπ μοίρας ἀπὸ τοῦ ἀπογείου μέχρι τοῦ περιγείου κεκινῆσθαι τὸν ἥλιον καὶ ἐπὶ τοῦ ζωδιακοῦ καὶ ἐπὶ τοῦ οἰκείου ἐκκέντρου κύκλου.

Καὶ πάλιν ἀπὸ τοῦ τοιούτου περιγείου προχωροῦντα τὸν ἥλιον εἰς τὸ ἐξ αὐτοῦ ἕτερον τεταρτημόριον, ἤτοι μέχρι σο μοιρῶν, πλείονα καταλαμβάνεσθαι ποιούμενον τὴν ἀνώμαλον αὐτοῦ κίνησιν, ἤτοι τὴν ἐπὶ τοῦ ζωδιακοῦ θεωρουμένην ἤπερ τὴν ἐπὶ τοῦ οἰκείου ἐκκέντρου κύκλου ὁμαλὴν αὐτοῦ κίνησιν, ὡς συνάγεσθαι πάλιν τὸ πλεονάζον τὸ πλεῖστον ἐν ταῖς σο μοίραις τοῦ οἰκείου κύκλου, ἤτοι ἐν τῇ περατώσει τῶν τριῶν τεταρτημορίων τῶν ἀπὸ τοῦ ἀπογείου μοίρας δύο ὡσαύτως καὶ λεπτὰ κγ΄, καθὼς εἶχε καὶ τὸ πλεῖον τῆς ἐλλείψεως. κἀντεῦθεν πάλιν προχωρῶν ὁ ἥλιος ἀπὸ τῶν σο μοιρῶν ὡς πρὸς τὸ ἀπόγειον αὐτοῦ, ἔλαττον καταλογίζεται κινεῖσθαι κατὰ τὴν ἀνώμαλον αὐτοῦ κίνησιν, ἤτοι ἐπὶ τοῦ ζωδιακοῦ, ἤπερ κατὰ τὴν ἐπὶ τοῦ οἰκείου ἐκκέντρου κύκλου ὁμαλὴν αὐτοῦ κίνησιν, ὡς συναποκαθίστασθαι αὖθις καὶ ἐξισάζειν ἀμφοτέρας τὰς κινήσεις κατὰ τὸ ἀπόγειον καὶ ἀναπληροῦσθαι μέχρι τῶν τξ μοιρῶν, ἤτοι τοῦ τέλους τῆς ὅλης περιόδου, τὸ ἔλλειμμα τῆς ὁμαλῆς κινήσεως πρὸς τὴν ἀνώμαλον, ἤτοι τὰς εἰρημένας μοίρας δύο καὶ λεπτὰ κγ΄.

Καὶ ἔστι λοιπὸν ἐν μὲν τῷ ἀπογειωτάτῳ καὶ περιγειωτάτῳ ἐξισάζουσα ἡ τοῦ ἡλίου κίνησις καὶ συναποκαθισταμένη ἐπί τε τοῦ ζωδιακοῦ καὶ τοῦ οἰκείου ἐκκέντρου κύκλου, ἐν δὲ τοῖς μέσοις τόποις, οἳ καὶ ἐκ διαμέτρου εἰσὶν ἀλλήλοις, ἤτοι κατὰ τὰς ϙ μοίρας ἀπὸ τοῦ ἀπογείου καὶ κατὰ τὰς σο, ἔστι τὸ πλεῖστον διάφορον τῆς ὁμαλῆς κινήσεως πρὸς τὴν ἀνώμαλον, ἤτοι αἱ β μοῖραι καὶ λεπτὰ κγ΄. καὶ κατὰ μὲν τὰς ϙ μοίρας ἔστι τὸ διάφορον ἀφαιρετικόν, κατὰ δὲ τὰς σο μοίρας προσθετικόν. καὶ ταῦτα δῆλα τοῖς ὁρῶσιν ἐν τῇ καταγραφῇ τοῦ κανόνος.

37 μέχρις ο C **55** διὰ μέτρου C

than the mean and its position is restored at the perigee where the maximum deficit of 2° 23′ is recovered and [the Sun] occupies the same position on the ecliptic and on its eccentric circle.

Then again from the perigee on, as the Sun advances into the next quadrant up to 270°, its anomalous motion on the ecliptic is larger than the uniform on its own eccentric. One again finds that the maximum excess occurs at 270°. The maximum occurs at the end of the third quadrant being again 2° and 23′, which is as large as the value of the deficit. And from here the Sun advances from 270° toward the apogee, with its anomalous rotation on the ecliptic being slower than on its eccentric. Both rotations are again restored and become equal at the apogee at 360°, that being the end of a complete period where the deficit of 2° and 23′ disappears.

In this manner the two motions of the Sun are restored and become equal at the apogees and perigees. At the intermediate places, that is at 90° and 270° from the apogee, which are diametrically opposite, appears the maximum difference between uniform and anomalous rotations, that of 2° and 23′. At 90° the difference is subtracted, and at 270° it is added. All these are evident to those who study the contents of the table.

25 παράδοσις μετὰ ἐπιλογισμοῦ χρονικῆς ἀρχῆς, ὅθεν δεῖ τὰς ψηφοφορίας τοῦ ἡλίου λογίζεσθαι

Καὶ τὰ μὲν τῶν ἡλιακῶν κανόνων περί τε τῶν ὁμαλῶν κατὰ τοὺς ὡρισμένους χρόνους αὐτοῦ κινήσεων, ἔτι δὲ καὶ περὶ τῶν ἀνωμάλων, τοῦτον ἔχει τὸν τρόπον καὶ ἱκανῶς ἡμῖν διηυκρίνηται καί, ὡς ἐχρῆν ἐμοὶ δοκεῖν, οἱ κανόνες αὐτοὶ ἐκτέθεινται. ἑπομένου δὲ λοιπὸν ἤδη τοῦ καὶ παράδοσιν καὶ διδασκαλίαν ποιήσασθαι πῶς ἐστι διὰ τῶν τοιούτων κανόνων τὰς τοῦ ἡλίου ψηφοφορίας ἑκάστοτε ποιεῖσθαι καὶ ἀνευρίσκειν τὰς ἐποχὰς καὶ τοὺς τόπους αὐτοῦ τοὺς ἐπὶ τοῦ ζωδιακοῦ, ἀναγκαῖόν ἐστι κατὰ τὴν ἡμετέραν πρόθεσιν ἐν χρόνῳ τινὶ ὡρισμένῳ σύνεγγυς τοῖς καθ᾽ ἡμᾶς τούτοις χρόνοις προϋποστήσασθαι καὶ προϋποθέσθαι τὴν ἀρχὴν τῶν κινήσεων αὐτοῦ τε τοῦ ἡλίου καὶ τῆς σελήνης καὶ τῶν ἄλλων ἀστέρων, ἀνευρόντας τίνας ἐποχὰς εἶχον ἐπὶ τοῦ ζωδιακοῦ κατ᾽ αὐτὸν τὸν χρόνον, ὡς ἂν ἐντεῦθεν εὐεπιλογίστως καὶ ῥᾷον ἀρχώμεθα ψηφοφορεῖν τὰς κινήσεις αὐτῶν διὰ τῶν ἐκτεθειμένων κανόνων. καὶ νῦν γε δὴ τοῦτο ποιητέον ἡμῖν προηγουμένως ἐν τοῖς περὶ ἡλίου λόγοις. καὶ γάρ, καθὼς προέφημεν ἐν τῇ ἀρχῇ τῆς παρούσης συντάξεως, ἐργῶδές ἐστι καὶ δυσμεταχειριστότερον τὰς ψηφοφορίας ἑκάστας ποιεῖσθαι ἢ ἀπὸ τῆς δεδομένης ἐποχῆς καὶ ἀρχῆς τῶν χρόνων κατὰ τὸν Πτολεμαῖον, ἤτοι ἀπὸ τῆς βασιλείας τοῦ Ναβονασάρου, ὡς προείρηται, ἢ ἀπὸ τῆς δεδομένης ἐποχῆς καὶ ἀρχῆς τῶν χρόνων παρὰ τοῦ Θέωνος, ἤτοι ἀπὸ τῆς ἀρχῆς τοῦ Φιλίππου. πλήθη γὰρ καὶ ἀπ᾽ ἀμφοτέρων τῶν ἀρχῶν, ὡς ἔστιν εὔδηλον, πλείστων ἐτῶν μέχρι καὶ εἰς ἡμᾶς συνάγονται. διὰ ταῦτα δὴ καὶ ἀνευρεῖν δοκιμάζομεν καὶ συστήσασθαι νέαν, ὡς ἔφην, ἐποχὴν καὶ χρονικὴν ἀρχὴν δι᾽ ἀκριβοῦς ἐπιλογισμοῦ ἐπὶ τῆς μοναρχίας τοῦ θειοτάτου βασιλέως ἡμῶν ἐν τοῖς πρώτοις ἔτεσι τῆς μοναρχίας αὐτοῦ, ὅπῃ παρήκει, καὶ πρῶτόν γε νῦν ἤδη μάλιστα, καθὼς προτιθέμεθα, περὶ τοῦ ἡλίου.

17-18 Cf. Ptol. *Alm.* 3.7 (vol. 1.1, p. 254 Heiberg) **19-20** Cf. Ptol., *Proch. Kan.* vol. 1, p. 1 Halma; Theon, *PC* p. 200 Tihon

3 διευκρίνηται C **13** προεκτεθειμένων C **25** ὅποι C **25** παροίκει V[ac]

25 Presentation with calculation for the origin where we begin measuring the time that will be used in computations

The tables for the uniform and anomalous motion of the Sun as a function of time are constructed in this manner. They are well understood and have been presented in a way that I consider useful. The next step is to communicate and teach how to calculate and determine the longitude and positions [of the Sun] on the ecliptic [at any time] using the tables. It is our intention to assume and introduce a reference time close to our current day for starting the motions of the Sun, the Moon and the other stars and compute the longitudes they had on the ecliptic at that time. From that time on we begin calculating reliably but simply their motion using the available tables. Now, we shall apply this [method] first for the Sun. As we mentioned at the beginning of this *Syntaxis*, it is a copious and difficult method to perform each computation, either from the assumed era and initial time introduced by Ptolemy, that is from the reign of Nabonassar, or from the era and initial time adopted by Theon, that is from the time of [King] Philip. It is evident that many years have passed from either of these initial times. For this reason we try to find and recommend, after careful consideration, a new era, as I stated, and an initial time for the calculations. As initial time we select the first year of the reign of our very divine King and begin measuring the time in any direction, as we intend to do this first for the rotation of the Sun.

Ἡ μὲν οὖν δεδομένη ἀρχὴ τῆς κινήσεως τῶν ἀστέρων ἐν τοῖς Προχείροις Κανόσιν, ἤτοι τὸ πρῶτον ἔτος τῆς ἀρχῆς Φιλίππου τοῦ Ἀριδαίου, τοῦ μετὰ Ἀλέξανδρον τὸν μέγαν βασιλέα Μακεδόνων, παραδέδοται ἡμῖν ἐκ διαδοχῆς τῶν ἠκριβωκότων ταῦτα, μετὰ ἔτη οὖσα ἀπὸ κτίσεως κόσμου, ὡς ἡμεῖς οἱ ὀρθῶς θεοσεβοῦντες δοκιμάζομεν, ͵ερπε. ταύτην μέντοι ἣν ἔσχομεν παράδοσιν ἐκ διαδοχῆς, ὡς ἔφημεν, ἀσφαλεστάτην καὶ ἀληθεστάτην πιστοῦνται πολλαὶ ψηφοφορίαι, οὐχ ἥκιστα καὶ παρ᾽ ἡμῶν μετ᾽ ἐπισκέψεως καὶ ἀκριβείας γενόμεναι ἐν τοῖς καθ᾽ ἡμᾶς τούτοις νῦν χρόνοις. ἐπὶ ταύτῃ γὰρ τῇ ἀρχῇ καὶ ὑποθέσει ἀπόπειραν πολλάκις ποιησάμενοι καὶ καταλογισάμενοι τοὺς ἀπὸ ταύτης τῆς ἀρχῆς χρόνους καὶ τὰ ἀνήκοντα αὐτοῖς κινήματα κατὰ τὴν περίληψιν τῶν εἰρημένων κανόνων, μέχρι καὶ τοῦ καιροῦ καθ᾽ ὃν προετιθέμεθα ψηφοφορίαν ἐνεργῆσαί τινα, εἴτε ἐποχὴν ἀνευρεῖν δηλαδὴ ἀστέρος μετά τινος ἐπισημειώσεως φαινομένης εἴτε σελήνης ἐκλειπτικὴν πανσεληνιακὴν συζυγίαν εἴτε ἡλίου ἐκλειπτικὴν συνοδικὴν συζυγίαν, πολλὴν τὴν συμφωνίαν τῆς εἰρημένης προλήψεως καὶ παραδόσεως πρὸς τὰ φαινόμενα καταλαμβάνομεν. κἀντεῦθεν ἐβεβαιωσάμεθα τὰ τῆς προλήψεως ταύτης καὶ πολὺ τὸ πιστὸν ἐκ τῶν ὁρωμένων αὐτῇ ἐπεψηφισάμεθα.

Τὸ δὲ πρῶτον ἔτος τῆς τοῦ θειοτάτου βασιλέως ἡμῶν μοναρχίας ἤρξατο μετὰ ἔτη ἀπὸ κτίσεως κόσμου ͵ϛψϛ πεπληρωμένα, παρελθόντων ἐπ᾽ αὐτοῖς καί τινων μηνῶν. κατὰ γὰρ τὸ ͵ϛψϛα ἔτος ἀπὸ κτίσεως κόσμου, ἑνδεκάτῃ τοῦ κατὰ Ῥωμαίους Δεκεμβρίου μηνός, γέγονεν ἡ καταρχὴ τῆς μοναρχίας αὐτοῦ. συνάγονται οὖν ἀπὸ τῆς παραδεδομένης

28 Cf. Ptol., *Proch. Kan.* vol. 1, p. 1 Halma; Theon, *PC* p. 200 Tihon **36-45** ἐπεὶ τοίνυν γέγονεν ἡ καταρχὴ τῆς μοναρχίας αὐτοῦ κατὰ τὸ ͵ϛψϛα ἔτος ἀπὸ κτίσεως κόσμου, ἑνδεκάτῃ ὀκτωβρίου μηνός, εὔδηλον ὅτι κατὰ τὸ ͵ϛψϛβ ἔτος ἀπὸ κτίσεως κόσμου ἕως ἑνδεκάτης ὀκτωβρίου μηνὸς ἐνιαυτὸν ἕνα ὁ βασιλεὺς ἐν τῇ ἀρχῇ διήγαγεν, ὥστε εἶναι τὸ ͵ϛψϛβ ἔτος α ἔτος τῆς αὐτοκρατορίας αὐτοῦ καὶ τὸ τέλος τοῦ ϛβ ἔτους τέλος εἶναι καὶ τοῦ α ἔτους τῆς ἀρχῆς αὐτοῦ πλὴν ὅσον μετὰ προσθήκης τινῶν ἡμερῶν sch. in mg. C (Chort.) **46-57** σημείωσαι ἐνταῦθα νέαν ἀρχὴν ψηφοφοριῶν ἐκτεθεῖσαν ὑπὸ τοῦ θαυμαστοῦ συγγραφέως, τοῦ μεγάλου λογοθέτου, εἰς ὄνομα τοῦ ἀοιδίμου βασιλέως κυροῦ Ἀνδρονίκου τοῦ δευτέρου Παλαιολόγου, υἱοῦ Μιχαὴλ τοῦ λατινόφρονος sch. in mg C (Chort.) ζήτει περὶ τούτου καὶ ἐν τῷ λε κεφαλαίῳ τοῦ α βιβλίου sch. in mg C (Chort.)

In the *Handy Tables* the assumed initial time for the motions of the stars is the first year of the reign of Philip Arrhidaios, the successor of Alexander the Great, King of the Macedonians. This has come down to us very accurately as the year 5185 after the creation of the world, which we adopt as faithful [people]. We consider this tradition, which has been delivered to us successively, as very certain, true and supported by many calculations, none the least those carried out by us with due consideration and accuracy. According to this principle and hypothesis we tried many times to compute for the intervening times the corresponding movements, in accordance with the mentioned tables, i.e. to find the longitude of a star associated with an outstanding event, either an eclipse of the Moon at a *syzygy* or an eclipse of the Sun at a synodic *syzygy*. [We found] excellent agreement between the predictions and the phenomena. From this we ascertained that the predictions accurately reproduce the observations.

The first year of the reign of our most divine King began 6790 complete years after the creation of the world, plus a few months. His reign began on the 11th day of the Roman month of December in the year 6791 from the creation of the world. Thus, from the initial time

36-45 (note by the hand of Chortasmenos in the margin of C): Since the beginning of his reign, is on the 11th of October in 6791 from the creation of the world (CW), it is evident that until the 11th of October of 6792 the king ruled for one year. Thus 6792 CW is the first year of his reign and at the end of the 92nd year he completed one year plus a few days.

46-57 (note by the hand of Chortasmenos in the margin of C): One must note here a new beginning of calculations by the admirable author and Grand Logothete, dedicated to the glorious Emperor Andronikos II Palaiologos, son of Michael the Latin sympathiser.

ἀρχῆς τῶν κινήσεων τῶν ἀστέρων τῆς ἐν τοῖς Προχείροις Κανόσι, τῆς κατὰ τὸ πρῶτον ἔτος δηλαδὴ τοῦ Φιλίππου μέχρι τῆς κατὰ τὴν ἕκτην τοῦ Ὀκτωβρίου μεσημβρίας τοῦ ͵ςψϛβ ἔτους, ἤτοι περὶ τὰ τέλη τοῦ πρώτου ἔτους τῆς μοναρχίας τοῦ αὐτοῦ θειοτάτου βασιλέως ἡμῶν, ἔτη κατ' Αἰγυπτίους ͵αχη τὸν τρόπον τοῦτον. ἀπὸ μὲν γὰρ τοῦ ͵ερπε ἔτους τοῦ ἀπ' ἀρχῆς κόσμου, ὅπερ δὴ καὶ πρῶτον ἔτος ἐστὶ τοῦ Φιλίππου καὶ ἀρχὴ τῆς παραδόσεως, ὡς ἔφημεν, ἐν τοῖς Προχείροις Κανόσι, μέχρι συμπληρώσεως τοῦ ͵ςψα ἔτους, καθ' ὃ γέγονε ἡ ἀρχὴ τῆς μοναρχίας τοῦ αὐτοῦ θειοτάτου βασιλέως ἡμῶν, ὡς ἔφημεν, εἰσὶν ἔτη ἐν μέσῳ ͵αχς. τούτοις προστίθεται καὶ τὸ κατὰ τὸ πέμπτον ἔτος τῆς Αὐγούστου μοναρχίας ἀποκαταστὰν ἀπὸ τῶν τετραετηρίδων, ὡς προέφημεν, αἰγυπτιακὸν ἔτος καὶ γίνονται ͵αχζ ἔτη αἰγυπτιακά. ἀπὸ δὲ τοῦ ἕκτου ἔτους τῆς Αὐγούστου μοναρχίας πάλιν ἤρξατο, ὡς ἔφημεν, ἀρχὴν λαμβάνειν ὁ παρὰ τοῖς Αἰγυπτίοις ἐπιλογιζόμενος ἀριθμὸς τῶν ἡμερῶν διὰ τῶν τετραετηρίδων.

Ἐπεὶ γοῦν ἀπὸ τῆς ἀρχῆς τοῦ κόσμου μέχρι συμπληρώσεως αὐτοῦ τοῦ πέμπτου ἔτους τῆς μοναρχίας τοῦ Αὐγούστου ἔτη παρῆλθον ͵ευπδ, ὡς καὶ τοῦτο ἀπὸ ἀσφαλοῦς καὶ ἀπλανοῦς παραδόσεως ἔχομεν, συνάγονται τὰ ἀπὸ ἕκτου ἔτους τῆς μοναρχίας Αὐγούστου μέχρι καὶ τοῦ εἰρημένου ͵ςψϛα ἔτους ἀπὸ κτίσεως κόσμου ἐν μέσῳ ἔτη ͵ατζ. ταῦτα δὴ ὡς ἔστι θεωρῆσαι καὶ συλλογίσασθαι ῥᾳδίως προσλαβόντα καὶ ἡμέρας λη διὰ τοῦ τετραετηρισμοῦ γίνονται ἔτη αἰγυπτιακὰ ͵ατη. προστίθενται γοῦν αἱ ἡμέραι αἱ λη ἀπὸ τοῦ ἑξῆς ἔτους, τοῦ ͵ςψϛβ δηλονότι, καὶ ἀπαρτίζονται ἐν αὐτῷ τῷ ͵ςψϛβ ἔτει τὰ δηλωθέντα αἰγυπτιακὰ ͵ατη ἔτη. ταῦτα δὴ τετραετηριζόμενα κατὰ τὴν τῶν Αἰγυπτίων τάξιν ποιοῦσιν ἡμέρας ἀπ' αὐτοῦ τοῦ τετραετηρισμοῦ τκζ, αἷς δηλονότι συντίθενται καὶ συναριθμοῦνται καὶ αἱ δηλωθεῖσαι λη

51 Cf. Ptol., *Proch. Kan.* vol. 1, p. 1 Halma; Theon, *PC* p. 200 Tihon **58-59** Cf. Ptol., *Proch. Kan.* vol. 1, p. 1 Halma; Theon, *PC* p. 200 Tihon

62 ἔτη αἰγυπτιακά V^{sl} C^{mg} **63-66** οὕτω γὰρ ἔχει ἡ παράδοσις τοῦ Θέωνος. συναγαγόντες τὰ ἀπ' ἀρχῆς Φιλίππου μέχρι τοῦ ἀναδιδομένου χρόνου, εἰσάγομεν εἰς τὸ (sic) τῶν εἰκοσαπενταετηρίδων κανόνι sch. in mg. C (Chort.) **72** διὰ τοῦ τετραετηρισμοῦ V^{pc} C^{pc} (in ras.) **72** ἔτη αἰγυπτιακὰ V^{sl} C^{mg} **77** συντίθενται καὶ συναριθμοῦνται καὶ V^{pc} C^{pc} (in ras. et mg.)

recorded in the *Handy Tables* for the motion of the stars, that is the first year of the reign of Philip, up to the midday of the sixth of October in the year 6792 — close to the end of the first year of our very divine King — elapsed 1608 Egyptian years. They are computed in the following manner. The reign of Philip, being the starting point for counting time, begins in the year 5185 after creation; from then on, up to 6791 intervened 1606 years. When on the fifth year of the reign of Augustus we add one year in order to correct for the four-year Egyptian cycles, we obtain 1607 Egyptian years. We begin again counting the four-year-cycles of the Egyptians from the sixth year of the reign of Augustus.

Since, from the creation of the world up to the fifth year of the reign of Augustus passed 5484 years, which we obtain from safe and reliable sources, 1307 years passed between the sixth year of Caesar and the year 6791 CW. Considering all facts, thinking simply and adding 38 days to complete the four-year-cycles, they become 1308 Egyptian years. The 38 days are added to the next year 6792, and to that year correspond the 1308 Egyptian years mentioned above. Then, dividing the 1308 years by four, according to the Egyptian custom, produces 327 days, to which we add 38 days

63-66 (note by the hand of Chortasmenos in the margin of C): This is the transmission of Theon, counting the years from the beginning of the reign of Philip up to a specific time in the table of 25-years.

ἡμέραι, αἱ ἀπὸ ͵ςψϛβ ἔτους, καὶ ἀπαρτίζουσιν ἕτερον ἔτος αἰγυπτιακόν. ὃ καὶ προστίθεται τοῖς εἰρημένοις ͵αχζ ἔτεσι καὶ γίνονται, ὡς ἀνωτέρω ἔφημεν, ͵αχη αἰγυπτιακὰ ἔτη ἀπὸ τῆς ἀρχῆς Φιλίππου, ἤτοι τῆς ὑποτιθεμένης ἐν τοῖς Προχείροις Κανόσιν ἀρχῆς τῆς κατὰ τὸ ͵ερπε ἔτος ἀπ᾽ ἀρχῆς κόσμου, μέχρι τῆς μεσημβρίας τῆς ἕκτης τοῦ κατὰ Ῥωμαίους Ὀκτωβρίου μηνὸς τοῦ δηλωθέντος ͵ςψϛβ ἔτους ἀπ᾽ ἀρχῆς κόσμου, ἤτοι τοῦ δευτέρου καθ᾽ Ἕλληνας μηνός.

Σημειωτέον γὰρ δὴ καὶ τοῦτο, ὅτι παραδέδοται ἡμῖν καὶ τοῦτο ἀσφαλῶς καὶ ἠκριβωμένως, ὡς οἱ Ἕλληνες ἀρχὴν ἔτους τὴν μεσημβρίαν τίθενται τῆς κθ τοῦ καθ᾽ ἡμᾶς Αὐγούστου μηνός, ὡς προλαμβάνειν τὴν ἀρχὴν τοῦ ἑλληνικοῦ ἔτους τῆς ἀρχῆς τοῦ καθ᾽ ἡμᾶς ῥωμαϊκοῦ ἔτους

81 Cf. Ptol., *Proch. Kan.* vol. 1, p. 1 Halma; Theon, *PC* p. 200 Tihon

82 sqq. τοῦ Χορτασμένου Ἰωάννου. λαβόντες γὰρ τὰ ἀπὸ τῆς τοῦ πρώτου ἐνιαυτοῦ παραγωγῆς μέχρι καὶ τοῦ παρόντος ἔτη πεπληρωμένα καὶ ἀπ᾽ αὐτῶν ἀφελόντες ἔτη ͵ερπε, τὰ λοιπὰ ἀπὸ τῆς ἀρχῆς Φιλίππου τοῦ Ἀριδαίου ἀπογράφομεν ἔτη μετὰ προσθήκης ἐνιαυτοῦ ἑνός. ἔπειτα τῶν συναγομένων ἐτῶν ἀπὸ τοῦ ἕκτου ἔτους τῆς Αὐγούστου βασιλείας μέχρι καὶ τοῦ νῦν ἀναδοθέντος χρόνου τὸ δ^ον λαβόντες, οὐκ ἀκριβῶς, ἀλλ᾽ ὅσον ἐμπίπτει, ἕξομεν τετραετηρίδας, ἤτοι ἐμβολίμους ἡμέρας, ὅσαις δηλονότι προείληφεν ὁ κατ᾽ Αἰγυπτίους χρόνος τὸν κατὰ Ἀλεξανδρέας, εἴτ᾽ οὖν Ἕλληνας· αἷς προσθέντες τὰς ἀπ᾽ ἀρχῆς Σεπτεβρίου τοῦ προκειμένου ἡμῖν ὑστάτου καὶ ἐνεστηκότος ἔτους ἡμέρας μέχρι καὶ αὐτῆς τῆς ἡμέρας καθ᾽ ἣν τὴν ψηφοφορίαν ποιούμεθα τῆς ἀπὸ τῆς παρελθούσης μεσημβρίας ἀρχομένης μετὰ προσθήκης νῦν μὲν δύο, νῦν δὲ τριῶν ἡμερῶν τοῦ αὐγούστου μηνός καὶ ἀπὸ τῶν συναγομένων ἀφελόντες, εἰ τύχοι αἰγύπτιον ἔτος ἕν, ἤτοι τξε ἡμέρας, καὶ προσθέντες αὐτὸ τοῖς ἀπογεγραμμένοις ἔτεσι, τὰ μὲν γεγονότα ἕξομεν ἔτη ἁπλᾶ αἰγύπτια, τὰς δὲ ὑπολειφθείσας ἡμέρας ἀπὸ τῆς τοῦ Θὼθ νεομηνίας εἰς τὰ τῶν αἰγυπτιακῶν μηνῶν ἑπόμενα ἀπολύσαντες διδόντες ἑκάστῳ μηνὶ ἡμέρας λ, εἰς ὃν ἂν καταντήσῃ μῆνα τὸ πέρας τοῦ ἀριθμοῦ, ἐκεῖνον ἀπογράφομεν αἰγυπτιακὸν μῆνα, τὰς ὑπολειφθείσας ἡμέρας τοῦ μηνὸς ἐπέχοντα. ἐὰν δὲ μὴ ὑπολειφθῶσιν ἐλάττονες τῶν λ, ἀλλ᾽ ἐκβαλλόμενον τὸ πλῆθος τῶν ἡμερῶν τὸ τριακονθήμερον ἀπαρτίσῃ, αὐτόν τε τὸν μῆνα εἰς ὃν ἀπήρτισεν καὶ τὰς λ ἡμέρας ἀπογραψόμεθα. καὶ ἂν δὲ καὶ πάλιν προσεκβαλλόμενον τὸ τῶν ἡμερῶν πλῆθος μετὰ τὸν ιβ μῆνα τὸν Μεσωρὶ εἰς τὸν ἐπαγόμενον καταντήσῃ αὐτόν τε καὶ τὰς ὑπολειφθείσας ἡμέρας ἀπογραψόμεθα β κεφάλαιον εἴτε μία ἐστίν, εἴτε δύο, καὶ ἕως τῶν ε. χρὴ δὲ γινώσκειν ὅτι κατὰ τὸν τοῦ βισέξτου ἐνιαυτόν (δηλονότι ὅτε τὸν ἀπὸ τοῦ ε ἔτους τῆς Αὐγούστου καίσαρος βασιλείας μέχρι καὶ τοῦ προκειμένου μεριζομένων παρὰ τὸν δ οὐδὲ ἓν ἔτος καταλιμπάνεται) τηνικαῦτα οὐκ ἀπὸ τῆς κθ τοῦ Αὐγούστου μηνός, ἀλλ᾽ ἀπὸ τῆς τούτου λ τὸ τῶν ἡμερῶν πλῆθος λαμβάνοντες, τὸν Φεβρουάριον κθ ἡμερῶν, ὅσων οὖν καὶ ἔστι τότε, μέχρι συμπληρώσεως λογιούμεθα τοῦ ἐνισταμένου ἔτους sch. in mg. C (Chort.)

from the year 6792, and they complete one additional Egyptian year. This added to the 1607, mentioned above, makes 1608 years. So many years passed from the beginning of the reign of Philip, which is also the starting point introduced in the *Handy Tables* and the 5185th year from the beginning of the world up to midday of the sixth day of the Roman month of October of 6792, [with October] being the second month for the Hellenes.

Let it be noted what has also been handed down to us as very certain and accurate, that the Hellenes begin their year at the midday of the 29th of our month of August. The beginning of the Hellenic year precedes our Roman

82ff. (note in the margin of C): By John Chortasmenos. Taking into account the number of years completed from the creation [of the world] up to this date and subtracting 5185 we obtain the difference, which denotes the years that elapsed since Philip Arrhidaios. To them we add one year to account for the adjustment of Egyptian years. Then we count the years from the sixth year of the reign of Augustus up to now and, by taking one-fourth of this number, not exactly, but including whatever [integer] comes out, we obtain the intercalary (*emvolimous*) days that are derived from the four-year periods. These are the additional days according to the Egyptian time in Alexandria or according to the Greeks. We add the [emvolimous] days to the accumulated days after the beginning of September for the year we carry out the computation. We count days from the noon of the previous day, adding sometimes two or three days from the month of August. In case that the total exceeds 365 days we subtract them and add one year to the recorded number of years. The final years are simple Egyptian years and there may be a remainder of days to be added to the beginning of the month Thoth or to the following months, attributing 30 days to each month. If there is no remainder with less than 30 days, but the resulting number completes whole months, we keep it as the number of months. If there are remaining days after the 12th month of Mesore, we assign them to the next intercalating (*epagomenon*) and carry the remaining one or two up to five days in the second Table.

It is also necessary to know that during a leap year (that is when the number of years from the fifth year of the reign of Caesar Augustus up to the present time is exactly divisible by four, without a remainder), we count the number of days not from the 29th but from the 30th of August, assigning 29 days to February and complete the computation from that time up to now.

ἡμέραις τρισίν, αἵτινές εἰσιν ἀπὸ τῆς μεσημβρίας τῆς κθ τοῦ Αὐγούστου μηνὸς μέχρι τῆς μεσημβρίας τῆς πρώτης τοῦ Σεπτεμβρίου μηνός. διὰ τοῦτο καὶ ἔφημεν κατὰ τὴν μεσημβρίαν τῆς ἕκτης ἡμέρας τοῦ Ὀκτωβρίου μηνὸς τοῦ ͵ϛψϟβ ἔτους ἀπὸ κτίσεως κόσμου ἀπαρτίζεσθαι καὶ ἀποκαθίστασθαι τὸ αἰγύπτιον ἔτος, ὡς τὰς λοιπαζομένας ἡμέρας λη εἰς ἀποπλήρωσιν ἔτους αἰγυπτιακοῦ ταῖς ἀπὸ τῶν τετραετηρίδων, ὡς ἔφημεν, ἡμέραις καταλογιζόμενοι ἀπὸ τῆς μεσημβρίας τῆς κθ τοῦ Αὐγούστου μηνὸς τοῦ ͵ϛψϟα ἔτους, ἀφ᾽ ἧς δηλονότι τὴν ἀρχὴν τοῦ ἔτους ποιοῦνται οἱ Ἕλληνες. ἐν γοῦν τοῖς τοιούτοις ͵αχη αἰγυπτιακοῖς ἔτεσι καταλαμβάνομεν, κατὰ τὴν παράδοσιν τῶν ἤδη ἐκτεθειμένων κανόνων ἐν τῷ προχείρῳ τῆς ὁμαλῆς τοῦ ἡλίου κινήσεως, ἐπικινηθῆναι τὸν ἥλιον ὁμαλῶς μετὰ πολλὰς δηλονότι κυκλικὰς περιόδους ἀπὸ τοῦ παραδεδομένου ἀπογείου αὐτοῦ τοῦ κατὰ τὰς ε μοίρας καὶ λ΄ τῶν Διδύμων μοίρας ρλα κη΄, ὥστε εἶναι τηνικαῦτα τὸν ἥλιον καὶ εὑρίσκεσθαι θεωρούμενον κατὰ τὴν ὁμαλὴν αὐτοῦ κίνησιν Ζυγοῦ μοίρας ιϛ καὶ νη΄.

Συμπεραίνομεν οὖν καὶ συλλογιζόμεθα ἀρχὴν αἰγυπτιακοῦ ἔτους ὑποθέσθαι, μᾶλλον δὲ τοῦ πρώτου παρ᾽ Αἰγυπτίοις μηνὸς τοῦ καλουμένου Θὼθ ἐν τῇ παρούσῃ συντάξει, καὶ ἀρχὴν τῆς ὁμαλῆς τοῦ ἡλίου κινήσεως· καθεξῆς δὲ καὶ τῆς τῶν ἄλλων ἀστέρων ὡσαύτως ὁμαλῆς κινήσεως, ἀπὸ τῆς μεσημβρίας τῆς ϛ τοῦ Ὀκτωβρίου τοῦ

93 λειπαζομένας V **94** ἀναπλήρωσιν C **100** μετὰ πολλὰς post δηλονότι transp. C
106 sqq. ὑπόδειγμα: ὑποκείσθω ἡμῖν χρόνος καθ᾽ ὃν τὴν ψηφοφορίαν ποιούμεθα ἀπὸ κτίσεως κόσμου ͵ϛψϟβ μετά γε τοῦ ἐνισταμένου μηνὸς Ὀκτωβρίου, ἡμέρα τούτου ϛ, ἀπὸ τῆς τοῦ πρώτου ἐνιαυτοῦ παραγωγῆς ἔτη πεπληρωμένα ͵ϛψϟα. ἄφες τούτων ͵ερπε. λείπονται ἔτη ἀπὸ τῆς Φιλίππου τοῦ Ἀριδαίου βασιλείας ͵αχϛ. πρόσθες τούτοις καὶ ἐνιαυτὸν τῆς ἀποκαταστάσεως α΄. γίνονται ὁμοῦ ͵αχζ. ἀπὸ τῆς τοῦ πρώτου ἐνιαυτοῦ παραγωγῆς μέχρι συμπληρώσεως τοῦ πέμπτου ἔτους τῆς Αὐγούστου βασιλείας ἔτη ͵ευπδ καὶ ἀπὸ τοῦ ἕκτου ἔτους τῆς Αὐγούστου βασιλείας μέχρι καὶ τοῦ νῦν ἐνεστηκότος χρόνου καὶ τρέχοντος ἔτη ͵ατη. τούτων τὸ $δ^{ον}$ τκζ ἡμέραι τετραετηρίδες. οὐδὲν καταλείπεται βίσεξτος ὁ ἐνιαυτός. πρόσθες ταύταις τὰς ἀπ᾽ ἀρχῆς Σεπτεμβρίου μέχρι καὶ τῆς $ϛ^{ης}$ Ὀκτωβρίου ἡμέρας λϛ καὶ τὰς πρὸ τοῦ Αὐγούστου ἡμέρας β· ἐπειδὴ βίσεξτος ἐστὶν ὁ ἐνιαυτὸς γίνονται ἡμέραι λη, ἤτοι ἔτος αἰγυπτιακὸν ὁλόκληρον. πρόσθες τοῦτο τοῖς ἀπογεγραμμένοις ἔτεσι τοῖς ἀπὸ Φιλίππου. γίνονται ἔτη ἀπὸ Φιλίππου μέχρι καὶ τοῦ νῦν ἐνεστηκότος ἡμῖν χρόνου ͵αχη μὴν Θὼθ ἡμέρα τούτου πρώτη. ἡ ἄρα τοῦ παρὰ Ῥωμαίοις Ὀκτωβρίου $ϛ^{η}$ εὑρέθη Θὼθ $α^{η}$ sch. in mg. C (Chort.)

year by three days, which are the days from the 29th of August to the noon of the first of September. For this reason we said that at midday on the sixth day of October of the year 6792 CW the Egyptian year is completed and restored, as the missing 38 days are added to supplement the four-year-cycles and complete one Egyptian year. These days are counted from the midday of the 29th of August of the year 6791, on which the Hellenes begin their year. In these 1608 years, according to the tables in the *Handy Book*, the Sun travelled in the uniform rotation and completed many periods from its accepted apogee at 5° and 30′ in the constellation of Gemini. In this time interval the Sun advanced by 131° and 28′ to find itself located at 16° and 58′ in Libra.

To conclude, we consider adopting in this book as the starting point for the uniform motion of the Sun the beginning of the Egyptian year, to be more precise, the first month of the Egyptian year called Thoth. [We also adopt], for the uniform motion of the Sun and the other stars the midday of the sixth of

106ff. (note by the hand of Chortasmenos in the margin of C): Example: Let us assume that the time we make the calculation is 6792 CW and on the sixth day of the month of October. On the first year of his reign 6791 full years passed plus a remainder. Subtracting from this 5185 for the reign of Philip Arrhidaios we obtain 1606 years.

Add to this one year for the first correction [in the Egyptian calendar] and makes altogether 1607. From the first year of creation up to the completion of the fifth year of the reign of Augustus there are 5484 years, and from the sixth year of the reign of Augustus up to the present year there are 1308 years. One fourth of them, for the four-year periods, gives 327 days without a remainder being a leap year. Add to them 36 days from the beginning of September to the sixth of October and two days from August. Because the year is a leap year, it makes 38 days, which is a complete Egyptian year. Add this to the calculated years from the epoch of Philip and we arrive at the 1st day of Thoth in the year 1608; thus we calculated that the Roman date sixth of October is the first day of Thoth.

κατὰ τὸ ͵ςψϛβ ἔτος ἀπ᾽ ἀρχῆς κόσμου, ἤτοι περὶ τὰ τέλη τοῦ πρώτου ἔτους τῆς μοναρχίας τοῦ θειοτάτου βασιλέως ἡμῶν Ἀνδρονίκου τοῦ Παλαιολόγου, καθ᾽ ὃν δὴ χρόνον εὑρίσκεται ὁ ἥλιος ἀπέχων κατὰ τὴν ὁμαλὴν αὐτοῦ κίνησιν ἀπὸ τοῦ ἀπογείου αὐτοῦ τὰς εἰρημένας ρλα μοίρας καὶ κη΄ καὶ ἐπέχων ις μοίρας τοῦ Ζυγοῦ νη΄. καὶ λοιπὸν ταύτην ἀρχὴν ὑποθέμενοι τῶν ἑξῆς ὁμαλῶν τοῦ ἡλίου κινήσεων, ἐξ αὐτῆς ποιησόμεθα τὰς ψηφοφορίας καὶ ἀνευρήσομεν ἑκάστοτε ἐν τοῖς δεδομένοις χρόνοις τὰς ὁμαλὰς τοῦ ἡλίου ἐπικινήσεις καὶ τοὺς ἐπὶ τοῦ ζῳδιακοῦ τόπους κατὰ τοὺς χρονικοὺς κανόνας, οὓς ἤδη ἐξεθέμεθα.

116-17 ἑκάστοτε post ἐν τοῖς δεδομένοις transp. C

October of the year 6792 CW, which is approximately the end of the first year of the reign of our very divine King Andronikos Palaiologos, at which time the Sun in its uniform motion is located at a distance of 131° and 28′ from its apogee, being at 16° and 58′ in the sign of Libra. Finally, we adopt this starting point for the uniform motion of the Sun, and we shall perform the calculations and compute for any given time the regular motion and its locations in the ecliptic using the tables that have been presented already.

26 παράδοσις ὅπως ἐστὶ καθ' ὁντιναοῦν δεδομένον χρόνον μετὰ τὴν ὑποτεθεῖσαν ἀρχὴν ψηφοφορεῖν καὶ ἐπιλογίζεσθαι τὰς ὁμαλὰς καὶ ἀνωμάλους κινήσεις καὶ ἐποχὰς τοῦ ἡλίου

Ἐπεὶ δὲ τὰ τοιαῦτα κατὰ τὸ εἰκὸς προδιελάβομεν, ἱκανῶς ἐμοὶ δοκεῖν καὶ τὴν τοιαύτην ἀρχὴν ὑπεθέμεθα τῶν ἑξῆς ἐκεῖθεν ἑκάστοτε ψηφοφοριῶν

1-27 sch. in mg. C (Chort.)

ψηφοφορία κατὰ τὸν πρόχειρον

		ἀποχῆς ἡλίου
εἰκοσιπενταετηρίδαι	͵α χα	ρλγ ι΄
ἔτη ἁπλᾶ	ζ	τνη ιη΄
μῆνες	Θώθ νεομηνία	~ ~
		ρλα κη΄

ταύτας ἀποληφθείσας ἀπὸ τῆς εης καὶ ἡμισείας μοίρας τῶν Διδύμων εἰς τὰ ἑπόμενα, ἐγένετο ἡλίου ὁμαλὴ ἐποχή Ζυγοῦ ις νη΄. εἴσαξον τὰς καταχθείσας μοίρας εἰς τὸν κανόνα τῆς τοῦ ἡλίου ἀνωμαλίας. παράκεινται αὐτῷ ἐξ ἀναλογίας α νδ΄ καί, ἐπεὶ ὁ καταχθεὶς ἀριθμὸς ἐλάττων ἐστὶν μοιρῶν ρπ, ἄφελε τὰ τῆς ἀνωμαλίας τῆς ἡλιακῆς ἐποχῆς· λείπεται ἡλίου ἀνώμαλος ἐποχὴ Ζυγοῦ ιε στ΄ νδ΄΄

ρλα α να΄
ρλα κη΄ πρόσθες α ν΄ δ΄΄
ρλβ α μθ΄

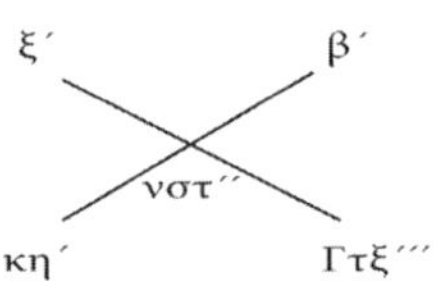

εἴρηται ἄρα ἐν τῷ ὑποκειμένῳ ἔτει ὅπερ ἐστὶ ἀπὸ κτίσεως κόσμου ͵ςψϚβ κατὰ τὴν ςην τοῦ Ὀκτωβρίου ὁ ἥλιος ὁμαλῶς μὲν ἐπέχων Ζυγοῦ ις νη΄, ἀνωμάλως δὲ Ζυγοῦ ιε στ΄ νδ΄΄

εἰκοσαπενταετηρίδες ἡλίου ἀπ' ἀρχῆς τοῦ ͵ςψϚβου ἔτους

α	ρλα κη΄
κστ	ρκε κγ΄
να	ριθ ιθ΄

ἢ πρόσθες τνγ νε΄ ἢ ἄφελε ς δ΄ μα΄΄
ὡς κἂν τοῖς προχείροις γίνεται

26 Description of how we calculate and determine, at a given moment of time, after the assumed starting time, the regular and anomalous movement and longitude of the Sun

Since all these [topics] were presented appropriately, it appears befitting for me to accept this starting point for [carrying out] the calculations.

1-27 (note by the hand of Chortasmenos in the margin of C):

	Handy Calculation		
25 years	1601	133°	10′
Single years	7	358°	18′
Months	beginning of Thoth	-	-
		131°	28′

Adding them to the 5 and half degrees of Gemini the mean longitude of the Sun becomes 16° and 58′ of Libra. Introduce the resulting degrees into the table of the Sun's anomaly. The number for the anomaly next to it, found from the proportionality, is 1° 54′ and, since the resulting degrees were less than 180°, we subtract it from the solar longitude leaving for the anomalous longitude 15° 6′ 54″.

for 131° anomaly	1° 51′
for 131° 28′ add	1° 50′ 6″
for 132° anomaly	1° 49′

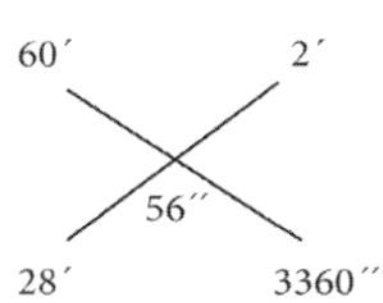

Thus we state that for the year 6792 from the creation of the World, on the sixth of October that we consider, the Sun was in the uniform rotation 16° 58′ of Libra and in the anomalous rotation 15° 6′ 54″ of Libra.

The 25-year intervals at the beginning of the year 6792

1st year	131° 28′
26 years	125° 23′
51 years	119° 19′

Either add 353° 53′ or subtract 6° 4′ 41″ obtained from handy computation.

καὶ αὐτοὺς τοὺς κανόνας προεξεθέμεθα, τούς τε περὶ τῶν κατὰ τοὺς ὡρισμένους χρόνους ὁμαλῶν τοῦ ἡλίου κινήσεων καὶ αὐτὸν τὸν περὶ τῆς ἀνωμαλίας τῆς κινήσεως αὐτοῦ σαφηνίσαντες καὶ διευκρινήσαντες καὶ τοὺς σκοποὺς καὶ τὰς μεθόδους αὐτῶν. ῥητέον ἤδη λοιπὸν καὶ ὅπως ἐστὶν ἑκάστοτε καθ᾽ ὁντιναοῦν δεδομένον ἐκεῖθεν χρόνον ἐξετάζοντας δι᾽ αὐτῶν τῶν κανόνων ἀνευρίσκειν τὴν τοῦ ἡλίου ἐποχήν, τήν τε ὁμαλὴν λεγομένην παρὰ τοῖς ἀστρονόμοις καὶ τὴν ἀνώμαλον, ταυτὸν δὲ εἰπεῖν καὶ ἀκριβῆ τε καὶ ἀληθῆ, ἤγουν κατὰ ποίου δωδεκατημορίου τοῦ ζωδιακοῦ εὑρίσκεται καὶ κατὰ ποίας μοίρας αὐτοῦ.

Ὅταν οὖν τοῦτο εὑρεῖν βουλώμεθα, ἐπισκεπτόμεθα πόσος ἐστὶν ὁ δεδομένος ἡμῖν χρόνος, ἀφ᾽ ἧς ὑπεθέμεθα χρονικῆς ἀρχῆς. καὶ ἀναζητοῦμεν αὐτόν, εἰ καταγράφεται ἐν τῷ προεκτεθειμένῳ κανόνι τῶν ὀκτωκαιδεκαετηρίδων, κατὰ τὸ πρῶτον σελίδιον· εἰ δὲ μή, ἀναζητοῦμεν τὸν καταγεγραμμένον ἐν τῷ τοιούτῳ κανόνι ἔγγιστα αὐτοῦ ἐλάσσονα καὶ ἀπογραφόμεθα τὰς παρακειμένας αὐτῷ ἐν τοῖς ἑξῆς σελιδίοις μοίρας ἢ λεπτά. εἶτα, εἰ ἐναπολιμπάνεται ἔτος ἢ ἔτη τινά, εἰσάγομεν αὖθις ταῦτα εἰς τὸν κανόνα τῶν ἁπλῶν ἐτῶν καὶ ἀπογραφόμεθα καὶ τὰ παρακείμενα καὶ αὐτοῖς ἐν τοῖς ἑξῆς σελιδίοις μοίρας καὶ λεπτά. εἶτα, εἰ πρὸς τούτοις εἰσὶ καὶ λείπονται ἐπὶ τοῦ δεδομένου ἡμῖν χρόνου καὶ μῆνές τινες καὶ ἡμέραι, ἔτι δὲ καὶ ὧραι ἀπὸ τῆς ἔγγιστα παρελθούσης μεσημβρίας, εἰσάγομεν καὶ ταῦτα εἰς τοὺς οἰκείους κανόνας, ἤτοι τοὺς μῆνας εἰς τὸν τῶν μηνῶν κανόνα καὶ τὰς ἡμέρας εἰς τὸν τῶν ἡμερῶν καὶ τὰς ὥρας εἰς τὸν τῶν ὡρῶν. καὶ εὑρίσκονται καὶ τὰ ἑκάστων τῶν εἰσαγομένων ἀριθμῶν ἐν τοῖς τοιούτοις κανόσι παρακείμενα, μοῖραι δηλονότι καὶ λεπτά, καὶ ἀπογραφόμεθα καὶ ταῦτα κατὰ μέρος ἰδίᾳ.

Εἶτα συνάπτομεν τὰς εὑρεθείσας κατὰ μέρος μοίρας καὶ λεπτὰ καὶ συνάγομεν ὁμοῦ. καὶ ἐπισυντιθέντες ταύτας τὰς συναχθείσας τῇ κατὰ

7 ἐξετάζοντας post κανόνων transp. C **12** ἐπισκεπτόμεθα C^{pc} (in mg.) **17** ἑξῆς C^{pc} (in mg.) **28** sqq. sch. in mg. C (Chort.) τὸ ὁμαλὸν ἡμερήσιον τοῦ ἡλίου κίνημά ἐστι μοίρας Ϥ νθ΄ η΄΄. τοσοῦτο γὰρ ἀναφαίνεται τῆς πρώτης ἀποκαταστάσεως, τουτέστι τῶν τξε ἡμερῶν τοῦ ἐνιαυτοῦ καὶ δ^{ον} ἔγγιστα μεριζομένων παρὰ τὰς τξ μοίρας τῶν δωδεκατημορίων τοῦ ζωδιακοῦ. τοῦτο οὖν τριακοντάκις συντιθέμενον μηνιαῖον ὁμαλὸν κίνημα ποιεῖ μοίρας κη λε΄ κατὰ τὸν Θέωνα, κατὰ δὲ Πτολεμαῖον πλέον τούτου διὰ τὴν ἀκριβεστέραν λῆψιν τῶν διαφόρων ἑξηκοστῶν. τοῦτο δ᾽ αὖθις τὸ τριακονθήμερον κίνημα δωδεκάκις συντιθέμενον μοίρας ποιεῖ τνδ μθ΄. ἀλλ᾽ ἐπείπερ ἡ μὲν δωδεκάκις

[In addition], we presented tables for the uniform movements of the Sun, clarified its anomalous motion and described the aims and the methods [involved]. We must describe next how at a specific time we can apply the tables in order to find and determine the precise and correct longitude of the Sun – including the one the astronomers call the regular and the anomalous – and give the precise and true position, that is the constellation of the ecliptic and the degrees within them where the Sun is located.

When we wish to determine [the longitude], we look at how much time has elapsed from the reference time. Then we look to see if this time appears somewhere in the first column of the table for 18 years. If it does not appear, then we search for the smallest number closest to it and record from the corresponding column the degrees or minutes. Then, if there remains a year or years we introduce them in the Table of Simple Years and record from the corresponding columns the degrees and minutes. Now, if there are missing toward [the time we are calculating] a few months and days, and even hours from the previous midday, we introduce them in the corresponding table, that is the months in the table of months, the days to that of days, the hours in the table of hours. Thus, we find for each entry of the table the degrees and minutes and record each of them separately.

Next, we collect the degrees and minutes that we found and add them up; we add the sum to the initial longitude that the Sun had at the assumed

1-27 (note by the hand of Chortasmenos in the margin of C): The uniform daily rotation of the Sun is 0° 59′ 8″ as it follows when the first revolution of the approximate 365 and one-quarter days of the year divide the 360° of the signs of the zodiac. Multiplying this with 30 we obtain the uniform motion in a month, being 28° 35′ according to Theon and more according to Ptolemy, because of a more precise calculation of the seconds. When the rotation of the thirty days is multiplied by 12 it produces 354° 49′. However, since the [product] [*continues to next page*]

τὴν ὑποτεθεῖσαν ἀρχὴν τοῦ χρόνου τῆς κινήσεως παρ᾽ ἡμῶν πρότερον ἐποχῇ τοῦ ἡλίου, ἤτοι ταῖς ρλα κη΄ μοίραις, ἅς ἀπεῖχε τηνικαῦτα ἀπὸ τοῦ οἰκείου ἀπογείου, ἤτοι τῶν ε μοιρῶν καὶ λ΄ τῶν Διδύμων, τὸν συναχθέντα ἐκ πάντων τούτων ἀριθμὸν καταλογιζόμεθα, εἴ ἐστι πλείων τξ μοιρῶν εἴτε μή. καὶ εἰ μέν ἐστι πλείων, ἀφαιροῦντες ἀπ᾽ αὐτοῦ κυκλικῆς περιόδου ἀριθμόν, ἤτοι αὐτὰς τὰς τξ μοίρας, ὅσαι ἂν ἔπειτα ἀπολειφθῶσι μοῖραι, τοσαύτας ἐροῦμεν προκεχωρηκέναι τηνικαῦτα τὸν ἀστέρα ἀπὸ τοῦ οἰκείου ἀπογείου, ἤτοι τῶν εἰρημένων ε μοιρῶν τῶν Διδύμων καὶ λ΄ ἑξηκοστῶν. εἰ δὲ ὁ ἀριθμὸς τῶν συναχθεισῶν μοιρῶν ἐντός ἐστι τῶν τξ, ὅσος ἐστί, τοσαύτας πάλιν φήσομεν μοίρας παροδεῦσαι τὸν ἥλιον ἀπὸ τοῦ τοιούτου ἀπογείου αὐτοῦ. ἵνα δὲ εὐεπιλόγιστος ἡ ψηφοφορία γένηται, προστίθενται τῷ τοιούτῳ ἀριθμῷ καὶ μοῖραι ξε καὶ λ΄ λεπτά.

Εἶτα τὸν συναγόμενον οὕτως ἀριθμὸν ἐκβάλλοντες ἀπὸ τῆς ἀρχῆς τοῦ Κριοῦ εἰς τὰ ἑπόμενα τῶν ζωδίων καὶ διδόντες ἑκάστῳ ζωδίῳ μοίρας λ, ἐὰν ἀπολειφθῶσι τινὲς μοῖραι μὴ ἀπαρτίζουσαι τὸν λ ἀριθμόν, τὰς τοσαύτας μοίρας τοῦ ἑξῆς ζωδίου φήσομεν τηνικαῦτα ἐπέχειν τὸν ἥλιον, ὁμαλῶς θεωροῦντες αὐτὸν κινούμενον καὶ εὑρίσκεσθαι

γινομένη σύνθεσις τῶν λ τξ΄ μόνον ἡμέρας ποιεῖ, ὁ δὲ αἰγύπτιος ἐνιαυτὸς τξε ἡμερῶν ἐστι μόνον, ἐὰν προσθῶμεν ταῖς τνδ μθ΄ μοίραις καὶ ε ἡμερῶν αἰγυπτίων κινήματα, τουτέστι δ νε΄, γενήσεται ὁ αἰγύπτιος ἐνιαυτὸς τνθ μοῖραι με΄ λεπτά.

Καὶ τοίνυν ἐπείπερ ἐν τῷ α^{ῳ} ἔτει τῆς Φιλίππου τοῦ Ἀριδαίου βασιλείας ἢ τῆς τοῦ Παλαιολόγου Ἀντωνίου—οὐδὲν γὰρ διαφέρει τό γε τοιοῦτον πρός γε τὸ ῥηθησόμενον—εὕρηται ὁ ἥλιος ἐκεῖ μὲν ἐπέχων ὁμαλῶς εἰς τὰ ἑπόμενα μοίρας ρξβ ι΄ ἀπὸ τῆς ε^{ης} καὶ ἡμισείας μοίρας τῶν Διδύμων, ἐνταῦθα δὲ ὁμοίως ὁμαλῶς ἀπέχων μοίρας ρλα κη΄. ἡ δὲ τῶν ἐνιαυτῶν ἐπισυναγωγὴ κατὰ εἰκοσαπενταετηρίδας γίνεται καὶ ἐν ἀμφοτέροις τοῖς κανόσι. καὶ εἰσὶν ἐκεῖσε τὰ τῷ α^{ῳ} ἔτει παρακείμενα, οὐ τοῦ α^{ου} ἔτους, ἀλλὰ τοῦ β^{ου} κινήματα, ὡς δὲ καὶ τὰ τοῦ β^{ου} τοῦ τρίτου καὶ ἐφεξῆς ἀκολούθως τὰ τοῦ κδ κινήματα τοῦ κε^{ου} ἔτους. τὰ δέ γε τοῦ κε^{ου} τῷ κστ^{ῷ} παράκειται. ἐὰν ἄρα τὸ ἐνιαύσιον τοῦ ἡλίου κίνημα, ὅπερ ἀποδέδεικται μοιρῶν τνθ με΄, εἰκοσάκις καὶ ἑξάκις πολλαπλασιάσωμεν, ποιήσωμεν ἀριθμὸν μοιρῶν ͵θ τνγ νε΄. ἐὰν δὲ καὶ αὐτὰ τὰ τξ εἰς ἃ τέμνεται ὁ ζωδιακὸς κύκλος εἰκοσάκις καὶ πεντάκις πολλαπλασιάσωμεν, ποιήσομεν ἀριθμὸν μοιρῶν ͵θ, ἅστινας ἐὰν ἀφέλωμεν τῶν ͵θ τνγ νε΄ μοιρῶν, καταλειφθήσονται ἡμῖν μοῖραι τνγ νε΄, ἅστινας ἂν προστιθέντες τῷ τε α^{ῳ} ἔτει τῆς τοῦ ἡλίου ὁμαλῆς εἰρημένης ἐποχῆς ὑπό τε Θέωνος καὶ ἑτέρου τῶν μετὰ ταῦτα σοφῶν εἰ τύχοι, καὶ ταῖς ὑπὸ τούτου γραφομέναις εἰκοσαπενταετηρίσι τὸν ἀριθμὸν αὐτῶν μέχρις οὗ βουλόμεθα παραυξήσομεν.

36 προκεχωρηκέναι post τηνικαῦτα transp. C

reference time, which is at 131° and 28′ away from the apogee , i.e. 5° and 30′ in Gemini. We examine the number collected this way whether it is greater or smaller than 360°. If it is greater, we subtract full revolutions and the degrees that remain denote the interval that the Sun advanced in that particular time from its apogee (the 5° and 30′ of Gemini). If the resulting number is less than 360°, then these are again the degrees that the Sun advanced from its apogee. In order to make the calculation more apparent (convenient), we add to it 65° and 30′.

This way we arrive at the number [that the Sun travelled], starting from the beginning of Aries and proceeding to the subsequent constellations, attributing 30° to each of them. [At the end it may happen that] the number obtained this way [measured from the beginning of Aries] does not complete 30°, but there is still a remainder. We carry the remainder to the next constellation and the degrees we find denote the location of the Sun, according to its uniform rotation.

[*continues from previous page*]
of the 30 days produces only 360 days but the Egyptian year has 365 days, the Egyptian year will get 359° and 43′ if we add to the 354° 49′ another 4° 55′ from the additional 5 days.

In the context of this discussion the computation for the first year of the reign Philip Arrhidaios, or for [the first year of] Antonios Palaiologos[1] are not very different, and one finds that the Sun, in its forward uniform rotation, was located at that time at 162° 10′ from the 5° and 10′ of Gemini, while now it is at 131° 28′ with both values computed using the 25-year cycles. In the calculation, the entry next to the first year corresponds to the second year of his reign; the entry for the second corresponds to the third year of his reign and so on up to 24th year, being the 25th year. Finally, the entry of the 25th corresponds to his 26th year. Thus, when we multiply 359° 45′ by 26 we obtain the number 9353° 55′; then, when we multiply the 360° of a circle by 25 we obtain 9000, which we subtract from 9353° 55′ and there remain 353° 55′; this result from the subtraction is added to the longitude reported by Theon and any other later expert, up to the date that we wish to carry out the calculation.

[1] Deposed by his grandson Andronikos III, Andronikos II was forced to become a monk under the name Antonios on 30 January 1330 (cf. *Prosopographisches Lexikon der Palaiologenzeit* 21436).

ἐν τοσῇδέ τινι μοίρᾳ τοῦδέ τινος ζωδίου. ἀκριβῶς δὲ καὶ ἀληθῶς, ὡς προέφημεν, εὑρήσομεν τὴν τότε ἐποχὴν αὐτοῦ τὸν τρόπον τοῦτον· τὸν συναχθέντα ἀριθμόν, ὡς προείρηται, τῆς κινήσεως αὐτοῦ, τῆς ἀπὸ τοῦ ἀπογείου αὐτοῦ, εἰσάξομεν εἰς τὸν κανόνα, ὃν προολίγου ἐξεθέμεθα, τῆς ἀνωμαλίας αὐτοῦ καὶ θεωρήσομεν τὰ παρακείμενα αὐτῷ ἐν τῷ τρίτῳ σελιδίῳ μοίρας καὶ λεπτά. καὶ ὅσα ἂν εὑρήσωμεν, εἰ μὲν ὁ ἀριθμὸς ἐν τῷ πρώτῳ σελιδίῳ τοῦ κανόνος εὑρίσκεται, ἤτοι τοῦ ἀπὸ μιᾶς μέχρι καὶ ρπ, ἀφαιρήσομεν ταῦτα ἐξ αὐτοῦ, εἰ δὲ εὑρίσκεται ἐν τῷ δευτέρῳ σελιδίῳ τῶν κοινῶν ἀριθμῶν, ἤτοι τοῦ ἀπὸ τῶν ρπα ἀνάπαλιν μέχρι καὶ τξ, προσθήσομεν αὐτῷ τὰ ὡς εἴρηται εὑρεθέντα παρακείμενα ἐν τῷ τρίτῳ σελιδίῳ. καὶ πάλιν διὰ τὸ εὐεπιλόγιστον, ὡς εἴρηται, ἐπισυνθέντες αὐτῷ ξε μοίρας καὶ λʹ λεπτά, τὸν συναχθέντα ἀριθμὸν ἐκβαλοῦμεν ἀπὸ τῆς ἀρχῆς τοῦ Κριοῦ. καὶ διδόντες ἑκάστῳ ζωδίῳ λ μοίρας, ὅπου ἂν καταλήξωμεν, ὡς ἔφημεν, ἐκεῖνο τὸ ζώδιον καὶ τὰς τοσάσδε μοίρας αὐτοῦ ἐπέχειν τηνικαῦτα τὸν ἥλιον, ἀκριβῶς τε καὶ ἀληθῶς ἐροῦμεν.

In order to find the precise and true longitude, we introduce the distance from Aries obtained this way in the Table of the Anomalies, which we described earlier, and record the degrees and minutes given in the third column, however many degrees they may be. If the degrees and minutes we found are in the first column of the table that is between 1° and 180°, we subtract it. If, on the other hand, it is in the second column of the common numbers [of the table], that is from 181° to 360°, we add what appears in the third column. Again, for simplicity, we add to the resulting number 65° and 30′ so that we measure the distance from the beginning of Aries. Finally, by assigning again 30° to each constellation, we end up at the final longitude, and we say that the Sun is precisely and truly located at this constellation and at so many degrees.

27 περὶ διαφορᾶς ὡρῶν, καὶ καθ' ὅσους τρόπους αἱ διαφοραὶ θεωροῦνται τῶν ὡρῶν, καὶ ὅτι ἀναγκαία ἡ περὶ τῶν ἀναφορῶν τῶν τοῦ ἰσημερινοῦ τμημάτων εἴτουν μοιρῶν κατάληψις

Καὶ τὴν μὲν ψηφοφορίαν καὶ εὕρεσιν ἑκάστοτε τῆς τοῦ ἡλίου παρόδου καὶ τῆς ἐποχῆς καὶ τοῦ κατὰ τὸν ζωδιακὸν τόπου, ἐν ᾧ εὑρίσκεται, οὕτω δὴ καταλογίζεσθαι ῥᾴδιον. ἐπεὶ δὲ τὰς ἐν τῷ κανόνι ἀπὸ μεσημβρίας ὥρας ὅ τε Πτολεμαῖος καὶ ὁ Θέων ὡς ἐν Ἀλεξανδρείᾳ ἐξέθεντο καὶ πρὸς τὸν ἐν αὐτῇ δηλονότι μεσημβρινὸν καὶ ἔτι ὡς ἰσημερινὰς πάσας καὶ οὐ καιρικάς, καὶ μὴν ἔτι ὡς ὁμαλῶν καὶ ἴσων πάντων τῶν νυχθημέρων ὄντων (τὰ τοιαῦτα δὲ οὐκ ὀλίγην ἔχει τὴν ἑτερότητα), ἀναγκαῖόν ἐστι διευκρινῆσαι τὰς περὶ τῶν ὡρῶν διαφορὰς διὰ τὸ ἀσφαλὲς τῶν ἑκάστοτε ψηφοφοριῶν, ὡς ἂν μὴ διὰ τὸ ἀνεξέταστον καὶ ἀνεπιλόγιστον τῶν ὡρῶν οὐκ ὀλίγη πλάνη περὶ τὴν εὕρεσιν καὶ κατάληψιν τῶν χρόνων ἐνίοτε τοῖς ψηφοφοροῦσι γίνηται καὶ δοκῇ τοῖς πολλοῖς τὸ τῆς ἐπιστήμης ἀσάλευτον προσπταίειν. ἐν πολλοῖς, ἡ μὲν οὖν ἔκθεσις τῶν κανόνων καὶ ἡ χρονικὴ ἀρχὴ ἐξετέθη Πτολεμαίῳ καὶ Θέωνι ἀπὸ μεσημβρίας, ὡς ἔφην, τῆς κατὰ τὴν Ἀλεξάνδρειαν, ταύτῃ δὲ τῇ ἐκθέσει καὶ ἡμεῖς ἠκολουθήσαμεν, διὰ τὸ ἀσύγχυτον τῶν ὑποθέσεων. δῆλον δὲ ὡς οἱ μεσημβρινοὶ παράλληλοι, τουτέστιν οἱ κατ' οὐρανὸν τόποι—ἐν οἷς ὁ ἥλιος γίνεται κατὰ τὸ μέσον τῆς ἡμέρας ἐφ' ἑκάστης οἰκήσεως ἐν τῇ πρώτῃ περιφορᾷ καὶ ταχυτάτῃ καὶ καθολικῇ, ὡς ἔφημεν, καὶ αὐτὸς μετὰ τοῦ παντὸς οὐρανίου σώματος συμπεριστρεφόμενος—διάφοροί εἰσι καὶ ἀλλήλων προηγούμενοι καὶ ὑστερίζοντες. καὶ ταῖς μὲν ἀνατολικωτέραις τῶν οἰκήσεων, ἐπειδὴ σφαιρικὸν νοεῖται τὸ σῶμα

4 Cf. Ptol. *Alm.* 3.2 (vol. 1.1, p. 212-213 Heiberg); Ptol. *Proch. Kan.* vol. 2, p. 76 Halma **13** Cf. Ptol. *Alm.* 2.13, 3.2 (vol. 1.1, p. 212-213 Heiberg), 4.4 (vol. 1.1, p. 286-287 Heiberg), 9.4 (vol. 1.2, p. 223, 229, 235, 241, 247 Heiberg); Ptol. *Proch. Kan.* vol. 2, p. 76, 130-132 Halma

tit. τῶν ante ὡρῶν[1] add. C   tit. μοιρῶν : ὡρῶν C[pc]   **2** τόπον C[ac]   **11** δοκῇ C[pc] : δοκοίη V

27 Various definitions of hours with a description of their differences, whose understanding is essential for describing the ascensions at various sections of the equator

It is now easy to carry out the calculation and determine the longitude of the Sun and its location on the ecliptic as a function of time. However, since the tables of Sun-risings, given by Ptolemy and Theon, were prepared for the meridian of Alexandria and since they used equinoctial instead of seasonal hours and their solar days were always equal — a feature that produces several peculiarities — it is necessary to clarify the differences between the various definitions of hours in order to be sure for the results of each calculation. [I do that] because not examining and computing the hours [precisely] leads sometimes to mistakes in the determination and recording of times, and then many [people] blame the unshakable nature of science. In general the exposition of the tables and the starting times adopted by Ptolemy and Theon refer to the midday of Alexandria. We also followed this presentation in order not to confuse [the reader by changing] the original assumptions. It is evident that the meridian circles, which define celestial locations, are the places where the Sun — as it rotates with the entire celestial globe on its first and very fast revolution — culminates at midday; they are different relative to each other, so that the Sun culminates at some places earlier and at others later. Since the

τῆς γῆς, καὶ προανατέλλει ὁ ἥλιος καὶ ταχύτερον ἐπὶ τοῦ μεσημβρινοῦ γίνεται καὶ ἐργάζεται τὸ μέσον τῆς ἡμέρας, ταῖς δὲ ἑξῆς δυτικωτέραις οἰκήσεσι καὶ ὑστερίζει ἐν ταῖς ἀνατολαῖς ὁ ἥλιος κατὰ ἀναλογίαν τῶν πρὸς ἀλλήλας διαστάσεων τῶν οἰκήσεων καὶ ὑστερίζει ὡσαύτως ἐν τοῖς μεσημβρινοῖς τόποις γινόμενος.

Προτέρα γάρ ἐστιν ἡ ἀρχὴ τῆς ἡμέρας καὶ πρότερον ἐν μεσημβρίᾳ γίνεται ὁ ἥλιος τοῖς κατὰ Βαβυλῶνα οἰκοῦσιν ἢ τοῖς κατὰ τὴν μεγαλόπολιν βασιλίδα ταύτην οἰκοῦσιν ἡμῖν καὶ ἔτι μᾶλλον ἢ τοῖς κατὰ τὴν Ῥώμην καὶ Ἰταλίαν οἰκοῦσι καὶ ἔτι πολλῷ μᾶλλον ἢ τοῖς κατὰ τὰς βρεττανικὰς νήσους οἰκοῦσι. καὶ μεθύστερον τοῖς κατὰ Βαβυλῶνα οἰκοῦσιν ὁ ἥλιος ἀνατέλλει καὶ ἀρχὴν ἡμέρας ποιεῖται καὶ αὖθις ἐν τοῖς μεσημβρινοῖς τόποις γίνεται ἢ τοῖς κατὰ τὸν ἰνδικὸν ποταμὸν Γάγγην οἰκοῦσι καὶ ἔτι πολλῷ μᾶλλον ἢ τοῖς κατὰ τὴν Χρυσῆν χερρόνησον καλουμένην οἰκοῦσιν ἢ τοῖς ἐν Σήραις. ἐπεὶ γοῦν ὡς εἴρηται ἡ κατὰ χρόνον ἀρχὴ τῆς κινήσεως τῶν ἀστέρων ἀπὸ τοῦ κατ' Ἀλεξάνδρειαν μεσημβρινοῦ θεωρεῖται (ἀπὸ γὰρ τῆς μεσημβρίας τῆς ἐν τῇ τοιαύτῃ πόλει ἐκτίθενται οἱ κανόνες), δῆλον πάντως ὡς ἐν ταῖς ἑκάστοτε ψηφοφορίαις, ἐὰν ἐν ἄλλαις πόλεσι γίνωνται, οὐ τὸν αὐτὸν εἶναι χρόνον φήσομεν, οἷος ἀναφαίνεται ἀπὸ τῶν κανόνων. οἷον ἐὰν ἀναφαίνηται τυχὸν ἀπὸ τῶν κανόνων τριῶν ὡρῶν ἰσημερινῶν χρόνος μετὰ μεσημβρίαν, ἐν Ἀλεξανδρείᾳ ἐστὶν ὁ τοσοῦτος χρόνος, ἐν δὲ ταῖς ἀνατολικωτέραις αὐταῖς πόλεσι πλείων ἐστὶ τριῶν καὶ ἡμισείας ὡρῶν ἴσως ἢ τεττάρων ἢ καὶ πλειόνων κατὰ ἀναλογίαν τῆς διαστάσεως. ἐὰν δὲ ἐν δυτικωτέραις χώραις ἢ πόλεσιν ἐν Ἰταλίᾳ τυχὸν ἢ Βρεττανίᾳ ὁ ἐπιλογισμὸς τοῦ χρόνου γένηται, δύο ὡρῶν ἢ μιᾶς ἢ ἡμισείας ἔσται ὁ χρόνος μετὰ μεσημβρίαν κατὰ ἀναλογίαν πάλιν τῆς διαστάσεως τῆς ἀπὸ τῆς πόλεως Ἀλεξανδρείας. τὰς δὲ διαστάσεις ἀπ' ἀλλήλων τῶν πόλεων τῶν ἐπιφανεστέρων κατὰ μέρος ἐν ἀριθμοῖς ἐκτίθεται ὁ Πτολεμαῖος ἐν τῇ βίβλῳ αὐτοῦ τῆς Γεωγραφίας, ἐκτίθεται δὲ τὰς τοιαύτας διαστάσεις καὶ ὁ Θέων ἐν τῷ Προχείρῳ κανονικῶς.

50-51 Cf. Ptol. *Geog.* 8; Ptol. *Proch. Kan.* vol. 1, p. 109-131 Halma

24 κατ' C **31** βρετανικὰς C **35** χερόνησον V **45** Βρετανίᾳ C

Earth is spherical, the Sun rises earlier in inhabited regions [located] further to the East and then reaches the meridian sooner as it travels its way to midday. For locations further to the West the Sun culminates and reaches the meridian at a later time, the delay being proportional to the relative distances.

The day begins and the Sun arrives at midday earlier for those living in Babylon than in the Great City — the Queen City [Byzantium] — where we live; and still earlier for us than those who live in Rome and Italy, and much earlier than those living in the British Isles. On the contrary, the Sun rises in the morning and again arrives at midday later for those living in Babylon than for those living in the river Ganges and much later than for those living in the Golden Peninsula or in Seres. Now since the initial time for counting the motion of the stars refers to the meridian of Alexandria (the tables are presented for the midday of that city), it is clear that for each calculation that refers to another city we should not quote the same time appearing in the tables. Thus, if the equinoctial time appearing in the tables of Alexandria is, for instance, three hours after midday, in the cities further to the East it is more than three and a half hours plus perhaps a quarter or more depending on their distance. If, on other hand, they are countries or cities further West in Italy or Britain, the correct time may be two or one and a half hours after midday, again depending on the distance from the city of Alexandria. Ptolemy in his book of *Geography* recorded numerically the distances between the major cities. Similarly, Theon presents the same distances in his *Handy Tables*.

Καὶ ἔξεστι τῷ μεταχειριζομένῳ ψηφοφορίαν τινὰ μετὰ τὴν εὕρεσιν τῆς χρονικῆς ἀποκαταστάσεως αὐτῆς ἀναζητεῖν καὶ ἀνευρίσκειν ῥᾳδίως ἀπὸ τοῦ εἰρημένου κανόνος τοῦ Θέωνος, ὁπόση τίς ἐστι διάστασις κατὰ μῆκος ἀπὸ τῆς Ἀλεξανδρείας εἰς τήνδε τινὰ τὴν πόλιν, ἐφ᾽ ἧς ἡ ψηφοφορία γίνεται. καὶ εἰ μέν ἐστιν ἀνατολικωτέρα ἡ πόλις, προστιθέναι ταῖς ἀναδοθείσαις ὥραις ἀπὸ τῆς κατὰ Ἀλεξάνδρειαν μεσημβρίας, ὅτι καὶ πρὸς τὸν κατὰ Ἀλεξάνδρειαν μεσημβρινόν, ὡς πολλάκις ἔφημεν, αἱ ἀρχαὶ τῶν ψηφοφοριῶν ἐξετέθησαν· εἰ δέ ἐστι δυτικωτέρα, ἀφαιρεῖν τῶν ἀναδοθεισῶν ὡρῶν ὅσαι ὧραι ἢ μέρος ὥρας καταλογίζονται ἀπὸ τῶν τῆς διαστάσεως μοιρῶν. ἰστέον δὲ ὅτι κατὰ ιε μοίρας τῆς κατὰ μῆκος διαστάσεως ὥρα μία ἀφαιρεῖται ἢ προστίθεται ἐν ταῖς ψηφοφορίαις, τὸν τρόπον ὃν ἔφημεν. εἰ δὲ καὶ ἥττονές εἰσιν αἱ μοῖραι τῶν ιε κατ᾽ ἀναλογίαν ὡς πρὸς τὰς ιε ἀπαρτιζούσας ὁλόκληρον ὥραν, ἐπιλογίζονται αἱ ἥττονες μοῖραι ἀπαρτίζειν μόριόν τι ὥρας. ἰστέον δὲ καὶ τοῦτο, ὅτι αἱ τοιαῦται μοῖραι χρόνοι παρὰ τοῖς ἀστρονόμοις καλοῦνται· τὸ γὰρ ἡμερονύκτιον κδ ὡρῶν ὄν, ὡς πολλάκις προέφημεν, χρόνος ἐστὶ τῆς πρώτης καὶ ταχυτάτης περιφορᾶς τῆς οὐρανίας σφαίρας, ἤτοι τοῦ μεγίστου ἰσημερινοῦ κύκλου, ὃς εἰς τξ μοίρας τέμνεται, ὡς ἐπιβάλλειν μιᾷ ἑκάστῃ ὥρᾳ ἀπὸ τῶν τοιούτων τξ μοιρῶν τοῦ ἰσημερινοῦ ἐν τῇ περιόδῳ ιε μοίρας (εἰκοσάκις γὰρ καὶ τετράκις τὰ ιε τξ εἰσί). καὶ καλοῦνται αἱ τοιαῦται ιε μοῖραι χρόνοι ὥρας, ὥστε ἐπεὶ καὶ ἡ γῆ σφαιρική ἐστιν, ὡς προέφημεν, καὶ ὁμόκεντρος τῷ παντί, τὰς ἐν αὐτῇ ιε μοίρας ὅμοια εἶναι τμήματα τῆς ὅλης κυκλικῆς περιφερείας αὐτῆς ταῖς ἰσαρίθμοις τῶν κατ᾽ οὐρανὸν κύκλων μοίραις. διὰ τοῦτο καὶ ἔφημεν τὰς κατὰ μῆκος διαστάσεις τῶν πόλεων κατὰ ιε μοίρας προστιθέναι ἢ ἀφαιρεῖν χρόνον ὥρας.

Ὅταν οὖν ποιούμενός τις ψηφοφορίαν τινὰ ἐπ᾽ ἄλλης πόλεως, οὐκ ἐπὶ τῆς Ἀλεξανδρείας, βουληθείη εὑρεῖν τοῦ χρόνου ἀκρίβειαν ἐπὶ τῆς τοιαύτης πόλεως, ὀφείλει ἀναζητεῖν τὴν τοιαύτην πόλιν ἐν τῷ εἰρημένῳ κανόνι τῶν ἐπιφανῶν πόλεων τῶν ἐν τῷ Προχείρῳ· καὶ

78-82 πρώτη διάκρισις τῶν ὡρῶν· πῶς ἂν τὰς ἀναδιδομένας ἐκ τῶν κανόνων ὥρας ἰσημερινὰς καὶ ὡς πρὸς τὸν δι᾽ Ἀλεξανδρείας μεσημβρινὸν μεταλάβοιμεν πρὸς τὰς ἀφ᾽ ἑτέρας πόλεως μεσημβρίας οἷον καθ᾽ ὑπόθεσιν τοῦ Βυζαντίου sch. in mg. C (Chort.)

This way it is possible for the one who is engaged in a calculation once, he determines the time of an event, then to find easily in Theon's tables how great is the distance in longitude of the city in question from the city of Alexandria. If the city is to the East he adds the appropriate hours to the midday of Alexandria, because as we stated many times, the initial time in the calculations is the meridian of Alexandria. If the city lies to the West, he subtracts from the given hours as many hours or fractions of an hour is obtained from the distance in degrees. It is known that a distance of 15° corresponds to one hour which is subtracted or added in the calculation and in the manner described already. If, furthermore, the degrees are less that 15°, then the correction to the calculation is proportional to the fraction of 15°, with 15° comprising an hour. Let it also be known, that astronomers call [these intervals of 15°] time-degrees. As we mentioned many times, the solar day consists of 24 hours, being the period for the first and very fast revolution of the major circle of the equator, which is subdivided in 360°; it follows from this definition that each hour consists of 15° and 360° are completed [in 24 hours]. These 15° are called time-degrees. As we mentioned, the Earth is spherical and homocentric to the universe, thus the 15° intervals are identical sections of the circular perimeter of the Earth and [stay] in one-to-one correspondence to the degrees of the celestial circles. For this reason, we said that for distances between cities of 15° we add or subtract one hour.

When one computes the time for a city other than Alexandria and wishes to find the precise time for it, he must find the city in the table of the famous cities in the *Handy* [*Tables*] and once he finds it,

ἀνευρίσκειν ταύτην τὲ καὶ τὰς παρακειμένας αὐτῇ μοίρας κατὰ μῆκος, καὶ ὁρᾶν ὅσον ἐστὶ τὸ διάφορον αὐτῶν πρὸς τὰς παρακειμένας κατὰ μῆκος μοίρας τῇ κατ' Αἴγυπτον Ἀλεξανδρείᾳ, καὶ κατὰ ιε μοίρας τῆς διαστάσεως ἐπιλογίζεσθαι χρόνον ὥρας, καὶ τοῦτον, ὡς ἀνωτέρω εἴρηται, προστιθέναι ἢ ἀφαιρεῖν. ἐπεὶ δὲ ὁ τοιοῦτος κανὼν τῶν ἐπιφανῶν πόλεων ὡς εἰκὸς πεπλατυσμένος ἐστί, καὶ ἡ νῦν κατ' ἐπιτομὴν αὕτη σύνταξις ἐκδίδεται πρὸς τὰς κατὰ τὴν μεγαλόπολιν τήνδε καθ' ἡμᾶς γενησομένας ψηφοφορίας. τοσοῦτο μόνον νῦν προσθέντες ἀρκεσθησόμεθα, ὅτι ἡ διάστασις τοῦ κατὰ τὴν Ἀλεξάνδρειαν μεσημβρινοῦ πρὸς τὸν καθ' ἡμᾶς μεσημβρινόν ἐστι τεσσάρων καὶ ἡμίσεος μοιρῶν, ἤτοι χρόνων. καὶ ἐπεὶ ἡ καθ' ἡμᾶς αὕτη μεγαλόπολις δυτικωτέρα ἐστὶ τῆς Ἀλεξανδρείας, ἐν ταῖς ἑκάστοτε ψηφοφορίαις ἀφαιρεῖν δεῖ τῶν ἀναδιδομένων ὡρῶν τὸ ἐπιβάλλον τοῖς τοιούτοις τέσσαρσι καὶ ἡμίσει χρόνοις, τρίτον ἔγγιστα ὥρας, καὶ οὕτω τὰς λοιπὰς ἐν τῇ ἑκάστοτε ψηφοφορίᾳ καταλογίζεσθαι ὥρας. καὶ τὸ μὲν διὰ τοὺς μεσημβρινοὺς διάφορον τῶν ὡρῶν ἱκανῶς ἡμῖν διευκρίνηται.

Δευτέρα δὲ ἡμῖν ἐπίσκεψίς ἐστι περὶ τῶν ὡρῶν αὕτη. πάσας τοῦ ἔτους ἡμέρας καὶ νύκτας εἰς δώδεκα ὥρας διαιροῦσιν οἱ ἀστρονόμοι, ὡς ἐντεῦθεν ἀκολουθεῖν ἐξανάγκης τὰς τοιαύτας ὥρας ἀνίσους ἀλλήλων εἶναι, ἤτοι τὰς ιβ ἡμερινὰς ταῖς ιβ νυκτεριναῖς. καὶ νῦν μὲν τὰς ἡμερινὰς μείζους, νῦν δὲ τὰς νυκτερινάς, καὶ οὐδέποτε ἴσας, εἰ μὴ μόνον δὶς τοῦ ἔτους, ὁπηνίκα ὁ ἥλιος ἐπὶ τῶν ἰσημερινῶν σημείων γινόμενος, ἤτοι ἐν τῇ ἀρχῇ τοῦ Κριοῦ καὶ τῇ ἀρχῇ τῶν Χηλῶν, ἰσημερίαν ἐμποιεῖ καὶ τὰς ὥρας ἴσας ὡς πρὸς αἴσθησιν τὰς ἡμερινὰς ταῖς νυκτεριναῖς. ἄνευ γοῦν τῶν εἰρημένων ἡμερῶν καὶ τῶν νυκτῶν αἱ πᾶσαι ἄλλαι ἡμέραι καὶ νύκτες τοῦ ἔτους ἀνίσους ἔχουσι τὰς ὥρας πρὸς ἀλλήλας τὲ καὶ τὰς ἰσημερινάς, καὶ καλοῦνται αἱ τοιαῦται ὧραι αἱ ἄνισοι καιρικαὶ ὧραι.

Ἐπεὶ γοῦν οἱ κανόνες τὰς ἐκτεθειμένας ἐν αὐτοῖς ὥρας ἰσημερινὰς ὑποτίθενται, χρεία πάντως ἐστὶν ἀναγκαία ἐν ταῖς ἑκάστοτε ψηφοφορίαις τὰς ἀναδιδομένας ὥρας ἰσημερινὰς ἐκ τῶν κανόνων

98-102 δευτέρα διάκρισις τῶν ὡρῶν· πῶς ἂν τὰς ἀναδεδομένας ἀπὸ τῶν κανόνων ὥρας ἰσημερινὰς μεταλάβοιμεν ἂν εἰς καιρικὰς εἴτ' οὖν ἀνίσους καὶ τὸ ἀνάπαλιν sch. in mg. C (Chort.)

and the adjacent distance in degrees from Alexandria of Egypt he calculates the difference in time, allocating 15° for each hour which is added or subtracted. Since the table of famous cities is a reasonable interpolation, and the present *Syntaxis* is published abridged with reference to the Great City [Byzantium], it suffices to add that the meridian of Alexandria relative to our meridian is four and a half time-degrees. Since our Great City is located to the West of Alexandria we must subtract from the time found in each calculation these four and a half time-degrees, i.e. approximately one third of an hour. So far we elucidated with sufficient clarity the time differences between meridians.

A second topic investigates the hours [of a day or a night]. Astronomers divide all days and nights of a year into twelve hours. It follows from this that the hours are unequal, i.e. the twelve hours of a day and night [are unequal]. Sometimes the hours of the day are longer and at other times the hours of the night are longer, and they are never equal except twice in a year, when the Sun is at the equinoctial points that are at the beginning of Aries and the beginning of Libra, at these times the days and nights are equal. Besides the days and nights just mentioned, all other days and nights of the year have unequal hours, also being different from the equinoctial, and are called seasonal hours.

Now, since the hours appearing in the tables are equinoctial, it is necessary in each calculation to understand how to convert for a specific day equinoctial to seasonal hours, or conversely seasonal to equinoctial

ὁπόσαι εἰσὶ καιρικαὶ ὧραι κατὰ τὴν τότε ἡμέραν ἢ νύκτα καταλαμβάνειν καὶ ἐπιλογίζεσθαι ἢ καὶ ἀνάπαλιν τὰς καιρικὰς ὥρας θεωρεῖν ὁπόσαι καταλαμβάνονται ἰσημεριναί, ἐν τῷ χρόνῳ δηλονότι τῆς τότε ἡμέρας ἢ τῆς τότε νυκτός, καθ' ἃς αἱ ψηφοφορίαι γίνονται. μεθοδεύεται δὲ τοῦτο τὸν τρόπον τοῦτον. γραμμικαῖς ἀποδείξεσιν ὁ Πτολεμαῖος ἀναζητήσας εὑρίσκει ὁπόσοις ἰσημερινοῖς χρόνοις, εἴτουν μοίραις τῶν τοῦ ἰσημερινοῦ μεγίστου κύκλου τξ, ἕκαστον τῶν τοῦ λοξοῦ καὶ ζωδιακοῦ δωδεκατημορίων ἀεὶ συναναφέρεται ἐπὶ τοῦ ὁρίζοντος τῆς τε ὀρθῆς νοουμένης σφαίρας καὶ ἐπὶ τοῦ ὁρίζοντος ἑκάστου κλίματος. οὐ γὰρ τοῖς αὐτοῖς χρόνοις ἐπὶ τοῦ ὁρίζοντος ἐπὶ τῆς ὀρθῆς σφαίρας καὶ ἐπὶ τοῦ ὁρίζοντος ἐφ' ἑκάστου τῶν κλιμάτων ἕκαστον δωδεκατημόριον συναναφέρεται, ἀλλὰ προηγουμένως μέν, ἐπειδὴ λοξός ἐστιν ὁ ζωδιακὸς πρὸς τὸν ἰσημερινὸν καὶ οὐ παράλληλος αὐτῷ, οὐχ' αἱ αὐταὶ καὶ ἶσαι μοῖραι τοῦ τε λοξοῦ καὶ τοῦ ἰσημερινοῦ ἐπὶ τοῦ ὁρίζοντος τῆς τε ὀρθῆς σφαίρας καὶ παντὸς ἄλλου συναναφέρονται, ἀλλὰ νῦν μὲν ἐπὶ τῶνδέ τινων δωδεκατημορίων πλείους ἀναφέρονται τοῦ ζωδιακοῦ ἐν τῷ αὐτῷ χρόνῳ ἐπὶ τοῦ ὁρίζοντος ἢ τοῦ ἰσημερινοῦ, νῦν δὲ πλείους τοῦ ἰσημερινοῦ ἢ τοῦ ζωδιακοῦ. ἔπειτα δὲ οὐδὲ ὅταν ἐπὶ ἄλλων ὁριζόντων τῶν διαφόρων κλιμάτων θεωρῶμεν ὅσοις χρόνοις, ἤτοι μοίραις τοῦ ἰσημερινοῦ, ἀναφέρεται ἕκαστον ζώδιον, καταλαμβάνεται ἴσοις χρόνοις, ἤτοι μοίραις τοῦ ἰσημερινοῦ, ὅσοις καὶ ἐπὶ τῆς ὀρθῆς σφαίρας συναναφέρεσθαι τὸ αὐτὸ ζώδιον, ἀλλὰ ἐπὶ πάντων κλιμάτων ἢ πλείοσιν ἢ ἐλάττοσι πρός τε αὐτὰς τὰς ἐπὶ τοῦ ὁρίζοντος τῆς ὀρθῆς σφαίρας ἀναφοράς, ἔτι γε μὴν καὶ πρὸς τὰς ἐπὶ τῶν ἄλλων κλιμάτων ἀναφοράς. καὶ παντελῶς εἰσιν ἄνισοι αἱ συναναφοραὶ τῶν τοῦ ἰσημερινοῦ χρόνων ἑνὸς ἑκάστου δωδεκατημορίου τοῦ ζωδιακοῦ, ἔν τε τῷ ὁρίζοντι τῆς ὀρθῆς σφαίρας καὶ ἐν τοῖς ὁρίζουσι πάντων τῶν κλιμάτων. τοῦτο δὲ γίνεται πάντως διά τε τὴν ἔγκλισιν τοῦ ζωδιακοῦ ὡς πρὸς τὸν ἰσημερινὸν καὶ διὰ τὴν διαφορὰν τῆς ἐγκλίσεως τῶν ὁριζόντων πρὸς τὸν ὁρίζοντα τῆς ὀρθῆς σφαίρας.

Ταῦτα μέντοι πάντα καὶ τὰς αὐτῶν ποικίλας ἑτερότητας ἀκριβέστατα ἐξετάσας καὶ ἀνευρὼν γραμμικαῖς δείξεσιν ὁ Πτολεμαῖος,

116 Cf. Ptol. *Alm.* 2.7 **143-144** Cf. Ptol. *Alm.* 2.7-8

for the day or the night at which the calculation takes place. This topic is studied by Ptolemy who investigated it with geometrical methods and determined the time-degrees, i.e. degrees of the equator that each sign of the zodiac always rises through the horizon of *sphaera recta* and the horizon of each zone. Each sign does not rise at the same time on the horizon of *sphaera recta* and on the horizon of each zone, but earlier, because the ecliptic is inclined relative to the equator and not parallel. Equal arcs of the inclined and of the equator do not correspond to equal arcs of the horizon of *sphaera recta* or any other horizon, but for some constellations the corresponding intervals are sometimes larger on the ecliptic than the equator, and for other times larger on the equator than the ecliptic. Furthermore, when we analyze the rising times for each constellation at the horizons of various zones — in terms of time-degrees — these rising times are either larger or smaller for each zone and also relative to *sphaera recta*. In general, the relative rising times for each sign of the zodiac are not the same on the horizon of *sphaera recta* and on the horizon of each zone. This is caused by the inclination of the ecliptic relative to the equator and the inclinations of [local] horizons relative to the horizon of *sphaera recta*.

In any case, all these and their peculiarities were studied carefully by Ptolemy, who found geometrical proofs and presented his results in

ἔπειτα καὶ διὰ κανόνων ἐν ἀριθμῶν ἐκθέσει παραδίδωσι. καὶ ποιεῖται μὲν πρῶτον κανόνα τῶν ἀναφορῶν ἐπὶ τῆς ὀρθῆς σφαίρας, ἔπειτα δὲ καὶ ἐπὶ τῆς ἐγκεκλιμένης κατὰ μέρος ἰδίους κανόνας, ἐπὶ τοῦ ὁρίζοντος ἑνὸς ἑκάστου τῶν κλιμάτων. καὶ ἐκτίθεται ἐν τοῖς τοιούτοις κανόσι κατὰ δεκαμοιρίαν ἑνὸς ἑκάστου ζωδίου, ὅσοι συναναφέρονται χρόνοι, ἤτοι τμήματα καὶ μοῖραι τοῦ ἰσημερινοῦ, ἐπί τε τοῦ ὁρίζοντος τῆς ὀρθῆς σφαίρας καὶ ἐπὶ τῶν ὁριζόντων ἑκάστου κλίματος. συναποδείκνυσι δὲ γραμμικαῖς καὶ τοῦτο δείξεσιν, ὅτι ὅσοις χρόνοις συναναφέρεται ἕκαστον δωδεκατημόριον τοῦ ζωδιακοῦ ἐπὶ τοῦ ὁρίζοντος τῆς ὀρθῆς σφαίρας, τοῖς ἴσοις χρόνοις διεξέρχεται καὶ τὸν μεσημβρινὸν ἑκάστου κλίματος. καὶ τούτου πρόδηλον καὶ παντὶ συνιδεῖν τὸ αἴτιον. ὅ τε γὰρ ὁρίζων τῆς ὀρθῆς σφαίρας μέγιστος κύκλος ὢν τῶν ἐν τῇ σφαίρᾳ καὶ διὰ τῶν πόλων αὐτῆς ἐστιν, ὡς ἔστι κατανοεῖν ῥᾴδιον· ὅ τε μεσημβρινὸς παντὸς κλίματος καὶ αὐτὸς μέγιστος κύκλος ὢν τῶν ἐν τῇ σφαίρᾳ, καὶ αὐτὸς κοινωνεῖ τῷ ὁρίζοντι τῆς ὀρθῆς σφαίρας, ὅτι διὰ τῶν πόλων αὐτῆς ἐστιν, ὡς ἔστι καὶ τοῦτο παντὶ δῆλον, τοῖς δὲ τῶν κλιμάτων ὁρίζουσι κατὰ τοῦτο οὐ κοινωνεῖ. οἱ γὰρ νοουμένοι ὁρίζοντες ἐφ᾽ ἑνὸς ἑκάστου τῶν κλιμάτων μέγιστοι μέν εἰσι καὶ αὐτοὶ κύκλοι τῶν ἐν τῇ σφαίρᾳ—εἰς ἶσα γὰρ αὐτὴν διαιροῦσι τμήματα, ἤτοι ἡμισφαίρια, τό τε ὑπὲρ γῆν καὶ τὸ ὑπὸ γῆν—οὐ μὴν καὶ διὰ τῶν πόλων αὐτῆς εἰσι. διὰ τοῦτο καὶ ἔφημεν ὅτι ἐπειδὴ ὁ ὁρίζων τῆς ὀρθῆς σφαίρας καὶ μέγιστός ἐστι κύκλος τῶν ἐν αὐτῇ καὶ διὰ τῶν πόλων αὐτῆς ἐστιν, ὡσαύτως δὲ καὶ ὁ μεσημβρινὸς ἑκάστου κλίματος καὶ μέγιστος κύκλος ἐστὶ τῶν ἐν τῇ σφαίρᾳ καὶ διὰ τῶν πόλων αὐτῆς ἐστιν ἐξανάγκης. καὶ τοῦτο κοινὸν ἔχουσι, ὅτι τὰ δι᾽ αὐτῶν διϊόντα τοῦ ζωδιακοῦ δωδεκατημόρια ἕκαστα τοῖς ἴσοις χρόνοις, ἤτοι τμήμασι καὶ μοίραις τοῦ ἰσημερινοῦ, συναναφέρεται καὶ συμπρόεισιν.

150 ἑνὸς ante ἑκάστου add. C **151** γραμμικαῖς post καὶ τοῦτο transp. C **151-55** σημείωσαι ἀναγκαῖον περὶ τοῦ ὁρίζοντος τῆς ὀρθῆς σφαίρας τοῦ καὶ ἐπὶ παντὸς γινομένου ἀεὶ καὶ νοουμένου μεσημβρινοῦ ὅτι ἄμφω διὰ τῶν πόλων τῆς ὀρθῆς σφαίρας εἰσίν Sch in mg. G

numerical tables. First he composed a table for the Sun-risings on *sphaera recta*. Subsequently, he prepared specific tables for the inclined and for the horizon of each zone, and recorded them in intervals of 10° of each sign. He expressed the rising times at the horizon of *sphaera recta* and the horizon of each zone in time-degrees.

He also proves and shows geometrically that the time it takes each sign of the zodiac to cross the horizon of *sphaera recta* is equal to the time it takes to cross the meridian of each zone. This is evident and also easy for everybody to understand, because the horizon of *sphaera recta* is a great circle which passes through the poles; similarly, the meridian of every zone is a major circle of the sphere related to the horizon of *sphaera recta* as it also passes through the poles; however, the [great circles that define] the horizons of the zones are different: the theoretical horizons for each zone are major circles of the sphere dividing it in two hemispheres, one above the Earth and the other below, but do not pass through the poles. For this [reason] we said that, because each horizon of *sphaera recta* is a meridian, there is the common property that each sign of the zodiac passes and advances through the meridian [of a location] in equal time-intervals with those measured in the corresponding horizon of the equator.

Ἀλλ', ὅπερ ἐλέγομεν, τοιούτους μὲν τοὺς κανόνας ἐκτίθεται ὁ Πτολεμαῖος ἐν τῇ Συντάξει τῶν ἀναφορῶν ἐπί τε τῆς ὀρθῆς σφαίρας καὶ ἐφ' ἑκάστου τῶν κλιμάτων κατὰ δεκαμοιρίαν ἑκάστου τῶν ζωδίων τοὺς ἀναφερομένους τοῦ ἰσημερινοῦ χρόνους ἐπιλογιζόμενος. ὁ μέντοι Θέων ἐν τοῖς Προχείροις καθ' ἑκάστην μοῖραν ἑνὸς ἑκάστου τῶν δωδεκατημορίων τοὺς ἐπιβάλλοντας χρόνους ἐπιλογίζεται καὶ συντίθησι καὶ οὕτως αὐτὸς τὴν ἔκθεσιν ποιεῖται τῶν κανόνων. ἐπεὶ γοῦν πρὸς τὸ εὐποριστότερον τῆς ἐργασίας καὶ ἀπονώτερον καὶ ῥᾷον καὶ ἑτοιμότερον ἡμῖν ἡ πραγματεία τῆς παρούσης Στοιχειώσεως προτίθεται, διὰ τοῦτο δὴ τοῖς τοῦ Θέωνος κανόσι χρήσασθαι δοκιμάζομεν καὶ αὐτοὺς νῦν ἐκθέσθαι, σαφηνείας ὡς ἐπὶ τὸ πλεῖστον πρόνοιαν ποιούμενοι ἄνευ μετρίας τινὸς μεταβολῆς, ἥτις ἐστὶν ἀφαίρεσις τοῦ τρίτου σελιδίου τοῦ κανόνος τῶν ἀναφορῶν τῆς ὀρθῆς σφαίρας διὰ σκοπὸν χρήσιμον, μᾶλλον δὲ ἀναγκαῖον καὶ ἀπαραίτητον, ὅτι οὐκ ἔστιν ἄλλως χρῆσθαι πρὸς τὴν καθ' ἡμᾶς δοθεῖσαν ἐποχὴν καὶ ἀρχὴν τοῦ χρόνου τῆς κινήσεως τοῦ ἡλίου, εἰ μὴ κατὰ τὴν παράδοσιν καὶ τὴν μέθοδον Πτολεμαίου, οὐ τοῦ Θέωνος. οὕτως γὰρ ἔσται καὶ ἡ ἐργασία ἀσφαλεστέρα δι' αἰτίαν καὶ λόγον, ὃν οὐκ εὐχερὲς νῦν προστιθέναι διὰ τὸ μῆκος τοῦ λόγου.

Εἰσὶ γοῦν οἱ τοῦ Θέωνος κανόνες οὕτως ἐκτεθειμένοι, ἐπί τε τοῦ ὁρίζοντος τῆς ὀρθῆς σφαίρας καὶ τῶν ἑπτὰ κλιμάτων. ῥᾷον δέ ἐστιν ἀπὸ τῶν παρακειμένων ἐφ' ἑκάστου τῶν ἑπτὰ κλιμάτων, κἂν ἐπί τινων πόλεων ἢ χωρῶν τὰς ψηφοφορίας ἑκάστοτε θεωρῶμεν, κατὰ τὴν ἐν μέσῳ διάστασιν δύο κλιμάτων διακειμένων ἐπιλογίζεσθαι ἀναλόγως τὲ καὶ καθ' ὁμαλὴν παραύξησιν ἢ ἴσως ἐλάττωσιν ἐκ τῆς διαφορᾶς τῶν παρακειμένων τοῖς ἑκατέρωθεν κλίμασι καὶ ἀνευρίσκειν τὰ ἐπιβάλλοντα αὐτῇ, οἱᾳδήτινι ἑκάστῃ χώρᾳ καὶ πόλει, ἐφ' ἧς ψηφοφορία γίνεται. ἔστι γοῦν, ὡς ἐλέγομεν, ἕκαστος ὁ κανὼν ἐκτεθειμένος κατὰ τρία ὁμοῦ ζώδια, διὰ τὸ συντομώτερον ἐν τῇ ἐπιγραφῇ τοῦ κανόνος προτιθεμένων τῶν τοιούτων τριῶν δωδεκατημορίων κατ' ὄνομα ἐν τοῖς οἰκείοις τόποις κατὰ τάξιν ἑξῆς.

172 Cf. Ptol. *Alm.* 2.8 (vol. 1.1, p. 134-141 Heiberg) **175** Cf. Ptol. *Proch. Kan.* p. 97-128 Tihon

As we were saying, Ptolemy presented such tables in the chapter of ascensions for *sphaera recta* and for each zone in steps of 10°, within each sign on the zodiac and uses equinoctial hours. Theon, on the other hand, computes, summarises and presents them in his *Handy Tables* for every degree of each sign on the zodiac. In order to enrich this treatise and to make it simpler, more convenient and easier to use, we decided to present Theon's tables. We shall present them now, paying attention to clarity and making a minor change, merely omitting the third column in the table for the Sun's risings on *sphaera recta*. This is necessary and unavoidable in view of the initial time and longitude we adopted for the movement of the Sun, which cannot be used in any other way but following the method of Ptolemy instead of Theon. This way the work becomes more reliable, for reasons and arguments which are not convenient to explain now since they are too long.[2]

We present Theon's tables for the horizon of *sphaera recta* and for the seven zones. It is rather convenient from [the results on] the seven zones to calculate the Sun risings for cities and countries which lie between, by proportionally computing the linear increase or perhaps the decrease from the difference of adjacent zones. This way one computes the appropriate rising times of the Sun for each country and city. As we mentioned, each table contains, for the sake of brevity, three signs of the zodiac, with their names appearing on the title of each table and presented in sequence of the zones.

---

2 Here Metochites means correction from the difference between uniform and anomalous rotations, discussed later in this chapter as the third topic. This column includes the correction from the equation (*prosthaphaeresis*).

Τὸ οὖν τοῦ κανόνος πρῶτον σελίδιον ἔχει τοὺς κοινοὺς ἀριθμοὺς τῶν μοιρῶν τῶν τριῶν δωδεκατημορίων, μίαν, δευτέραν, τρίτην, μέχρι καὶ τῆς τριακοστῆς· εἶτα ὑφ᾽ ἑκάστην ἐπιγραφὴν ὀνόματος δωδεκατημορίου ἕτερα δύο σελίδια, ἄνευ μόνου τοῦ κανόνος τῶν ἀναφορῶν τῆς ὀρθῆς σφαίρας. ἐνταῦθα γὰρ ἀφαιρεῖται τὸ τρίτον σελίδιον, ὡς ἔφημεν, ὡς γίνεσθαι τὰ τοῦ ὅλου κανόνος σελίδια ἑπτά. ἀπὸ τῶν δύο δὲ σελιδίων καθ᾽ ἑκάστην μοῖραν ἑκάστου τῶν τριῶν δωδεκατημορίων, ἐν μὲν τῷ ἑνὶ παράκειται ὁ ἀριθμὸς τῶν συναναφερομένων αὐτῇ χρόνων, ἤτοι μοιρῶν τοῦ ἰσημερινοῦ, ἐνίοτε δὲ ἅμα καὶ λεπτῶν, ἐν δὲ τῷ ἑτέρῳ σελιδίῳ τῶν δύο παράκειται ὁ ἀριθμός, ὅς ἐπιβάλλει τῇ αὐτῇ μοίρᾳ ἐπιγραφόμενος ὡριαίων χρόνων καὶ ἑξηκοστῶν ἐπὶ τῶν κλιμάτων μόνων. ἐπὶ δὲ τῆς ὀρθῆς σφαίρας ἀφῄρηται παρ᾽ ἡμῶν τὸ τοιοῦτον σελίδιον, ὡς ἔφημεν. ἔχει δὲ τὸ τοιοῦτον σελίδιον σκοπὸν τόνδε. ἐφ᾽ ἑκάστου τῶν κλιμάτων, οἵτινες νοοῦνται τὴν οἴκησιν ἔχοντες ἐπὶ τῆς γῆς ὑπ᾽ αὐτὸν τὸν ἰσημερινὸν μέγιστον παράλληλον τῆς ὀρθῆς σφαίρας, οἱ τοιοῦτοι ἐν παντὶ τμήματι τοῦ ζωδιακοῦ περιοδεύοντος καὶ εὑρισκομένου τοῦ ἡλίου ἰσημερίαν ἔχουσι. τοῖς γὰρ ὑπὸ τὸν ἰσημερινὸν οἰκεῖν νοουμένοις οἱ παράλληλοι κύκλοι—ὧν ἅπτεται ὁ ἥλιος ἐφ᾽ ἑκάτερα τοῦ ἰσημερινοῦ περιϊών, ἐπὶ τὰ βόρεια δηλαδὴ καὶ τὰ νότια—οἱ πάντες οὗτοι παράλληλοι τοῖς ὑπὸ τὸν ἰσημερινόν, ὡς εἴρηται, οἰκοῦσιν εἰς δύο ἡμικύκλια ὑπὸ τοῦ ὁρίζοντος τῆς ὀρθῆς σφαίρας ἀκριβέστατα τέμνονται. διὰ τοῦτο καὶ ἐπὶ πάντων αὐτῶν ὁ ἥλιος εὑρισκόμενος ἐν τῇ καθόλου περιστροφῇ τοῦ παντὸς ἴσας περιφερείας ἀπολαμβάνει ὑπὲρ γῆν τε καὶ ὑπὸ γῆν, καὶ ἐν ἴσοις χρόνοις τοῦ ἰσημερινοῦ περιστρέφεται τὸ ὑπὲρ γῆν ἡμικύκλιον καὶ τὸ ὑπὸ γῆν. ὅμοια γὰρ τὰ ἡμικύκλια τά τε ὑπὲρ γῆν καὶ ὑπὸ γῆν τῶν παραλλήλων ἀλλήλοις τὲ καὶ τοῖς ἡμικυκλίοις τοῦ ἰσημερινοῦ, τῷ τε ὑπὲρ γῆν τε καὶ ὑπὸ γῆν, εἰ καὶ ἥττονα αὐτοῦ τε, ἀλλὰ δὴ καὶ ἀλλήλων ἔνια ἐνίων καὶ ἰσοχρονίως συμπεριοδεύοντα καὶ συναποκαθιστάμενα ἐπὶ τῶν σημείων ἀφ᾽ ὧν θεωροῦνται ἀρξάμενα. διὰ τοῦτο καὶ ὡς ἔφημεν ἀείποτε τοῖς ὑπὸ τὸν ἰσημερινὸν οἰκεῖν νοουμένοις ἶσαι ἂν νοοῖντο αἱ ἡμέραι πᾶσαι ταῖς νυξὶν ἐν οἱῳδήτινι τῶν παραλλήλων κύκλων, ὧν

214 τοιοῦτο C **229** εἰ καὶ : κἂν C

The first column of the table contains the numbers of degrees common for the three signs of the zodiac, [being enumerated as] one, two, three up to the thirtieth subdivision for each sign. Then, below the title for each sign there are two columns — except for the table of the risings for *sphaera recta* where the third column is omitted — thus making the entire table to have seven columns. In the two columns adjacent to each degree of the three signs, the first one records the rising times in degrees of the equator and sometimes with minutes. Only in the tables for the zones next to each degree there is a second column with the title hourly-times expressed in degrees and minutes. For [the table of] *sphaera recta* we omit this column, which serves a purpose only for each zone of the Earth [where days and nights are different]. In the regions below the equator — [below] the major circle of *sphaera recta* — the days and nights are always equal, no matter where the Sun is located on its trajectory along the ecliptic. All inhabited parallel circles below the equator, which the Sun crosses [twice] as it travels to the north and to the south, are divided by the horizon of *sphaera recta* into two semicircles. For this reason in all horizons [of *sphaera recta*] the Sun, as it travels its diurnal rotation, covers equal arcs and spends equal time above and below the Earth. The semicircles of the parallels above and below the horizon are similar to each other and to the semicircles of the equator. Even though they are smaller, there is a one to one correspondence, they rotate with the same speed and restore themselves together to their initial positions. For all these reasons, those living below the equator, on any of the parallel circles close to it, experience their days to be

ἅπτεται ὁ ἥλιος, εὑρισκομένου αὐτοῦ εἴτε εἰς τὰ βόρεια εἴτε εἰς τὰ νότια αὐτοῦ. ὅσαι δὲ οἰκήσεις ἐκεῖθεν ἀρξαμένοις ἡμῖν καταλαμβάνονται πρὸς τὰ βόρεια τοῦ ἰσημερινοῦ ἑξῆς, ἐπὶ τούτων συνεχῶς κατ᾽ ὀλίγον ἀεὶ ἄνισοι καταλαμβάνονται αἱ ἡμέραι ταῖς νυξί.

Καὶ τὰ μὲν βορειότερα τοῦ ζωδιακοῦ περιϊόντος τοῦ ἡλίου αἱ ἡμέραι μείζους εἰσὶ τῶν νυκτῶν καὶ κατ᾽ ὀλίγον ἀλλήλων ἔτι μείζους, καθ᾽ ὅσον πρόεισιν ἡ οἴκησις ἐπὶ τὰ βορειότερα. ἀπολαμβάνονται γὰρ ἀπὸ τῶν παραλλήλων κύκλων, ὧν ἐφάπτεται ὁ ἥλιος, μείζους ἀεὶ συνεχῶς περιφέρειαι, αἱ ὑπὲρ γῆν τῶν ὑπὸ γῆν, καὶ διὰ τοῦτο ἐξανάγκης πλείονα χρόνον ὁ ἥλιος ὢν ὑπὲρ γῆν ἢ ὑπὸ γῆν ποιεῖ τοῖς βορειοτέρας ἔχουσιν οἰκήσεις μείζονας τὰς ἡμέρας τῶν νυκτῶν, καὶ τοῦτο παντάπασιν ἀνίσως, τοῖς μὲν κατὰ τὸ εἰκὸς πλεῖον πάντως, τοῖς δὲ ἔλαττον. ὅταν δὲ ἐπὶ τὰ νότια τοῦ ἰσημερινοῦ περιοδεύῃ ὁ ἥλιος, ἀνάπαλιν γίνεται· τῇ γὰρ καθ᾽ ἡμᾶς οἰκουμένῃ ταύτῃ πρὸς τὰ βορειότερα μέρη ἐγκλινομένῃ, οἱ μετὰ τὸν ἰσημερινὸν ὡς πρὸς τὰ νότια αὐτῷ παράλληλοι κύκλοι τὰς ὑπὲρ γῆν περιφερείας αὐτῶν ἐλάττονας ἡμῖν ἔχουσιν ἢ τὰς ὑπὸ γῆν. καὶ οἱ καθ᾽ ἕκαστον κλίμα τῆς καθ᾽ ἡμᾶς τοιαύτης οἰκουμένης ὁρίζοντες ἐγκεκλιμένοι ὄντες πρὸς τὸν καθόλου τῆς ὀρθῆς σφαίρας ὁρίζοντα εἰς ἄνισα τέμνουσι τοὺς παραλλήλους κύκλους, ὡς πρὸς τὰ νότια τοῦ ἰσημερινοῦ. καὶ αἱ ὑπὲρ γῆν ὑπ᾽ αὐτῶν ἀποτεμνόμεναι περιφέρειαι ἥττονές εἰσι τῶν ὑπὸ γῆν καὶ κατ᾽ ὀλίγον συνεχῶς ἀεὶ ἥττονες, καθ᾽ ὅσον ἂν αἱ οἰκήσεις πρὸς τὰ βορειότερα μέρη καταλαμβάνωνται. καὶ τοῦτό ἐστιν, ὅπερ ἔφημεν, ἀνάπαλιν γίνεσθαι, ὅτι δὴ καὶ ἶσαι αἱ περιφέρειαι γραμμικαῖς δείξεσιν ἀποδείκνυνται, αἱ ἀπολαμβανόμεναι ὑπὲρ γῆν ἐπὶ τὰ νότια μέρη ταῖς ἀπολαμβανομέναις ὑπὸ γῆν ὑπὸ τῶν αὐτῶν ὁριζόντων ἐπὶ τὰ βόρεια μέρη. καὶ πάλιν ἶσαι αἱ ὑπὸ γῆν περιφέρειαι ἐπὶ τὰ νότια μέρη ταῖς ὑπὸ τῶν αὐτῶν ὁριζόντων ἀπολαμβανομέναις ὑπὲρ γῆν ἐπὶ τὰ βόρεια μέρη, ὡς ἐντεῦθεν ἐξανάγκης ἀκολουθεῖν ἐν τῇ περιόδῳ τοῦ ἡλίου γενομένου αὐτοῦ ἐπί τε τὰ βόρεια καὶ νότια καὶ ἁπτομένου τῶν ἑκατέρωθεν τοῦ ἰσημερινοῦ παραλλήλων αὐτῶν κύκλων τοῖς πρὸς τὰ βορειότερα οἰκοῦσι μέρη ἴσας γίνεσθαι τὰς

242 συνεχῶς post περιφέρειαι transp. C **259-61** καὶ πάλιν ... βόρεια μέρη om. C[ac] (homoeotel.)

equal to the nights, independently of whether the Sun is located to the north or to the south. For the locations starting close to the equator and extending to the north with increasing latitude, the days become gradually and continuously unequal relative to the nights.

As the Sun travels the northern [section] of the ecliptic, the days become longer than the nights and increase gradually for the zone further to the north. This is understood from the parallel circles where the Sun is positioned, because the arcs of the parallels above [the horizon] become continuously larger than the one below. Consequently, the Sun spends more time above the Earth than below and makes the days in the northern zones longer. They are always unequal, with the difference being sometimes larger and for others smaller. When the Sun travels to the south of the equator, the opposite takes place. Our ecumene is inclined to the north and the parallel circles in the south of the equator have smaller arcs above the Earth than below. The horizons for each zone in our ecumene are inclined relative to the horizon of *sphaera recta* and cut the parallel circles of the south in unequal parts, with the arc [of the parallel] above being smaller than the arc below, becoming continuously smaller for locations further to the north. As we said, the converse indeed happens in the south, because we can show geometrically that the arc of the parallel above the Earth in the southern hemisphere equals the arc of the parallel below the horizon for a zone in the northern hemisphere. Similarly, the arc below the horizon in the south equals the arc above the horizon in the north. It now follows that during the movement of the Sun, when it is located on parallels which are at equal [latitudes] from the equator to the north or to the south, those living in the north have the largest days equal

μείζονας τοῦ ἔτους ἡμέρας ταῖς μείζοσι τοῦ ἔτους νυξὶ καὶ τὰς ἥττονας νύκτας τοῦ ἔτους ἴσας ταῖς ἥττοσιν ἡμέραις τοῦ ἔτους. ὅσῳ δὲ μείζονες, γραμμικαῖς ὡσαύτως δείξεσιν ἀνευρὼν ὁ Πτολεμαῖος καθ᾽ ἕκαστον κλίμα ἐξ αὐτῆς τῆς ἀρχῆς τῆς ἐγκλίσεως εἰς τὰ βορειότερα καθεξῆς, ἔπειτα κανονικῶς ἐκτίθεται τὰς ἐν ἑκάστῳ κλίματι διαφοράς.

Ἐπεὶ γοῦν καθ᾽ ἕκαστον, ὡς λέγομεν, κλίμα ὡς πρὸς τὰ βορειότερα προϊόντες θεωροῦμεν τὰς ὑπὲρ γῆν περιφερείας τῶν κύκλων μείζους τῶν ὑπὸ γῆν καὶ συνεχῶς ἀεὶ μείζους, κἀντεῦθεν ἐξανάγκης καὶ αἱ παρ᾽ αὐταῖς ἡμέραι μείζους εἰσὶ τῶν νυκτῶν, ὅταν τῶν ὑπὲρ τὸν ἰσημερινὸν βορειοτέρων ὁ ἥλιος ἅπτηται. πάντως καὶ καθ᾽ ἕκαστον κλίμα ἀναλόγως ἀπ᾽ ἀνατολῆς εἰς δύσιν τοῦ ἡλίου πλεῖον διάστημα χρόνου ἐστὶν ἐν ταῖς ἡμέραις κατὰ τὰς τοιαύτας βορειοτέρας οἰκήσεις καὶ βραχύτερον ἐν ταῖς νυξί. καὶ ἀνάπαλιν, ὅταν εἰς τὰ κατὰ διάμετρον ᾖ τοῦ ζῳδιακοῦ μέρη καὶ τμήματα ὁ ἥλιος καὶ τῶν νοτιωτέρων τοῦ ἰσημερινοῦ παραλλήλων ἐφάπτηται ταῖς βορειοτέραις οἰκήσεσιν, ὀλιγοχρονιώτερον τὸ διάστημα τῆς ἡμέρας γίνεται καὶ πολυχρονιώτερον τὸ τῆς νυκτός. διὰ τοῦτο καὶ ἐπεὶ πᾶσαν ἡμέραν καὶ νύκτα εἰς ιβ ὥρας, ἃς καὶ καιρικὰς καλοῦσιν, ὁποῖαί ποτ᾽ ἂν ὦσιν, οἱ ἀστρονόμοι διαιροῦσι—καὶ εἰσὶ λοιπὸν αἱ τοιαῦται ὧραι, νῦν μὲν ἰσημεριναί, νῦν δὲ μείζους τῶν ἰσημερινῶν, νῦν δὲ ἐλάττους—, οὐ πᾶσαι ὧραι πάσης ἡμέρας ἰσαρίθμοις χρόνοις, ἤτοι τμήμασι τοῦ ἰσημερινοῦ, συγκαταλογίζονται καὶ συγκαταριθμοῦνται ἐν τῇ περιόδῳ καὶ περιστροφῇ τοῦ παντός, ἀλλὰ διαφόροις καὶ ἀνίσοις. καὶ εἰς μὲν ἑκάστην ἰσημερινὴν ὥραν συγκαταλογίζονται χρόνοι, ἤτοι τμήματα τοῦ ἰσημερινοῦ, ιε, ἐπειδὴ καὶ κδ ἐστὶ τὰ ιε τμήματα τῶν τξ, εἰς ἃ διαιρεῖται ὁ ἰσημερινός, καθὼς καὶ ἄλλος ὁστισοῦν κύκλος—καὶ ἡ ὥρα γὰρ ἡ μία κδ´ ἐστὶ τοῦ ἡμερονυκτίου. πάλιν δὲ καὶ εἰς ἑκάστην καιρικὴν ὥραν πρὸς τὸ διάστημα τοῦ χρόνου τῆς τηνικαῦτα ἡμέρας ἢ νυκτὸς ἀναλόγως καταριθμοῦνται οἱ χρόνοι, ἤτοι τὰ τμήματα τοῦ ἰσημερινοῦ, ἐν τῇ περιστροφῇ τοῦ παντός, εἴτε ις ἢ ιζ ἢ ιη, ὅταν ἐπὶ τὸ πλεῖον τῆς ἰσημερίας θεωρῶνται, καὶ τὸ ἀνάπαλιν ιδ ιγ, ὅταν ἐπὶ τὸ ἔλαττον.

267 Cf. Ptol. *Alm.* 2.3 (vol. 1.1, p. 93-96 Heiberg)

275 ἐστὶν ante χρόνου habet C **280** τὸ om. C **294** πλεῖστον C

to the largest nights in the south and the largest nights of the north equal to the largest days in the south. Ptolemy proved geometrically the increase as a function of the distance of each zone from the equator and presented the difference in tables for each zone.

As we advance to regions to the north, we find the arcs of the parallel circles above the Earth being larger than those below and always growing. It follows that the days for those regions are longer than the nights when the Sun is located to the north of the equator. In any case, for each zone the time from sunrise to sunset is longer in the northern zones with the nights shorter. The opposite happens when the Sun is located at the diametrically opposite regions of the ecliptic at the parallels of the southern hemisphere. Then in the northern regions the length of the day becomes shorter and the night longer. No matter how different they may be, the astronomers divide each day and night in 12 hours which they call seasonal. These hours are sometimes equal to equinoctial hours, sometimes longer and at other times shorter. All hours of a day are not considered and counted to correspond to equal intervals of the equator or of the period produced by the revolution of the celestial sphere, but [they are] different and unequal. To each equinoctial hour correspond 15° of the equator, because 24 fractions make 360°, into which we divide the equator or any other circle. Each [equinoctial] hour is 1/24 of a *nychthemeron* (solar day). Again, a seasonal hour of a day (or a night) is expressed in equinoctial hours, which could be 16°, 17° or 18°, when the days are longer than the nights; or conversely, 14° or 13° when they are shorter.

Διὰ ταῦτα δὴ κανόνες πεποίηνται τῷ Πτολεμαίῳ καὶ Θέωνι καθ' ἕκαστον κλίμα, ἐπειδήπερ καὶ καθ' ἕκαστον κλίμα διαφορὰς ἔφημεν ἔχειν πρὸς τὴν ἀπόστασιν δηλονότι τοῦ ἰσημερινοῦ τὰς ἡμέρας τὲ καὶ τὰς νύκτας. οἱ τοιοῦτοι οὖν κανόνες καθ' ἕκαστον κλίμα καὶ καθ' ἕκαστον δωδεκατημόριον τοῦ ζωδιακοῦ ἐκτέθεινται δηλοποιοῦντες ὁπόσων χρόνων ἐστὶν ἡ καιρικὴ ὥρα τῆς ἡμέρας ἐν ἐκείνῳ τῷ κλίματι καὶ τῷ δωδεκατημορίῳ. καὶ οὗτοί εἰσιν οἱ κανόνες, περὶ ὧν πρὸ ὀλίγου ἐλέγομεν, ὅτι ἐν τῷ πρώτῳ σελιδίῳ ἐκτίθενται οἱ κοινοὶ ἀριθμοὶ τῶν τριῶν δωδεκατημορίων, ὧν ἐστιν ὁ κανών. τῶν δὲ ἑξῆς δύο σελιδίων τῶν ὑφ' ἕκαστον δωδεκατημόριον, τὸ μὲν ἓν ἔχει τὰς ἀναφορὰς τοῦ ἰσημερινοῦ συντιθεμένας ἑξῆς καθ' ἑκάστην μοῖραν ἀπὸ τῆς ἀρχῆς τοῦ Κριοῦ, τὸ δὲ ἕτερον ἐζητοῦμεν τὶ βούλεται. καὶ ἔστιν ἡ ἔκθεσις τῶν ἐν αὐτῷ ἀριθμῶν δηλοποιοῦσα αὐτὸ τοῦτο περὶ οὗ νῦν ἡμῖν ὁ λόγος, πόσων χρόνων ἐστὶ δηλονότι ἡ καιρικὴ ὥρα ἐπὶ τοῦδέ τινος τοῦ κλίματος τῆς ἡμέρας ἐκείνης, ἐν ᾗ ὁ ἥλιος εὑρίσκεται, ἐν τῇδέ τινι τῇ μοίρᾳ τοῦδέ τινος ζωδίου.

Ὅταν οὖν βουλώμεθα τὰς καιρικὰς ὥρας τὰς δεδομένας ἡμῖν ἐφ' ἑκάστης ψηφοφορίας—τοῦτο γὰρ ἦν ἡμῖν ἡ πρόθεσις ἐν τοῖς νῦν λόγοις—εἰς ἰσημερινὰς μεταβάλλειν ἢ τὸ ἀνάπαλιν, τὰς δεδομένας ἡμῖν ἰσημερινὰς ὥρας ἐπὶ τῆς προκειμένης ψηφοφορίας μεταβάλλειν εἰς καιρικὰς ἔχοντες δεδομένον τὸ κλίμα, ἐφ' οὗ ψηφοφοροῦμεν, καὶ τὴν μοῖραν τοῦ δωδεκατημορίου, ἐν ᾧ εὑρίσκεται ὁ ἥλιος, ὅταν ποιῶμεν τὴν ψηφοφορίαν, θεωρῶμεν τὸν κανόνα τοῦ κλίματος καὶ τοῦ δωδεκατημορίου καὶ εὑρίσκομεν ὅσος παράκειται ἀριθμὸς ὡριαίων χρόνων τῇ δεδομένῃ μοίρᾳ τοῦ δωδεκατημορίου, ἣν ἐπέχει τηνικαῦτα ὁ ἥλιος ἐν τῷ τρίτῳ σελιδίῳ τοῦ αὐτοῦ δωδεκατημορίου, ὃ οὕτω πως ἐπιγράφεται ὡριαίων χρόνων, καὶ ὁ τοιοῦτος ἀριθμός ἐστι τῶν χρόνων τῆς κατ' ἐκείνην τὴν ἡμέραν καιρικῆς ὥρας. καὶ ἂν ταύτας τὰς καιρικὰς ὥρας, αἵ εἰσι πάντοτε ἑκάστης ἡμέρας ιβ, βουλώμεθα μεταβαλεῖν εἰς ἰσημερινάς, δωδεκαπλασιάζομεν τὸν ἀριθμὸν τῶν

296 Cf. Ptol. *Alm.* 2.8

299 τὰς om. C **303-4** τῶν δωδεκατημορίων τῶν τριῶν C **308-9** ἡμῖν post ὁ λόγος transp. C **311-12** ἡ δευτέρα τῶν ὡρῶν διάκρισις sch. in mg. C (Chort.)

For all these, Ptolemy and Theon prepared tables for each zone, because the days and nights are different depending on their distance from the equator. The tables are presented for each zone and for each sign of the zodiac, expressing the length of each seasonal hour [in terms of equinoctial hours] of the day at that specific location and for that sign. These are the tables for which we mentioned earlier that in the first column are written the degrees that are common for the three signs of the table. The next two columns under each sign of the zodiac give the following: the first one begins at Aries and tabulates for each degree the rising times. The other column tabulates what we wish to find, that is how long is the seasonal hour [in units of equinoctial hours] at that location and for that day, when the Sun is located at a specific degree within a sign of the zodiac.

When in a calculation we are given seasonal hours and wish to convert them to equinoctial — this is indeed our intention — or the opposite, given equinoctial hours of a previous calculation to convert them to seasonal, provided we are given the zone of the calculation and the degree of the sign where the Sun is located, we look at the table for the zone and the degrees within the sign where the Sun is located. Then we find in the third column, which has the title hourly-times, the value of hourly-times; this is the length of a seasonal hour at that moment. If now we wish to convert these seasonal hours of a day, which are always 12, to equinoctial, we multiply the number of hourly-times by 12

ὡριαίων χρόνων τὸν εὑρισκόμενον τηνικαῦτα καὶ τὸν συναγόμενον ἐκ τοῦ τοιούτου πολλαπλασιασμοῦ ὅλον ἀριθμὸν μερίζομεν ἐπὶ τὸν ιε. καὶ ὅσος ἐστὶν ὁ γινόμενος ἐκ τοῦ μερισμοῦ ἀριθμός, τοσαῦταί εἰσιν αἱ ὧραι ἰσημεριναὶ τῆς ἡμέρας ἐκείνης. ἐὰν δὲ ἀπολειφθῶσι καί τινες χρόνοι μὴ ἀπαρτίζοντες τὸν πεντεκαίδεκα ἀριθμόν, ὅσον μέρος ἐστὶ τῶν πεντεκαίδεκα οἱ τοιοῦτοι χρόνοι, ἢ τρίτον τυχὸν ἢ ἥμισυ ἢ δίτριτον καὶ ἄλλο ὁτιοῦν, καὶ τοῦτο τὸ μέρος προσθήσομεν ταῖς ἀνευρεθείσαις ὥραις ἰσημεριναῖς.

Οἷον ὑποκείσθω εἶναι τοὺς ὡριαίους χρόνους ἑπτακαίδεκα. αἱ ιβ οὖν καιρικαὶ ὧραι πολλαπλασιαζόμεναι ἐπὶ τὸν ἑπτακαίδεκα ἀριθμὸν ποιοῦσιν ἀριθμὸν σδ. τούτου οὖν μεριζομένου παρὰ τὸν ιε ἀναφαίνεται ἐκ τοῦ μερισμοῦ ἀριθμὸς ιγ. ἐναπολιμπάνονται καὶ θ χρόνοι, ἤτοι δίτριτον ὥρας ἰσημερινῆς ἔγγιστα. καὶ φήσομεν τηνικαῦτα εἶναι τὴν ἡμέραν ιγ ὡρῶν ἰσημερινῶν καὶ διτρίτου ὥρας. καὶ τοίνυν ὅταν ἔχωμεν ἐπί τινος ψηφοφορίας ὥρας τινὰς δεδομένας ἢ πρὸ μεσημβρίας ἢ μετὰ μεσημβρίαν, ἢ τρεῖς ἢ τέτταρας, ἂν μὲν ὦσι δεδομέναι καιρικαὶ τυχὸν τρεῖς καὶ βουλώμεθα ταύτας ποιῆσαι ἰσημερινάς, τριπλασιάζομεν τοὺς ὡριαίους χρόνους, οἵτινες καί εἰσιν ἐν τῷ κανόνι κατά τε τὸ κλίμα, ἐν ᾧ ψηφοφοροῦμεν, καὶ κατὰ τὴν μοῖραν τοῦ δωδεκατημορίου, ἐν ᾗ εὑρίσκεται ὁ ἥλιος, καὶ τὸν γινόμενον ἐκ τοῦ τοιούτου πολλαπλασιασμοῦ ἀριθμὸν μερίζομεν παρὰ τὸν ιε. καὶ ὁ γινόμενος ἀριθμὸς ἐκ τοῦ μερισμοῦ, ὡς εἴρηται, ἔσται ὁ ἀριθμὸς τῶν τηνικαῦτα ἰσημερινῶν ὡρῶν, καὶ μέρος τι ὥρας ἐνίοτε ἐπὶ ταῖς ὥραις, ὡς καὶ τοῦτ᾽ ἔφημεν.

Ἐὰν δὲ αἱ δεδομέναι ὧραι τυχὸν αἱ τρεῖς εἰσιν ἰσημεριναὶ καὶ βουλώμεθα εὑρεῖν πόσαι αὗταί εἰσι καιρικαί, τριπλασιάζομεν τοὺς ιε χρόνους τῆς ἰσημερινῆς ὥρας, καὶ τὸν συναγόμενον ἐκ τοῦ τοιούτου πολλαπλασιασμοῦ ἀριθμὸν μερίζομεν παρὰ τοὺς ὡριαίους χρόνους, ὅσοι ποτ᾽ ἂν ὦσιν, ἐν τῷ κανόνι τοῦ κλίματος ἐν ᾧ ψηφοφοροῦμεν, κατὰ τὴν μοῖραν τοῦ δωδεκατημορίου, εἰς ἣν εὑρίσκεται τηνικαῦτα ὁ ἥλιος. καὶ ὅσος ἐστὶν ὁ ἐκ τοῦ μερισμοῦ ἀριθμός, τοσαῦταί εἰσι καιρικαὶ ὧραι, τυχὸν δὲ ἐνίοτε καὶ μέρος ὥρας, αἱ δεδομέναι τρεῖς ἰσημεριναί. καὶ οὕτω μὲν εὑρίσκονται ἐφ᾽ ἡστινοσοῦν ἡμέρας αἱ δεδομέναι ἰσημεριναὶ ὧραι πόσαι εἰσὶ καιρικαί, καὶ ἀνάπαλιν αἱ δεδομέναι καιρικαὶ ὧραι

and the number obtained from the multiplication we divide by 15. The number resulting from the division denotes the number of equinoctial hours in that day. In case there is a remainder which is not an integer, but is fraction of these 15 times, that is 1/3 or 1/2 or 2/3 or something else, then we add the fraction to the equinoctial hours that we found.

Let us assume the hourly-time to be 17. When we multiply 12 seasonal hours by 17 we obtain 204 and this divided by 15 produces the number 13 with the remainder of 9, which is approximately 2/3 of an equinoctial hour. We say that the particular day has 13 equinoctial hours and 2/3 of an hour. Thus, when in a calculation we are given some hours, either before or after midday, which are either three or four, let us say they happen to be three seasonal hours and we wish to convert them to equinoctial, we multiply the number of hourly-times that is in the table (for the specific zone and for the sign of the zodiac where the Sun is located) by three, and the product of the multiplication we divide by 15. As we said, the number that comes out from the division is the number of equinoctial hours and sometimes contains a fraction of an hour.

If the three hours are equinoctial and we wish to find how many seasonal hours correspond to them, we multiply three by the 15° equinoctial hours and the result of the product is divided by the hourly-times, as many as they appear in the table of the zone where we are calculating and for the degree of the sign in which the Sun is located. The number that results from the division is equal to the seasonal hours corresponding to three equinoctial hours. This way one finds for a specific day how many seasonal hours correspond to equinoctial hours and vice-versa: given the seasonal hours, how many are the corresponding equinoctial.

πόσαι εἰσὶν ἰσημεριναί.

Τὰ αὐτὰ δὲ καὶ παραπλήσια ῥᾴδιόν ἐστιν ἐνεργεῖν καὶ ἐπὶ τῆς νυκτὸς τῆς μετὰ τὴν τοιαύτην ἡμέραν καὶ ἀνευρίσκειν οἵτινες καὶ ἐν τῇ τοιαύτῃ νυκτὶ ἐν ψηφοφορίᾳ τινὶ ἀναδίδονται ὧραι εἴτε καιρικαὶ εἴτε ἰσημεριναί. οἱ γὰρ ἐν τῷ αὐτῷ κλίματι τῇ κατὰ διάμετρον μοίρᾳ τῆς μοίρας, εἰς ἣν εὑρίσκεται τηνικαῦτα ὁ ἥλιος, παρακείμενοι ἐν τῷ αὐτῷ κλίματι ὡριαῖοι χρόνοι, οἱ αὐτοί εἰσι καὶ ὡριαῖοι χρόνοι τῆς δηλωθείσης νυκτός, οἵτινες ἐπιβάλλουσι δηλαδὴ μιᾷ ὥρᾳ καιρικῇ τῶν ιβ τῆς τοιαύτης νυκτός. τούτου δὲ οὕτω λαμβανομένου, ῥᾴδιον τὰ αὐτὰ ποιεῖν, ὡς εἴρηται, καὶ ἐπὶ τῶν νυκτερινῶν ὡρῶν καὶ μεταποιεῖν τὰς καιρικὰς εἰς ἰσημερινὰς καὶ τὰς ἰσημερινὰς εἰς καιρικάς. ὁ μέντοι Πτολεμαῖος καὶ ἄλλως ῥᾷστα παραδίδωσιν ἀνευρίσκειν ἐκ τοῦ κανόνος διὰ τῶν ἀναφορῶν μόνων ὁπόσαι εἰσὶν αἱ ἰσημεριναὶ ὧραι ἑκάστης ἡμέρας καὶ νυκτός, καὶ πόσων χρόνων ἐστὶν ἡ καιρικὴ ὥρα ἑκάστης ἡμέρας καὶ νυκτός. λαμβάνοντες γὰρ ἐπὶ τοῦ κλίματος, ἐφ᾽ οὗ ψηφοφοροῦμεν, πόσος ἐστὶν ἀριθμὸς ἀναφορῶν ἀπὸ τῆς μοίρας τοῦ δωδεκατημορίου, καθ᾽ ἣν εὑρίσκεται τηνικαῦτα ὁ ἥλιος, μέχρι τῆς κατὰ διάμετρον αὐτῇ μοίρας, τὸν τοιοῦτον ἀριθμὸν μερίζομεν παρὰ τὸν ιε. καὶ ὅσος ἐστὶν ὁ γινόμενος ἐκ τοῦ μερισμοῦ ἀριθμός, τοσαῦταί εἰσιν ἐν ἐκείνῃ τῇ ἡμέρᾳ αἱ ἰσημεριναὶ ὧραι. τοῦ τοιούτου πάλιν ἀριθμοῦ τῶν ἀναφορῶν τὸ ιβ λαμβάνοντες ἔχομεν πάντως αὐτὸ τὸ ιβ ὡς ὡριαίους χρόνους τῆς κατ᾽ ἐκείνην τὴν ἡμέραν καιρικῆς ὥρας. πάλιν τὸν ἀριθμὸν τῶν ἀναφορῶν τὸν ἀπὸ τῆς διαμετρούσης, ὡς εἴρηται, μοίρας τῇ μοίρᾳ καθ᾽ ἣν τηνικαῦτα ὁ ἥλιος εὑρίσκεται μέχρις αὐτῆς τῆς μοίρας, καθ᾽ ἣν εὑρίσκεται ὁ ἥλιος, μερίζοντες ἐπὶ τὸν ιε, ὅσος ἐστὶν ἐκ τοῦ μερισμοῦ ἀριθμός, τοσαῦταί εἰσιν αἱ ἰσημεριναὶ ὧραι τῆς μετὰ τὴν ἡμέραν ἐκείνην νυκτός. καὶ αὐτῶν τῶν ἀναφορῶν αὖθις τὸ ιβ´ ἔστιν ὁ ἀριθμὸς τῶν ὡριαίων χρόνων τῆς καιρικῆς ὥρας τῆς κατ᾽ ἐκείνην νυκτός. ἀλλ᾽ ἱκανῶς ἡμῖν καὶ τὰ περὶ τοῦ δευτέρου τρόπου τοῦ περὶ τῆς διαφορᾶς τῶν ὡρῶν διευκρίνηται.

371 Cf. Ptol. *Alm.* 2.9 (vol. 1.1, p. 142 Heiberg)

367-68 οἵτινες ... νυκτός om. C[ac] (homoeotel.) **379-80** τῶν ἀναφορῶν post τὸ ιβ λαμβάνοντες transp. C

When in a computation we deal with seasonal or equinoctial hours of a night that follows a day, the same or very similar methods can be easily used. For a specific zone we find the hourly-times for the night under consideration next to the degrees diametrically opposite to [the location of] the Sun at that time. The hourly-times define one of the 12 seasonal hours of the night. Once we obtain this number it is easy to convert seasonal to equinoctial and equinoctial to seasonal. At any rate, Ptolemy describes clearly how to find from the table of ascensions how many equinoctial hours are contained in each day and night and the length of each seasonal hour. He does this for each day and night [in the following way]. [In the table] of the zone for a calculation we compute the increase in the entries for ascensions from the location, where the Sun is located, up to the diametrically opposite location. We divide the increase by 15 and the result gives the length of the day in equinoctial hours. Dividing next the equinoctial hours by 12 we obtain the hourly-times of that day. Finally, counting the Sun risings [in the same sense of rotation], from the diametrically opposite degree up to the degree where the Sun is located now and, dividing [the increase] by 15, we obtain the equinoctial hours for the night which follows that day. One twelfth [of the increase] of the rising-times is the value of hourly-times for the seasonal hour of that night. Thus, we described with sufficient clarity the second aspect of differences between the hours.

Καὶ δὴ λοιπὸν καὶ περὶ τοῦ τρίτου ῥητέον ἡμῖν ἔκθεσις τῶν κανόνων, καὶ παρὰ τῷ Πτολεμαίῳ καὶ παρὰ τῷ Θέωνι, ὡς περὶ ὁμαλῶν νυχθημέρων ἐστὶ καὶ ἴσας ἐχόντων ἰσημερινὰς ὥρας κδ· ἀλλὰ μὴν τὰ νυχθήμερα οὐκ εἰσὶν ἀεὶ τὰ αὐτά, ἀλλὰ διαπαντός εἰσιν ἀνώμαλα. προδιευκρινητέον δὲ τί ἐστι τὸ νυχθήμερον καὶ ποῖόν ἐστι τὸ ὁμαλὸν καὶ ποῖον τὸ ἀνώμαλον· νυχθήμερον τοίνυν ἐστὶν ὅταν ἀπὸ τοῦ αὐτοῦ τοῦδέ τινος τόπου τοῦ ὁρίζοντος τυχὸν ἢ τοῦ μεσημβρινοῦ ἢ ἄλλου του συγκινηθεὶς καὶ συμπεριοδεύσας τῷ παντὶ καὶ οὐρανίῳ σώματι ὁ ἥλιος ἐπ' αὐτοῦ πάλιν γένηται ἢ τοῦ ὁρίζοντος δηλονότι ἢ τοῦ μεσημβρινοῦ—ἐπεὶ γοῦν ὁ ἥλιος καὶ ἰδίαν ποιεῖται κίνησιν εἰς τὰ ἐναντία τῷ παντὶ καὶ ἔστιν ἡ κατ' αὐτὸν τοιαύτη κίνησις ἡμερησία οὐδὲν μοίρας νθ ἔγγιστα ἐπὶ τοῦ μέσου τοῦ ζῳδιακοῦ—ὁμαλῶς νοοῖτ' ἂν ὁμαλὸν ἡμερονύκτιον ὁ χρόνος τῆς τοῦ παντὸς περιστροφῆς, ἤτοι τῶν τξ τμημάτων, ἃ καὶ χρόνοι καλοῦνται, ἡ περίοδος τοῦ ἰσημερινοῦ μεγίστου κύκλου, ὥσπερ καὶ ἄλλου του τῶν παραλλήλων, καὶ ἔτι ὅσον τμῆμα τοῦ ἰσημερινοῦ συναναφέρεται τῇ εἰρημένῃ μεταβάσει καὶ προχωρήσει τοῦ ἡλίου τῇ οὐδὲν νθ ἔγγιστα καὶ ποιεῖ ὁτιποτοῦν χρονικὸν ὥρας μέρος διάστημα.

Ἐπεὶ δὲ τὸ τοιοῦτο προχωρητικὸν καὶ μεταβατικὸν ἐπικίνημα τοῦ ἡλίου, ὡς πολλάκις ἔφημεν, οὐκ ἔστιν ἴσον οὐδὲ ὁμαλόν, ἀλλ' ὁτὲ μὲν πλεῖον, ὁτὲ δὲ ἔλαττον διὰ τὴν ἐκκεντρότητα τοῦ κύκλου, ἐφ' οὗ φέρεται ὁ ἥλιος, τὴν πρὸς τὸν ζῳδιακὸν καὶ τὴν διαφορὰν τῶν ἐν τοῖς ἀπογείοις αὐτοῦ κινημάτων καὶ τῶν ἐν τοῖς περιγείοις, περὶ οὗ φθάσαντες πλατυκώτερον διεξήλθομεν, διὰ τοῦτο καὶ τὰ νυχθήμερα πάντα τοῦ ἔτους εἰσὶν ἀνώμαλα καὶ τὸ χρονικὸν αὐτῶν διάστημα ὁτὲ μὲν πλεῖον ὁτὲ δὲ ἔλαττον. συνεπινοεῖται δὲ καὶ τοῦτο εἰς τὴν αἰτίαν τῆς ἀνωμαλίας τῶν νυχθημέρων, ὅτι αὐτὰ δὴ τὰ ἡμερήσια ἐπικινήματα, ἃ καὶ ἀνώμαλα καὶ ἄνισα ἔφημεν, τοῦ ἡλίου, κἂν ἴσα ἦν, οὐ τοῖς αὐτοῖς τμήμασιν, ἤτοι χρόνοις, τοῦ ἰσημερινοῦ συναναφέρονται. προέφημεν γὰρ καὶ τοῦτο, ὡς τὰ τοῦ ζῳδιακοῦ τμήματα οὐκ ἐν ἴσοις χρόνοις τοῦ ἰσημερινοῦ συναναφέρεται, κἂν καὶ ἴσα ᾖ ἐπί τε τῆς ὀρθῆς σφαίρας καὶ πάντων τῶν κλιμάτων. διὰ ταύτας γοῦν τὰς δύο αἰτίας, τήν τε

394 ἐστι om. C **394-95** σημείωσαι τοῦτο περὶ τοῦ νυχθημέρου πάνυ καλῶς εἰρημένον sch. in mg. C

Finally we come to the third topic. In the exposition of the tables, Ptolemy and Theon [assumed] the solar days to be regular and have 24 equinoctial hours. However, the solar days are not always equal but in many respects anomalous. From the start we must investigate how the solar day (*nychthemeron*) is defined, which solar days are regular and which are anomalous. When the Sun, starting at an arbitrary place of the horizon, or at the meridian, or somewhere else, rotates with everything else on the celestial sphere and returns to the same location either on the horizon or the meridian, at the same time it performs its own revolution through the middle of the zodiac, but in opposite sense, with its daily angular velocity being approximately 0° and 59′. In our approach so far, we visualized the rotation of any parallel and of the equator, which contain the 360 subdivisions, to be uniform and to include the second rotation of 0° 59′ for any segment of the equator and for any fraction of an hour.

The additional revolution of the Sun, as we said many times, is neither regular nor uniform but sometimes faster and other times slower, due to the eccentricity of the circle on which the Sun rotates relative to the zodiac and produces the difference of its motion at the apogees and perigees, as we described extensively at the appropriate place. For this reason, throughout a year all solar days are irregular, their length being sometimes longer and at other times shorter. The reason for this is the anomalous motion of the Sun on its trajectory. But even if the rotation of the Sun on the ecliptic were uniform, equal arcs of the ecliptic do not correspond to equal time intervals of the equator. We also mentioned that even when arcs of the ecliptic correspond to equal segments of *sphaera recta* or of any zone, they do not rise simultaneously

ἀνωμαλίαν τοῦ ἡλιακοῦ ἡμερησίου ἐπικινήματος καὶ τὴν ἀνισότητα τῶν συναναφερομένων αὐτῷ χρόνων τοῦ ἰσημερινοῦ μεγίστου κύκλου, ἀνώμαλα καὶ ἄνισα τὰ νυχθήμερα καταλαμβάνονται καὶ οὐ τῶν αὐτῶν ἰσημερινῶν ὡρῶν, μᾶλλον δὲ οὐ τῶν ἴσων χρόνων ἐν τῇ περιστροφῇ τοῦ παντὸς καὶ ἀποκαταστάσει αὐτῶν. τοῦτο μέν γε τὸ διάφορον καὶ ἄνισον αὐτῶν ἐν ὀλίγῳ μὲν ἀριθμῷ νυχθημέρων βραχύτατόν ἐστι καὶ σχεδὸν ἀνεπιλόγιστον καὶ ἄρρητον καὶ οὐκ ἄξιον λόγου τινός, κατ' ὀλίγον δὲ προστιθέμενον ὑποσημαίνει καὶ διαφορὰν ἐμποιεῖ λόγου καὶ διακρίσεως ἀξίαν.

Τοιγαροῦν ἐν ταῖς ψηφοφορίαις τῶν ἀναφαινομένων καὶ δεδομένων νυχθημέρων, ὡς ὁμαλῶν καταλογιζομένων ἐκ τῶν κανόνων, χρεία ἀπαραίτητος ἀποκαθιστάνειν αὐτὰ ὡς καιρικὰ καὶ ἀνώμαλα, ἃ καὶ ἁπλᾶ καλοῦσιν οἱ ἀστρονόμοι, καὶ ἀνευρίσκειν καὶ διευκρινεῖν τὸ διάφορον, ὃ κέκτηνται πρὸς τὰ ὁμαλά. τοῦτο τοίνυν ἐσπουδάσθη Πτολεμαίῳ καὶ τοῖς ἀστρονόμοις σὺν τοῖς ἄλλοις ἀκριβέστατα καὶ ἀποδεικτικώτατα καὶ ἀναντίρρητα. καὶ εὕρηται καὶ μεθοδεύεται οὕτως. ὅταν γὰρ ἐν ταῖς ἑκάστοτε ψηφοφορίαις τὰ ἀναδιδόμενα νυχθήμερα καιρικὰ καὶ ἁπλᾶ ὄντα βουλώμεθα ἀποκαθιστάνειν καὶ καταλογίζεσθαι ὡς ὁμαλά, θεωρῶμεν τίνα τόπον ἐπεῖχεν ὁ ἥλιος ἐπὶ τοῦ ζωδιακοῦ, ἤγουν τίνα μοῖραν τίνος δωδεκατημορίου ὁμαλῶς, καὶ τίνα ἀνωμάλως κατὰ τὸν προειρημένον χρόνον, ὃν ὑποτιθέμεθα ἐν τῇ παρούσῃ πραγματείᾳ ἀρχὴν τῆς αὐτοῦ κινήσεως. ἐπεῖχε δέ, ὡς ἔφημεν, ὁμαλῶς Ζυγοῦ μοίρας ις νη΄, ἀνωμάλως δὲ ιε στ΄ νδ΄΄. εἶτα θεωροῦμεν καὶ τίνα τόπον, ἤτοι μοῖραν δωδεκατημορίου τινός, ἐπέχει κατ' αὐτὸν τὸν καιρόν, καθ' ὃν ποιούμεθα τὴν ψηφοφορίαν, ὁμαλῶς τε καὶ ἀνωμάλως. εἶθ' οὕτως ἐπιλογιζόμεθα τὴν διάστασιν τῶν μεταξὺ τόπων ἀμφοτέρων, τῶν τε δηλονότι ἀπὸ τῆς πρώτης ὁμαλῆς ἐποχῆς μέχρι τῆς ὁμαλῆς αὖθις ἐποχῆς, καθ' ὃν καιρὸν ψηφοφοροῦμεν, καὶ τὴν ἀπὸ τῆς πρώτης φαινομένης καὶ ἀνωμάλου ἐποχῆς μέχρι πάλιν τῆς ἀνωμάλου ἐποχῆς τῆς κατὰ τὸν καιρόν, ὃν ψηφοφοροῦμεν. εἶτα εἰσάγομεν τοὺς τόπους τῆς διαστάσεως τῆς ἀπὸ τῆς πρώτης ἀνωμάλου καὶ φαινομένης ἐποχῆς εἰς τὴν δευτέραν ἀνώμαλον καὶ φαινομένην ἐποχὴν ἐν τῷ κανόνι τῶν

444 τόπον : τρόπον C[ac]

on the equator. For these two reasons, the anomaly in the daily rotation and the difference in the Sun risings on the equator, the solar days are anomalous and contain a different number of equinoctial hours. To say it better, a complete revolution of the universe and its return to the same configuration is not [always] completed at equal time intervals. This difference and discrepancy over an interval of a few solar days is extremely small, incalculable, undetermined and not worth mentioning. However, adding the small parts becomes noticeable and produces a difference worth mentioning.

Indeed, the solar days (*nychthemera*) that are included and appear in the calculations were computed from the tables of the uniform motion. It is necessary to convert them to seasonal and anomalous, which the astronomers call simple, and in addition to elucidate the difference between them. Ptolemy and other astronomers studied this topic very accurately and found the following reliable method. Whenever in a calculation the solar days are seasonal and simple and we wish to convert them and replace them with mean days, we consider the longitude of the Sun on the ecliptic, that is the degree within a sign, at the initial time of its uniform and its anomalous rotation. In this treatise the predetermined position at the initial time for the uniform motion[3] is at Libra 16° 58′ and for the anomalous at Libra 15°, 6′ and 54″.

Next we compute the longitudes that the Sun occupies at the time of the calculation for both the uniform and the anomalous rotations. Then we compute the increase of longitude for the uniform motion from the initial position to the present time and also the increase from the initial anomalous position to the present time. We introduce the increases of longitudes into the

[3] These are initial values of the calendar proposed by Metochites.

ἀναφορῶν τῆς ὀρθῆς σφαίρας. καὶ καταλογιζόμεθα ὅσαι ἀναφοραί εἰσι προστεθειμέναι καὶ καταγεγραμμέναι ἐν τῷ κανόνι ἀπὸ τοῦ πρώτου τόπου τῆς διαστάσεως τῆς ἀνωμάλου μέχρι τοῦ δευτέρου τόπου τῆς αὐτῆς ἀνωμάλου διαστάσεως. ὡσαύτως ποιοῦμεν καὶ ἐπὶ τῆς ὁμαλῆς διαστάσεως. ἔπειτα ἀναλογιζόμεθα πόση ἐστὶν ἡ διαφορὰ τούτων πρὸς τὴν ὁμαλὴν διάστασιν, καὶ ὅση ἐστὶν ἡ διαφορά, λογιζόμεθα αὐτὴν ὡς χρόνους ὥρας. καὶ ποιοῦμεν αὐτὰς μόριον ὥρας κατὰ ἀναλογίαν ὡς προσήκει, ὡς τῶν ιε ἀναφορῶν, ἤτοι χρόνων, συμπληρούντων ὡριαῖον χρόνον, καὶ τὸ τοιοῦτο μόριον τῆς ὥρας, ὅπερ ἔφημεν, γίνεσθαι ἀπὸ τῆς διαφορᾶς τῶν εἰρημένων ἀναφορῶν.

Ὅταν μὲν ἀποκαθιστάνειν βουλώμεθα τὰ καιρικὰ καὶ ἁπλᾶ καὶ ἀνώμαλα νυχθήμερα εἰς ὁμαλά, ἐὰν μὲν πλεονάζῃ ἐν ταῖς ἀναφοραῖς τῆς διαστάσεως τῶν ἀναμεταξὺ ἀνωμάλων ἐποχῶν, προστιθέαμεν ταῖς ὥραις, ἃς καταλογιζόμεθα ἀπὸ τῶν κανόνων ἐν ταῖς ψηφοφορίαις, καὶ ποιοῦμεν τὰς ἀναδοθείσας καιρικῶν νυχθημέρων ὥρας ὁμαλῶν νυχθημέρων τοσαύτας καὶ μόριον ὥρας· ἐὰν δὲ αὐτὸ τὸ διάφορον πλεονάζῃ ἐν τῇ μεταξὺ διαστάσει τῶν ὁμαλῶν ἐποχῶν, τὸ γινόμενον μέρος ὥρας ἀφαιροῦμεν ἀπὸ τῶν ὡρῶν, ἃς καταλογιζόμεθα ἀπὸ τῶν κανόνων ἐν ταῖς ψηφοφορίαις. καὶ ποιοῦμεν τὰς ἀναδοθείσας καιρικῶν νυχθημέρων τόσας ὥρας ὁμαλῶν νυχθημέρων, τοσαύτας ὥρας παρά τι μέρος ὥρας.

Ὅταν δὲ τὰ ὁμαλὰ νυχθήμερα ἀποκαθιστάνειν βουλώμεθα εἰς καιρικὰ καὶ ἁπλᾶ καὶ ἀνώμαλα—καὶ τοῦτο γὰρ ἔστιν ὅτε γενήσεται—, ἀνάπαλιν ποιήσομεν. τοῦ μὲν εἰρημένου διαφόρου πλεονάζοντος ἐν τῇ διαστάσει τῶν δύο ὁμαλῶν ἐποχῶν, αὐτὸ τοῦτο τὸ διάφορον ἀφαιρήσομεν ἀπὸ τῶν ὡρῶν τῶν εὑρισκομένων ἐν ταῖς ψηφοφορίαις ἀπὸ τῶν κανόνων, καὶ ποιοῦμεν τὰ ὁμαλὰ νυχθήμερα καιρικά. τοῦ δὲ αὐτοῦ διαφόρου πλεονάζοντος ἐν ταῖς ἀναφοραῖς ταῖς μεταξὺ τῆς διαστάσεως τῶν ἀνωμάλων ἐποχῶν, προστίθεμεν τὸ εἰρημένον διάφορον ταῖς εὑρεθείσαις ὥραις ἐκ τῶν κανόνων ἐν ταῖς ἑκάστοτε ψηφοφορίαις, καὶ οὕτως πάλιν ἀποκαθιστῶμεν ἀπὸ τῶν ὁμαλῶν νυχθημέρων τὰ καιρικά. ἀλλ' οὕτω μὲν καὶ ὁ τρίτος τρόπος τῆς τῶν

465 μεταξὺ C

table of right ascensions and read off and record how many Sun-risings took place between the initial anomalous position and the present position. We do the same for the uniform rotation. Finally, we compute the difference between the increments of Sun-risings in the two cases and convert the difference to a fraction of an equinoctial hour dividing by 15°.

When we wish to convert seasonal and anomalous (simple) hours to mean, we look at the increment of anomalous Sun-risings between the two longitudes; if it is larger [than the uniform], then we add the fraction to the value computed from the tables. This way we convert given seasonal to mean solar-hours expressed in so many hours and fractions of an hour. If, on the other hand, the increment of the mean Sun-risings is larger, then we subtract it from the value computed from the tables. This way we convert seasonal to mean solar days, containing so many hours and fraction of an hour.

When we wish to convert mean solar days to seasonal, because it will also happen some times, then we simply reverse the steps. Whenever the mentioned increment between the mean longitudes is larger, then we subtract the difference of increments from the hours computed from the tables. Thus we convert mean solar days to seasonal. If, however, the anomalous increment exceeds the mean, then we add the increment of hours found from the tables; this way we again convert mean to seasonal solar days. This is the third method for

ὡρῶν διακρίσεως μεθοδεύεται, ὅπως δηλονότι ἔξεστι τὰς τῶν καιρικῶν νυχθημέρων ὥρας ἀποκαθιστᾶν ὁμαλῶν νυχθημέρων ὥρας καί, τὸ ἀνάπαλιν, τὰς τῶν ὁμαλῶν νυχθημέρων ὥρας ἀποκαθιστᾶν καιρικῶν νυχθημέρων ὥρας.

Ἰστέον μέντοι ὅτι ἡ τοιαύτη μέθοδος τοῦ τρίτου τρόπου τῆς διακρίσεως τῶν ὡρῶν παρὰ τοῦ Πτολεμαίου ἐν τῇ Συντάξει παραδέδοται. οἱ δέ γε Πρόχειροι Κανόνες τοῦ Θέωνος ἄλλως τὰ περὶ τούτου μεθοδεύουσι, διὰ τρίτου σελιδίου προστιθεμένου ἐν τῷ κανόνι τῶν ἀναφορῶν τῆς ὀρθῆς σφαίρας, περὶ οὗ προείπομεν. ἀλλ᾽ ἡμῖν γε βελτίων καὶ ἀσφαλεστέρα ἐπιμελῶς σκεψαμένοις ἔδοξεν ἡ παράδοσις τοῦ Πτολεμαίου. διὰ ταῦτα δὴ καὶ χρῆσθαι ταύτῃ δοκιμάζομεν καὶ παραδιδόαμεν ἐν τῇ παρούσῃ Στοιχειώσει. τὸ δ᾽ ὅπως καὶ διὰ τίνας τοὺς λόγους τὸ μῆκος τοῦ λόγου φεύγοντες νῦν ἐῶμεν. ἴσως δὲ τοῖς ἐπιστατικώτερον περὶ τούτων σκεψαμένοις ἐξέσται καταλαβεῖν, ὅπως ταύτην ἡμεῖς τὴν μέθοδον προειλόμεθα καὶ ὅτι βελτίων καὶ ἀσφαλεστέρα ἐστὶν αὕτη ἢ κατὰ τὴν τῶν Προχείρων Κανόνων.

490 Cf. Ptol. *Alm*. 3.9 (vol. 1.1, p. 262-263 Heiberg) **491** Cf. Ptol. *Proch. Kan*. p. 97-100 Tihon

499 ἔφοδον V

handling the distinction among the hours, which provides a method for converting seasonal hours of a day to mean and vice-versa the mean to seasonal hours.

It should be noted that the third method for distinguishing the various definitions of hours appears in the *Syntaxis* of Ptolemy. The *Handy Tables* of Theon include a different procedure — they add a third column to the table of Sun-risings in *sphaera recta*, as was mentioned earlier. However, thinking carefully and to be safer, we selected the method of Ptolemy and decided to use and present it in this work. How and for which reasons we arrived at this decision we shall omit for reasons of brevity. Perhaps it is more appropriate to remark that we prefer this method as being better and more reliable than the one in the *Handy Tables*.

28 ἔκθεσις κανόνος ἀναφορῶν ἐπὶ τῆς ὀρθῆς νοουμένης οὐρανίου σφαίρας, ἔτι δὲ ἔκθεσις κανόνων ἀναφορῶν ἐπὶ τῶν ἑπτὰ κλιμάτων καὶ ἐπ' αὐτοῦ τοῦ διὰ Βυζαντίου, εἴτουν τῆς βασιλίδος πόλεως, παραλλήλου

Καὶ τοίνυν τὴν ἔκθεσιν ποιούμενοι τῶν κανόνων ἤδη τῶν ἀναφορῶν, τοὺς μὲν ἄλλους κανόνας τῶν ἀναφορῶν ἑκάστου κλίματος αὐτοὺς ἐκείνους τοὺς ἐν τοῖς Προχείροις Κανόσι διὰ τὸ εὐμεταχείριστον τῆς χρήσεως, ὡς προειρήκαμεν, ἀμεταποιήτους ἐκτιθέμεθα· μόνον δὲ τὸν κανόνα τῶν ἀναφορῶν τῆς ὀρθῆς σφαίρας τοσοῦτον μεταποιοῦμεν ἐκτιθέντες ἐνταῦθα, ὅσον ὑφαιρεῖν αὐτὸ μόνον τὸ ἐν αὐτῷ τρίτον, ὡς ἔφημεν, σελίδιον. ἐπειδὴ τὴν αὐτοῦ χρῆσιν καὶ τὴν δι' αὐτοῦ διόρθωσιν τῶν ὡρῶν τῶν ὁμαλῶν καὶ ἀνωμάλων νυχθημέρων παρεάσαντες προειλόμεθα τὴν μικρῷ πρόσθεν παραδεδομένην ἕνεκεν τούτου μέθοδον τοῦ Πτολεμαίου ἐν τῇ Συντάξει. καὶ ἔχουσιν οἱ κανόνες οὕτως·

3 Cf. Ptol. *Proch*. Kan. p. 97-134 Tihon

28 Presentation of the table for the ascensions of the Sun on sphaera recta; in addition tables for the ascensions of the Sun on the seven zones and on the parallel of Byzantium — the Queen City

Now we present the tables for the ascensions. For the tables of each zone we present those that appear in the *Handy Tables* because they are easier to use. We present them without changes, except for the table of ascensions of *sphaera recta*, where we omit the third column, because, as we described a little earlier, we decided to present the corrections from regular solar hours to anomalous using the method of Ptolemy from the *Syntaxis*. Next follow the Tables

Κανὼν αναφορῶν τῆς ὀρθῆς σφαίρας

ἡλί-ου	Αἰγοκέρου		Ὑδροχόου		Ἰχθύων	
μοῖ-ραι	μοῖ-ραι	λε-πτά	μοῖ-ραι	λε-πτά	μοῖ-ραι	λε-πτά
Α	α	ϛ	λγ	ιη	ξγ	ζ
β	β	ιβ	λδ	κ	ξδ	δ
γ	γ	ιη	λε	κβ	ξε	α
δ	δ	κδ	λϛ	κδ	ξε	νη
ε	ε	λ	λζ	κϛ	ξϛ	νε
ϛ	ϛ	λε	λη	κη	ξζ	να
ζ	ζ	μ	λθ	κθ	ξη	μζ
η	η	με	μ	λ	ξθ	μγ
θ	θ	ν	μα	λα	ο	λθ
ι	ι	νε	μβ	λβ	οα	λε
ια	ιβ	ō	μγ	λβ	οβ	λα
ιβ	ιγ	ε	μδ	λβ	ογ	κζ
ιγ	ιδ	ι	με	λβ	οδ	κγ
ιδ	ιε	ιε	μϛ	λβ	οε	ιθ
ιε	ιϛ	κ	μζ	λβ	οϛ	ιε
ιϛ	ιζ	κε	μη	λβ	οζ	ι
ιζ	ιη	λ	μθ	λβ	οη	ε
ιη	ιθ	λδ	ν	λβ	οθ	ō
ιθ	κ	λη	να	λα	οθ	νε
κ	κα	μβ	νβ	λ	π	ν
κα	κβ	μϛ	νγ	κη	πα	με
κβ	κγ	ν	νδ	κϛ	πβ	μ
κγ	κδ	νδ	νε	κδ	πγ	λε
κδ	κε	νη	νϛ	κβ	πδ	λ
κε	κζ	α	νζ	κ	πε	κε
κϛ	κη	δ	νη	ιη	πϛ	κ
κζ	κθ	ζ	νθ	ιϛ	πζ	ιε
κη	λ	ι	ξ	ιδ	πη	ι
κθ	λα	ιγ	ξα	ιβ	πθ	ε
λ	λβ	ιϛ	ξβ	ι	ϟ	ō

ἡλί-ου	Κριοῦ		Ταύρου		Διδύμων	
μοῖ-ραι	μοῖ-ραι	λε-πτά	μοῖ-ραι	λε-πτά	μοῖ-ραι	λε-πτά
α	ϟ	νε	ριη	μη	ρμη	μζ
β	ϟα	ν	ριθ	μϛ	ρμθ	ν
γ	ϟβ	με	ρκ	μδ	ρν	νγ
δ	ϟγ	μ	ρκα	μβ	ρνα	νϛ
ε	ϟδ	λε	ρκβ	μ	ρνβ	νθ
ϛ	ϟε	λ	ρκγ	λη	ρνδ	γ
ζ	ϟϛ	κε	ρκδ	λϛ	ρνε	ϛ
η	ϟζ	κ	ρκε	λδ	ρνϛ	ι
θ	ϟη	ιε	ρκϛ	λβ	ρνζ	ιδ
ι	ϟθ	ι	ρκζ	λ	ρνη	ιη
ια	ρ	ε	ρκη	κθ	ρνθ	κβ
ιβ	ρα	ō	ρκθ	κη	ρξ	κϛ
ιγ	ρα	νε	ρλ	κη	ρξα	λ
ιδ	ρβ	ν	ρλα	κη	ρξβ	λε
ιε	ργ	με	ρλβ	κη	ρξγ	μ
ιϛ	ρδ	μα	ρλγ	κη	ρξδ	με
ιζ	ρε	λζ	ρλδ	κη	ρξε	ν
ιη	ρϛ	λγ	ρλε	κη	ρξϛ	νε
ιθ	ρζ	κθ	ρλϛ	κη	ρξη	ō
κ	ρη	κε	ρλζ	κη	ρξθ	ε
κα	ρθ	κα	ρλη	κθ	ρο	ι
κβ	ρι	ιζ	ρλθ	λ	ροα	ιθ
κγ	ρια	ιγ	ρμ	λα	ροβ	κ
κδ	ριβ	θ	ρμα	λβ	ρογ	κε
κε	ριγ	ε	ρμβ	λδ	ροδ	λ
κϛ	ριδ	β	ρμγ	λϛ	ροε	λϛ
κζ	ριδ	νθ	ρμδ	λη	ροϛ	μβ
κη	ριε	νϛ	ρμε	μ	ροζ	μη
κθ	ριϛ	νγ	ρμϛ	μβ	ροη	νδ
λ	ριζ	ν	ρμζ	μδ	ρπ	ō

Table of ascensions of *sphaera recta*

D = Degrees, M = Minutes

Sun	Capricorn		Aquarius		Pisces			Sun	Aries		Taurus		Gemini	
D	D	M	D	M	D	M		D	D	M	D	M	D	M
1	1	6	33	18	63	7		1	90	55	118	48	148	47
2	2	12	34	20	64	4		2	91	50	119	46	149	50
3	3	18	35	22	65	1		3	92	45	120	44	150	53
4	4	24	36	24	65	58		4	93	40	121	42	151	56
5	5	30	37	26	66	55		5	94	35	122	40	152	59
6	6	35	38	28	67	51		6	95	30	123	38	154	3
7	7	40	39	29	68	47		7	96	25	124	36	155	6
8	8	45	40	30	69	43		8	97	20	125	34	156	10
9	9	50	41	31	70	39		9	98	15	126	32	157	14
10	10	55	42	32	71	35		10	99	10	127	30	158	18
11	12	0	43	32	72	31		11	100	5	128	29	159	22
12	13	5	44	32	73	27		12	101	0	129	28	160	26
13	14	10	45	32	74	23		13	101	55	130	28	161	30
14	15	15	46	32	75	19		14	102	50	131	28	162	35
15	16	20	47	32	76	15		15	103	45	132	28	163	40
16	17	25	48	32	77	10		16	104	41	133	28	164	45
17	18	30	49	32	78	5		17	105	37	134	28	165	50
18	19	34	50	32	79	0		18	106	33	135	28	166	55
19	20	38	51	31	79	55		19	107	29	136	28	168	0
20	21	42	52	30	80	50		20	108	25	137	28	169	5
21	22	46	53	28	81	45		21	109	21	138	29	170	10
22	23	50	54	26	82	40		22	110	17	139	30	171	19
23	24	54	55	24	83	35		23	111	13	140	31	172	20
24	25	58	56	22	84	30		24	112	9	141	32	173	25
25	27	1	57	20	85	25		25	113	5	142	34	174	30
26	28	4	58	18	86	20		26	114	2	143	36	175	36
27	29	7	59	16	87	15		27	114	59	144	38	176	42
28	30	10	60	14	88	10		28	115	56	145	40	177	48
29	31	13	61	12	89	5		29	116	53	146	42	178	54
30	32	16	62	10	90	0		30	117	50	147	44	180	0

Κανὼν αναφορῶν τῆς ὀρθῆς σφαίρας

ἡλίου	Καρκίνου		Λέοντος		Παρθένου		ἡλίου	Ζυγοῦ		Σκορπίου		Τοξότου	
μοῖραι	μοῖραι	λεπτά	μοῖραι	λεπτά	μοῖραι	λεπτά	μοῖραι	μοῖραι	λεπτά	μοῖραι	λεπτά	μοῖραι	λεπτά
α	ρπα	ς	σιγ	ιη	σμγ	ζ	α	σο	νε	σϙη	μη	τκη	μζ
β	ρπβ	ιβ	σιδ	κ	σμδ	δ	β	σοα	ν	σϙθ	μς	τκθ	ν
γ	ρπγ	ιη	σιε	κβ	σμε	α	γ	σοβ	με	τ	μδ	τλ	νγ
δ	ρπδ	κδ	σις	κδ	σμε	νη	δ	σογ	μ	τα	μβ	τλα	νς
ε	ρπε	λ	σιζ	κς	σμς	νε	ε	σοδ	λε	τβ	μ	τλβ	νθ
ς	ρπς	λε	σιη	κη	σμζ	να	ς	σοε	λ	τγ	λη	τλδ	β
ζ	ρπζ	μ	σιθ	κθ	σμη	μζ	ζ	σος	κε	τδ	λς	τλε	ς
η	ρπη	με	σκ	λ	σμθ	μγ	η	σοζ	κ	τε	λδ	τλς	ι
θ	ρπθ	ν	σκα	λα	σν	λθ	θ	σοη	ιε	τς	λβ	τλζ	ιδ
ι	ρϙ	νε	σκβ	λβ	σνα	λε	ι	σοθ	ι	τζ	λ	τλη	ιη
ια	ρϙβ	ō	σκγ	λβ	σνβ	λα	ια	σπ	ε	τη	κθ	τλθ	κβ
ιβ	ρϙγ	ε	σκδ	λβ	σνγ	κζ	ιβ	σπα	ō	τθ	κη	τμ	κς
ιγ	ρϙδ	ι	σκε	λβ	σνδ	κγ	ιγ	σπα	νε	τι	κη	τμα	λ
ιδ	ρϙε	ιε	σκς	λβ	σνε	ιθ	ιδ	σπβ	ν	τια	κη	τμβ	λε
ιε	ρϙς	κ	σκζ	λβ	σνς	ιε	ιε	σπγ	με	τιβ	κη	τμγ	μ
ις	ρϙζ	κε	σκη	λβ	σνζ	ι	ις	σπδ	μα	τιγ	κη	τμδ	με
ιζ	ρϙη	λ	σκθ	λβ	σνη	ε	ιζ	σπε	λζ	τιδ	κη	τμε	ν
ιη	ρϙθ	λδ	σλ	λβ	σνθ	ō	ιη	σπς	λγ	τιε	κη	τμς	νε
ιθ	σ	λη	σλα	λα	σνθ	νε	ιθ	σπζ	κθ	τις	κη	τμη	ō
κ	σα	μβ	σλβ	λ	σξ	ν	κ	σπη	κε	τιζ	κη	τμθ	ε
κα	σβ	μς	σλγ	κη	σξα	με	κα	σπθ	κα	τιη	κθ	τν	ι
κβ	σγ	ν	σλδ	κς	σξβ	μ	κβ	σϙ	ιζ	τιθ	λ	τνα	ιε
κγ	σδ	νδ	σλε	κδ	σξγ	λε	κγ	σϙα	ιγ	τκ	λα	τνβ	κ
κδ	σε	νη	σλς	κβ	σξδ	λ	κδ	σϙβ	θ	τκα	λβ	τνγ	κε
κε	σζ	α	σλζ	κ	σξε	κε	κε	σϙγ	ε	τκβ	λδ	τνδ	λ
κς	ση	δ	σλη	ιη	σξς	κ	κς	σϙδ	β	τκγ	λς	τνε	λς
κζ	σθ	ζ	σλθ	ιζ	σξζ	ιε	κζ	σϙδ	νθ	τκδ	λη	τνς	μβ
κη	σι	ι	σμ	ιδ	σξη	ι	κη	σϙε	νς	τκε	μ	τνζ	μη
κθ	σια	ιγ	σμα	ιβ	σξθ	ε	κθ	σϙς	νγ	τκς	μβ	τνη	νδ
λ	σιβ	ις	σμβ	ι	σο	ō	λ	σϙζ	ν	τκζ	μδ	τξ	ō

Table of ascensions of *sphaera recta*

Sun	Cancer		Leo		Virgo		Sun	Libra		Scorpio		Sagittarius	
D	D	M	D	M	D	M	D	D	M	D	M	D	M
1	181	6	213	18	243	7	1	270	55	298	48	328	47
2	182	12	214	20	244	4	2	271	50	299	46	329	50
3	183	18	215	22	245	1	3	272	45	300	44	330	53
4	184	24	216	24	245	58	4	273	40	301	42	331	56
5	185	30	217	26	246	55	5	274	35	302	40	332	59
6	186	35	218	28	247	51	6	275	30	303	38	334	2
7	187	40	219	29	248	47	7	276	25	304	36	335	6
8	188	45	220	30	249	43	8	277	20	305	34	336	10
9	189	50	221	31	250	39	9	278	15	306	32	337	14
10	190	55	222	32	251	35	10	279	10	307	30	338	18
11	192	0	223	32	252	31	11	280	5	308	29	339	22
12	193	5	224	32	253	27	12	281	0	309	28	340	26
13	194	10	225	32	254	23	13	281	55	310	28	341	30
14	195	15	226	32	255	19	14	282	50	311	28	342	35
15	196	20	227	32	256	15	15	283	45	312	28	343	40
16	197	25	228	32	257	10	16	284	41	313	28	344	45
17	198	30	229	32	258	5	17	285	37	314	28	345	50
18	199	34	230	32	259	0	18	286	33	315	28	346	55
19	200	38	231	31	259	55	19	287	29	316	28	348	0
20	201	42	232	30	260	50	20	288	25	317	28	349	5
21	202	46	233	28	261	45	21	289	21	318	29	350	10
22	203	50	234	26	262	40	22	290	17	319	30	351	15
23	204	54	235	24	263	35	23	291	13	320	31	352	20
24	205	58	236	22	264	30	24	292	9	321	32	353	25
25	207	1	237	20	265	25	25	293	5	322	34	354	30
26	208	4	238	18	266	20	26	294	2	323	36	355	36
27	209	7	239	17	267	15	27	294	59	324	38	356	42
28	210	10	240	14	268	10	28	295	56	325	40	357	48
29	211	13	241	12	269	5	29	296	53	326	42	358	54
30	212	16	242	10	270	0	30	297	50	327	44	360	0

Κανὼν ἀναφορῶν τοῦ διὰ Μερόης πρώτου κλίματος, ὡρῶν ιγ, μοιρῶν ις, λεπτῶν κζ΄

Ἡλίου	Κριοῦ				Ταύρου				Διδύμων			
Μ	Ἀνα	φοραί	Ὥρ χρ	Λ	Ἀνα	φοραί	Ὥρ χρ	Λ	Ἀνα	φοραί	Ὥρ χρ	Λ
α	ō	μζ	ιε	α	κε	ια	ιε	λς	νβ	κγ	ις	δ
β	α	λδ	ιε	β	κς	β	ιε	λζ	νγ	κβ	ις	ε
γ	β	κα	ιε	δ	κς	νγ	ιε	λη	νδ	κα	ις	ε
δ	γ	ι	ιε	ε	κζ	μδ	ιε	λθ	νε	κ	ις	ς
ε	γ	νη	ιε	ς	κη	λς	ιε	μ	νς	κ	ις	ς
ς	δ	μς	ιε	ζ	κθ	κη	ιε	μα	νζ	κ	ις	ζ
ζ	ε	λδ	ιε	η	λ	κ	ιε	μβ	νη	κ	ις	η
η	ς	κβ	ιε	ι	λα	ιβ	ιε	μδ	νθ	κ	ις	η
θ	ζ	ι	ιε	ια	λβ	δ	ιε	με	ξ	κ	ις	θ
ι	ζ	νη	ιε	ιβ	λβ	νς	ιε	μς	ξα	κ	ις	ι
ια	η	μς	ιε	ιγ	λγ	ν	ιε	μζ	ξβ	κβ	ις	ι
ιβ	θ	λδ	ιε	ιδ	λδ	μδ	ιε	μη	ξγ	κδ	ις	ι
ιγ	ι	κβ	ιε	ιε	λε	λη	ιε	μθ	ξδ	κς	ις	ια
ιδ	ια	ι	ιε	ιζ	λς	λβ	ιε	ν	ξε	κη	ις	ια
ιε	ια	νη	ιε	ιη	λζ	κς	ιε	να	ξς	λ	ις	ιβ
ις	ιβ	μζ	ιε	ιθ	λη	κ	ιε	να	ξζ	λβ	ις	ιβ
ιζ	ιγ	λς	ιε	κ	λθ	ιδ	ιε	νβ	ξη	λδ	ις	ιβ
ιη	ιδ	κε	ιε	κα	μ	η	ιε	νγ	ξθ	λς	ις	ιγ
ιθ	ιε	ιδ	ιε	κβ	μα	β	ιε	νδ	ο	λθ	ις	ιγ
κ	ις	γ	ιε	κδ	μα	νς	ιε	νε	οα	μγ	ις	ιγ
κα	ις	νβ	ιε	κε	μβ	νγ	ιε	νς	οβ	μζ	ις	ιδ
κβ	ιζ	μα	ιε	κς	μγ	μθ	ιε	νζ	ογ	να	ις	ιδ
κγ	ιη	λ	ιε	κζ	μδ	με	ιε	νη	οδ	νε	ις	ιδ
κδ	ιθ	κ	ιε	κη	με	μβ	ιε	νθ	ος	ō	ις	ιδ
κε	κ	ι	ιε	κθ	μς	λθ	ιε	νθ	οζ	ε	ις	ιδ
κς	κα	ō	ιε	λ	μζ	λς	ις	ō	οη	ι	ις	ιδ
κζ	κα	ν	ιε	λβ	μη	λγ	ις	α	οθ	ιε	ις	ιε
κη	κβ	μ	ιε	λγ	μθ	λ	ις	β	π	κ	ις	ιε
κθ	κγ	λ	ιε	λδ	ν	κζ	ις	γ	πα	κε	ις	ιε
λ	κδ	κ	ιε	λε	να	κδ	ις	γ	πβ	λ	ις	ιε
	κδ	κ			κζ	δ			λα	ς		

Table of the ascensions on the parallel of Meroë,
13 equinoctial hours, 16 degrees, 27 minutes

H-t = Hourly times

Sun	Aries				Taurus				Gemini			
D	Ascension		H-t	M	Ascensions		H-t	M	Ascensions		H-t	M
1	0	47	15	1	25	11	15	36	52	23	16	4
2	1	34	15	2	26	2	15	37	53	22	16	5
3	2	21	15	4	26	53	15	38	54	21	16	5
4	3	10	15	5	27	44	15	39	55	20	16	6
5	3	58	15	6	28	36	15	40	56	20	16	6
6	4	46	15	7	29	28	15	41	57	20	16	7
7	5	34	15	8	30	20	15	42	58	20	16	8
8	6	22	15	10	31	12	15	44	59	20	16	8
9	7	10	15	11	32	4	15	45	60	20	16	9
10	7	58	15	12	32	56	15	46	61	20	16	10
11	8	46	15	13	33	50	15	47	62	22	16	10
12	9	34	15	14	34	44	15	48	63	24	16	10
13	10	22	15	15	35	38	15	49	64	26	16	11
14	11	10	15	17	36	32	15	50	65	28	16	11
15	11	58	15	18	37	26	15	51	66	30	16	12
16	12	47	15	19	38	20	15	51	67	32	16	12
17	13	36	15	20	39	14	15	52	68	34	16	12
18	14	25	15	21	40	8	15	53	69	36	16	13
19	15	14	15	22	41	2	15	54	70	39	16	13
20	16	3	15	24	41	56	15	55	71	43	16	13
21	16	52	15	25	42	53	15	56	72	47	16	14
22	17	41	15	26	43	49	15	57	73	51	16	14
23	18	30	15	27	44	45	15	58	74	55	16	14
24	19	20	15	28	45	42	15	59	76	0	16	14
25	20	10	15	29	46	39	15	59	77	5	16	14
26	21	0	15	30	47	36	16	0	78	10	16	14
27	21	50	15	32	48	33	16	1	79	15	16	15
28	22	40	15	33	49	30	16	2	80	20	16	15
29	23	30	15	34	50	27	16	3	81	25	16	15
30	24	20	15	35	51	24	16	3	82	30	16	15
	24	20			27	4			31	6		

Κανὼν ἀναφορῶν τοῦ διὰ Μερόης πρώτου κλίματος, ὡρῶν ιγ, μοιρῶν ις, λεπτῶν κζʹ

Ἡλίου	Καρκίνου		Καρκίνου		Λέοντος		Λέοντος		Παρθένου		Παρθένου	
Μ	Ἀνα	φοραί	Ὡρ χρ	Λ	Ἀνα	φοραί	Ὡρ χρ	Λ	Ἀνα	φοραί	Ὡρ χρ	Λ
α	πγ	λς	ις	ιε	ριζ	γ	ις	γ	ρμθ	μδ	ιε	λδ
β	πδ	μβ	ις	ιε	ριη	ι	ις	β	ρν	μη	ιε	λγ
γ	πε	μη	ις	ιε	ριθ	ιζ	ις	α	ρνα	νβ	ιε	λβ
δ	πς	νδ	ις	ιδ	ρκ	κδ	ις	ō	ρνβ	νε	ιε	λ
ε	πη	ō	ις	ιδ	ρκα	λα	ιε	νθ	ρνγ	νη	ιε	κθ
ς	πθ	ς	ις	ιδ	ρκβ	λζ	ιε	νθ	ρνε	α	ιε	κη
ζ	Ϙ	ιβ	ις	ιδ	ρκγ	μγ	ιε	νη	ρνς	δ	ιε	κζ
η	Ϙα	ιθ	ις	ιδ	ρκδ	μθ	ιε	νζ	ρνζ	ζ	ιε	κς
θ	Ϙβ	κς	ις	ιδ	ρκε	νε	ιε	νς	ρνη	ι	ιε	κε
ι	Ϙγ	λγ	ις	ιγ	ρκζ	α	ιε	νε	ρνθ	ιγ	ιε	κδ
ια	Ϙδ	μ	ις	ιγ	ρκη	ζ	ιε	νδ	ρξ	ις	ιε	κβ
ιβ	Ϙε	μζ	ις	ιγ	ρκθ	ιγ	ιε	νγ	ρξα	ιθ	ιε	κα
ιγ	Ϙς	νδ	ις	ιβ	ρλ	ιθ	ιε	νβ	ρξβ	κβ	ιε	κ
ιδ	Ϙη	α	ις	ιβ	ρλα	κε	ιε	να	ρξγ	κε	ιε	ιθ
ιε	Ϙθ	η	ις	ιβ	ρλβ	λα	ιε	να	ρξδ	κη	ιε	ιη
ις	ρ	ιε	ις	ια	ρλγ	λς	ιε	ν	ρξε	λα	ιε	ιζ
ιζ	ρα	κβ	ις	ια	ρλδ	μα	ιε	μθ	ρξς	λδ	ιε	ιε
ιη	ρβ	κθ	ις	ι	ρλε	μς	ιε	μη	ρξζ	λς	ιε	ιδ
ιθ	ργ	λς	ις	ι	ρλς	να	ιε	μζ	ρξη	λη	ιε	ιγ
κ	ρδ	μδ	ις	ι	ρλζ	νς	ιε	μς	ρξθ	μ	ιε	ιβ
κα	ρε	νβ	ις	θ	ρλθ	α	ιε	με	ρο	μβ	ιε	ια
κβ	ρζ	ō	ις	η	ρμ	ς	ιε	μδ	ροα	μδ	ιε	ι
κγ	ρη	ζ	ις	η	ρμα	ια	ιε	μγ	ροβ	μς	ιε	η
κδ	ρθ	ιδ	ις	ζ	ρμβ	ις	ιε	μβ	ρογ	μη	ιε	ζ
κε	ρι	κα	ις	ς	ρμγ	κ	ιε	μα	ροδ	ν	ιε	ς
κς	ρια	κη	ις	ς	ρμδ	κδ	ιε	μ	ροε	νβ	ιε	ε
κζ	ριβ	λε	ις	ε	ρμε	κη	ιε	λθ	ρος	νδ	ιε	δ
κη	ριγ	μβ	ις	ε	ρμς	λβ	ιε	λη	ροζ	νς	ιε	β
κθ	ριδ	μθ	ις	δ	ρμζ	λς	ιε	λς	ροη	νη	ιε	α
λ	ριε	νς	ις	γ	ρμη	μ	ιε	λε	ρπ	ō	ιε	ō
	λγ	κς			λβ	μδ			λα	κ		

Table of the ascensions on the parallel of Meroë,
13 equinoctial hours, 16 degrees, 27 minutes

Sun	Cancer		Cancer		Leo		Leo		Virgo		Virgo	
D	Ascensions		H-t	M	Ascensions		H-t	M	Ascensions		H-t	M
1	83	36	16	15	117	3	16	3	149	44	15	34
2	84	42	16	15	118	10	16	2	150	48	15	33
3	85	48	16	15	119	17	16	1	151	52	15	32
4	86	54	16	14	120	24	16	0	152	55	15	30
5	88	0	16	14	121	31	15	59	153	58	15	29
6	89	6	16	14	122	37	15	59	155	1	15	28
7	90	12	16	14	123	43	15	58	156	4	15	27
8	91	19	16	14	124	49	15	57	157	7	15	26
9	92	26	16	14	125	55	15	56	158	10	15	25
10	93	33	16	13	127	1	15	55	159	13	15	24
11	94	40	16	13	128	7	15	54	160	16	15	22
12	95	47	16	13	129	13	15	53	161	19	15	21
13	96	54	16	12	130	19	15	52	162	22	15	20
14	98	1	16	12	131	25	15	51	163	25	15	19
15	99	8	16	12	132	31	15	51	164	28	15	18
16	100	15	16	11	133	36	15	50	165	31	15	17
17	101	22	16	11	134	41	15	49	166	34	15	15
18	102	29	16	10	135	46	15	48	167	36	15	14
19	103	36	16	10	136	51	15	47	168	38	15	13
20	104	44	16	10	137	56	15	46	169	40	15	12
21	105	52	16	9	139	1	15	45	170	42	15	11
22	107	0	16	8	140	6	15	44	171	44	15	10
23	108	7	16	8	141	11	15	43	172	46	15	8
24	109	14	16	7	142	16	15	42	173	48	15	7
25	110	21	16	6	143	20	15	41	174	50	15	6
26	111	28	16	6	144	24	15	40	175	52	15	5
27	112	35	16	5	145	28	15	39	176	54	15	4
28	113	42	16	5	146	32	15	38	177	56	15	2
29	114	49	16	4	147	36	15	36	178	58	15	1
30	115	56	16	3	148	40	15	35	180	0	15	0
	33	36			32	44			31	20		

Κανὼν ἀναφορῶν τοῦ διὰ Μερόης πρώτου κλίματος, ὡρῶν ιγ, μοιρῶν ις, λεπτῶν κζ΄

Ἡλίου	Ζυγοῦ		Ζυγοῦ		Σκορπίου		Σκορπίου		Τοξότου		Τοξότου	
Μ	Ἀνα	φοραί	Ὥρ χρ	Λ	Ἀνα	φοραί	Ὥρ χρ	Λ	Ἀνα	φοραί	Ὥρ χρ	Λ
α	ρπα	β	ιδ	νθ	σιβ	κδ	ιδ	κδ	σμε	ια	ιγ	νς
β	ρπβ	δ	ιδ	νη	σιγ	κη	ιδ	κγ	σμς	ιη	ιγ	νε
γ	ρπγ	ς	ιδ	νς	σιδ	λβ	ιδ	κβ	σμζ	κε	ιγ	νε
δ	ρπδ	η	ιδ	νε	σιε	λς	ιδ	κα	σμη	λβ	ιγ	νδ
ε	ρπε	ι	ιδ	νδ	σις	μ	ιδ	κ	σμθ	λθ	ιγ	νδ
ς	ρπς	ιβ	ιδ	νγ	σιζ	μδ	ιδ	ιθ	σν	μς	ιγ	νγ
ζ	ρπζ	ιδ	ιδ	νβ	σιη	μθ	ιδ	ιη	σνα	νγ	ιγ	νβ
η	ρπη	ις	ιδ	ν	σιθ	νδ	ιδ	ις	σνγ	ō	ιγ	νβ
θ	ρπθ	ιη	ιδ	μθ	σκ	νθ	ιδ	ιε	σνδ	η	ιγ	να
ι	ρϙ	κ	ιδ	μη	σκβ	κδ	ιδ	ιδ	σνε	ις	ιγ	ν
ια	ρϙα	κβ	ιδ	μζ	σκγ	θ	ιδ	ιγ	σνς	κδ	ιγ	ν
ιβ	ρϙβ	κδ	ιδ	μς	σκδ	ιδ	ιδ	ιβ	σνζ	λα	ιγ	ν
ιγ	ρϙγ	κς	ιδ	με	σκε	ιθ	ιδ	ια	σνη	λη	ιγ	μθ
ιδ	ρϙδ	κη	ιδ	μγ	σκς	δ	ιδ	ι	σνθ	με	ιγ	μθ
ιε	ρϙε	λ	ιδ	μβ	σκζ	κθ	ιδ	θ	σξ	νβ	ιγ	μη
ις	ρϙς	λβ	ιδ	μα	σκη	λε	ιδ	θ	σξα	νθ	ιγ	μη
ιζ	ρϙζ	λδ	ιδ	μ	σκθ	μα	ιδ	η	σξγ	ς	ιγ	μη
ιη	ρϙη	λη	ιδ	λθ	σλ	μζ	ιδ	ζ	σξδ	ιγ	ιγ	μζ
ιθ	ρϙθ	μγ	ιδ	λη	σλα	νγ	ιδ	ς	σξε	κ	ιγ	μζ
κ	σ	μζ	ιδ	λς	σλβ	νθ	ιδ	ε	σξς	κζ	ιγ	μζ
κα	σα	ν	ιδ	λε	σλδ	ε	ιδ	δ	σξζ	λδ	ιγ	μς
κβ	σβ	νγ	ιδ	λδ	σλε	ια	ιδ	γ	σξη	μα	ιγ	μς
κγ	σγ	νς	ιδ	λγ	σλς	ιζ	ιδ	β	σξθ	μη	ιγ	μς
κδ	σδ	νθ	ιδ	λβ	σλζ	κγ	ιδ	α	σο	νδ	ιγ	μς
κε	σς	β	ιδ	λα	σλη	κθ	ιδ	α	σοβ	ō	ιγ	με
κς	σζ	ε	ιδ	λ	σλθ	λς	ιδ	ō	σογ	ς	ιγ	με
κζ	ση	η	ιδ	κη	σμ	μγ	ιγ	νθ	σοδ	ιβ	ιγ	με
κη	σθ	ιβ	ιδ	κζ	σμα	ν	ιγ	νη	σοε	ιη	ιγ	με
κθ	σι	ις	ιδ	κς	σμβ	νζ	ιγ	νζ	σος	κδ	ιγ	με
λ	σια	κ	ιδ	κε	σμδ	δ	ιγ	νζ	σοζ	λ	ιγ	με
	λα	κ			λβ	μδ						

Table of the ascensions on the parallel of Meroë,
13 equinoctial hours, 16 degrees, 27 minutes

Sun	Libra		Libra		Scorpio		Scorpio		Sagittarius		Sagittarius	
D	Ascensions		H-t	M	Ascensions		H-t	M	Ascensions		H-t	M
1	181	2	14	59	212	24	14	24	245	11	13	56
2	182	4	14	58	213	28	14	23	246	18	13	55
3	183	6	14	56	214	32	14	22	247	25	13	55
4	184	8	14	55	215	36	14	21	248	32	13	54
5	185	10	14	54	216	40	14	20	249	39	13	54
6	186	12	14	53	217	44	14	19	250	46	13	53
7	187	14	14	52	218	49	14	18	251	53	13	52
8	188	16	14	50	219	54	14	16	253	0	13	52
9	189	18	14	49	220	59	14	15	254	8	13	51
10	190	20	14	48	222	4	14	14	255	16	13	50
11	191	22	14	47	223	9	14	13	256	24	13	50
12	192	24	14	46	224	14	14	12	257	31	13	50
13	193	26	14	45	225	19	14	11	258	38	13	49
14	194	28	14	43	226	24	14	10	259	45	13	49
15	195	30	14	42	227	29	14	9	260	52	13	48
16	196	32	14	41	228	35	14	9	261	59	13	48
17	197	34	14	40	229	41	14	8	263	6	13	48
18	198	38	14	39	230	47	14	7	264	13	13	47
19	199	43	14	38	231	53	14	6	265	20	13	47
20	200	47	14	36	232	59	14	5	266	27	13	47
21	201	50	14	35	234	5	14	4	267	34	13	46
22	202	53	14	34	235	111	14	3	268	41	13	46
23	203	56	14	33	236	17	14	2	269	48	13	46
24	204	59	14	32	237	23	14	1	270	54	13	46
25	206	2	14	31	238	29	14	1	272	0	13	45
26	207	5	14	30	239	36	14	0	273	6	13	45
27	208	8	14	28	240	43	13	59	274	12	13	45
28	209	12	14	27	241	50	13	58	275	18	13	45
29	210	16	14	26	242	57	13	57	276	24	13	45
30	211	20	14	25	244	4	13	57	277	30	13	45
	31	20			32	44						

Κανὼν ἀναφορῶν τοῦ διὰ Μερόης πρώτου κλίματος, ὡρῶν ιγ, μοιρῶν ις, λεπτῶν κζ΄

Ἡλίου	Αἰγοκέρωτος				Ὑδροχόου				Ἰχθύων			
Μ	Ἀναφοραί		Ὥρ χρ	Λ	Ἀναφοραί		Ὥρ χρ	Λ	Ἀναφοραί		Ὥρ χρ	Λ
α	σοη	λε	ιγ	με	τθ	λγ	ιγ	νζ	τλς	α	ιδ	κς
β	σοθ	μ	ιγ	με	τι	λ	ιγ	νη	τλζ	κ	ιδ	κζ
γ	σπ	με	ιγ	με	τια	κζ	ιγ	νθ	τλη	ι	ιδ	κη
δ	σπα	ν	ιγ	μς	τιβ	κδ	ιδ	ō	τλθ	ō	ιδ	λ
ε	σπβ	νε	ιγ	μς	τιγ	κα	ιδ	α	τλθ	ν	ιδ	λα
ς	σπδ	ō	ιγ	μς	τιδ	ιη	ιδ	α	τμ	μ	ιδ	λβ
ζ	σπε	ε	ιγ	μς	τιε	ιε	ιδ	β	τμα	λ	ιδ	λγ
η	σπς	θ	ιγ	μς	τις	ια	ιδ	γ	τμβ	ιθ	ιδ	λδ
θ	σπζ	ιγ	ιγ	μς	τιζ	ζ	ιδ	δ	τμγ	η	ιδ	λε
ι	σπη	ιζ	ιγ	μζ	τιη	γ	ιδ	ε	τμγ	νζ	ιδ	λς
ια	σπθ	κ	ιγ	μζ	τιη	νη	ιδ	ς	τμδ	μς	ιδ	λη
ιβ	ρϞ	κγ	ιγ	μζ	τιθ	νβ	ιδ	ζ	τμε	λε	ιδ	λθ
ιγ	ρϞα	κς	ιγ	μη	τκ	μς	ιδ	η	τμς	κδ	ιδ	μ
ιδ	ρϞβ	κη	ιγ	μη	τκα	μ	ιδ	θ	τμζ	ιγ	ιδ	μα
ιε	ρϞγ	λ	ιγ	μη	τκβ	λδ	ιδ	θ	τμη	β	ιδ	μβ
ις	ρϞδ	λβ	ιγ	μθ	τκγ	κη	ιδ	ι	τμη	ν	ιδ	μγ
ιζ	ρϞε	λδ	ιγ	μθ	τκδ	κβ	ιδ	ια	τμθ	λη	ιδ	με
ιη	ρϞς	λς	ιγ	ν	τκε	ις	ιδ	ιβ	τν	κς	ιδ	μς
ιθ	ρϞζ	λη	ιγ	ν	τκς	ι	ιδ	ιγ	τνα	ιδ	ιδ	μζ
κ	ρϞη	μ	ιγ	ν	τκζ	δ	ιδ	ιδ	τνβ	β	ιδ	μη
κα	ρϞθ	μβ	ιγ	να	τκζ	νς	ιδ	ιε	τνβ	ν	ιδ	μθ
κβ	τ	μ	ιγ	νβ	τκη	μη	ιδ	ις	τνγ	λη	ιδ	ν
κγ	τα	μ	ιγ	νβ	τκθ	μ	ιδ	ιζ	τνδ	κς	ιδ	νβ
κδ	τβ	μ	ιγ	νγ	τλ	λβ	ιδ	ιη	τνε	ιδ	ιδ	νγ
κε	τγ	μ	ιγ	νδ	τλα	κδ	ιδ	ιθ	τνς	β	ιδ	νδ
κς	τδ	μ	ιγ	νδ	τλβ	ις	ιδ	κ	τνς	ν	ιδ	νε
κζ	τε	λθ	ιγ	νε	τλγ	ζ	ιδ	κα	τνζ	λη	ιδ	νς
κη	τς	λη	ιγ	νε	τλγ	νη	ιδ	κβ	τνη	κς	ιδ	νη
κθ	τζ	λζ	ιγ	νς	τλδ	μθ	ιδ	κδ	τνθ	ιγ	ιδ	νθ
λ	τη	λς	ιγ	νζ	τλε	μ	ιδ	κε	τξ	ō	ιε	ō
	λα	ς			κζ	δ			κδ			

Table of the ascensions on the parallel of Meroë,
13 equinoctial hours, 16 degrees, 27 minutes

Sun	Capricorn				Aquarius				Pisces			
D	Ascension		H-t	M	Ascensions		H-t	M	Ascensions		H-t	M
1	278	35	13	45	309	33	13	57	336	1	14	26
2	279	40	13	45	310	30	13	58	337	20	14	27
3	280	45	13	45	311	27	13	59	338	10	14	28
4	281	50	13	46	312	24	14	0	339	0	14	30
5	282	55	13	46	313	21	14	1	339	50	14	31
6	284	0	13	46	314	18	14	1	340	40	14	32
7	285	5	13	46	315	15	14	2	341	30	14	33
8	286	0	13	46	316	11	14	3	342	19	14	34
9	287	13	13	46	317	7	14	4	343	8	14	35
10	288	17	13	47	318	3	14	5	343	57	14	36
11	289	20	13	47	318	58	14	6	344	46	14	38
12	290	23	13	47	319	52	14	7	345	35	14	39
13	291	26	13	48	320	46	14	8	346	24	14	40
14	292	28	13	48	321	40	14	9	347	13	14	41
15	293	30	13	48	322	34	14	9	348	2	14	42
16	294	32	13	49	323	28	14	10	348	50	14	43
17	295	34	13	49	324	22	14	11	349	38	14	45
18	296	36	13	50	325	16	14	12	350	26	14	46
19	297	38	13	50	326	10	14	13	351	14	14	47
20	298	40	13	50	327	4	14	14	352	2	14	48
21	299	42	13	51	327	56	14	15	352	50	14	49
22	300	40	13	52	328	48	14	16	353	38	14	50
23	301	40	13	52	329	40	14	17	354	26	14	52
24	302	40	13	53	330	32	14	18	355	14	14	53
25	303	40	13	54	331	24	14	19	356	2	14	54
26	304	40	13	54	332	16	14	20	356	50	14	55
27	305	39	13	55	333	7	14	21	357	38	14	56
28	306	38	13	55	333	58	14	22	358	26	14	58
29	307	37	13	56	334	49	14	24	359	13	14	59
30	308	36	13	57	335	40	14	25	360	0	15	0
	31	6			27	4			24			

24 tables that follow here (before the one below) are given in an Appendix

Κανὼν ἀναφορῶν τοῦ διὰ Βυζαντίου κλίματος, ὡρῶν ἰσημερινῶν ιε δ, μοιρῶν μγ, λεπτῶν ε΄

Ἡλίου	Κριοῦ				Ταύρου				Διδύμων			
Μοῖραι	Ἀναφοραί		Ὡριαῖοι χρόνοι	Λεπτά	Ἀναφοραί		Ὡριαῖοι χρόνοι	Λεπτά	Ἀναφοραί		Ὡριαῖοι χρόνοι	Λεπτά
α	ō	λβ	ιε	δ	ιζ	κ	ις	νδ	λη	θ	ιη	κς
β	α	δ	ιε	η	ιζ	νζ	ις	νη	λη	νθ	ιη	κθ
γ	α	λς	ιε	ια	ιη	λδ	ιζ	α	λθ	μθ	ιη	λα
δ	β	θ	ιε	ιε	ιθ	ιβ	ιζ	ε	μ	μ	ιη	λγ
ε	β	μβ	ιε	ιθ	ιθ	μθ	ιζ	η	μα	α	ιη	λε
ς	γ	ιδ	ιε	κγ	κ	κζ	ιζ	ιβ	μβ	κα	ιη	λζ
ζ	γ	μς	ιε	κς	κα	δ	ιζ	ιε	μγ	ιβ	ιη	λθ
η	δ	ιθ	ιε	λ	κα	μβ	ιζ	ιθ	μδ	γ	ιη	μα
θ	δ	να	ιε	λδ	κβ	ιθ	ιζ	κβ	μδ	νδ	ιη	μγ
ι	ε	κδ	ιε	λη	κβ	μζ	ιζ	κε	με	με	ιη	με
ια	ε	νς	ιε	μα	κγ	λζ	ιζ	κθ	μς	μα	ιη	μζ
ιβ	ς	κθ	ιε	με	κδ	ιη	ιζ	λβ	μζ	λη	ιη	μθ
ιγ	ζ	α	ιε	μθ	κδ	νη	ιζ	λε	μη	λδ	ιη	ν
ιδ	ζ	λδ	ιε	νγ	κε	λθ	ιζ	λη	μθ	λα	ιη	να
ιε	η	ζ	ιε	νς	κς	ιθ	ιζ	μα	ν	κζ	ιη	νβ
ις	η	μ	ις	ō	κζ	ō	ιζ	μδ	να	κδ	ιη	νδ
ιζ	θ	ιδ	ις	δ	κζ	μα	ιζ	μη	νβ	κ	ιη	νε
ιη	θ	μζ	ις	η	κη	κγ	ιζ	να	νγ	ιζ	ιη	νς
ιθ	ι	κα	ις	ια	κθ	δ	ιζ	νδ	νδ	ιδ	ιη	νη
κ	ι	νδ	ις	ιε	κθ	μς	ιζ	νζ	νε	ια	ιη	νθ
κα	ια	κθ	ις	ιθ	λ	λα	ιη	ō	νς	ιγ	ιθ	ō
κβ	ιβ	γ	ις	κβ	λα	ις	ιη	β	νζ	ιε	ιθ	ō
κγ	ιβ	λη	ις	κς	λβ	δ	ιη	ε	νη	ιη	ιθ	α
κδ	ιγ	ιβ	ις	λ	λβ	μς	ιη	η	νθ	κ	ιθ	α
κε	ιγ	μζ	ις	λγ	λγ	λα	ιη	ια	ξ	κγ	ιθ	β
κς	ιδ	κβ	ις	λζ	λδ	ιζ	ιη	ιγ	ξα	κε	ιθ	β
κζ	ιδ	νζ	ις	μ	λε	β	ιη	ις	ξβ	κη	ιθ	γ
κη	ιε	λβ	ις	μδ	λε	μη	ιη	ιθ	ξγ	λα	ιθ	γ
κθ	ις	η	ις	μζ	λς	λγ	ιη	κα	ξδ	λδ	ιθ	γ
λ	ις	μγ	ις	να	λζ	ιθ	ιη	κδ	ξε	λζ	ιθ	δ

δ δ δ δ δ δ

ις μγ κ λς κη ιη

24 tables that follow here (before the one below) are given in an Appendix

Table of the ascensions on the parallel of Byzantium,
15 ¼ equinoctial hours, 43 degrees, 5 minutes

Sun	Aries				Taurus				Gemini			
D	Ascensions		H-t	M	Ascensions		H-t	M	Ascensions		H-t	M
1	0	32	15	4	17	20	16	54	38	9	18	26
2	1	4	15	8	17	57	16	58	38	59	18	29
3	1	36	15	11	18	34	17	1	39	49	18	31
4	2	9	15	15	19	12	17	5	40	40	18	33
5	2	42	15	19	19	49	17	8	41	1	18	35
6	3	14	15	23	20	27	17	12	42	21	18	37
7	3	46	15	26	21	4	17	15	43	12	18	39
8	4	19	15	30	21	42	17	19	44	3	18	41
9	4	51	15	34	22	19	17	22	44	54	18	43
10	5	24	15	38	22	47	17	25	45	45	18	45
11	5	56	15	41	23	37	17	29	46	41	18	47
12	6	29	15	45	24	18	17	32	47	38	18	49
13	7	1	15	49	24	58	17	35	48	34	18	50
14	7	34	15	53	25	39	17	38	49	31	18	51
15	8	7	15	56	26	19	17	41	50	27	18	52
16	8	40	16	0	27	0	17	44	51	24	18	54
17	9	14	16	4	27	41	17	48	52	20	18	55
18	9	47	16	8	28	23	17	51	53	17	18	56
19	10	21	16	11	29	4	17	54	54	14	18	58
20	10	54	16	15	29	46	17	57	55	11	18	59
21	11	29	16	19	30	31	18	0	56	13	19	0
22	12	3	16	22	31	16	18	2	57	15	19	0
23	12	38	16	26	32	4	18	5	58	18	19	1
24	13	12	16	30	32	46	18	8	59	20	19	1
25	13	47	16	33	33	31	18	11	60	23	19	2
26	14	22	16	37	34	17	18	13	61	25	19	2
27	14	57	16	40	35	2	18	16	62	28	19	3
28	15	32	16	44	35	48	18	19	63	31	19	3
29	16	8	16	47	36	33	18	21	64	34	19	3
30	16	43	16	51	37	19	18	24	65	37	19	4
		4		4		4		4		4		4
	16	43			20	36			28	18		

Κανὼν ἀναφορῶν τοῦ διὰ Βυζαντίου κλίματος, ὡρῶν ἰσημερινῶν ιε δ, μοιρῶν μγ, λεπτῶν ε΄

Ἡλίου	Καρκίνου				Λέοντος				Παρθένου			
Μοῖραι	Ἀναφοραί		Ὡριαῖοι χρόνοι	Λεπτά	Ἀναφοραί		Ὡριαῖοι χρόνοι	Λεπτά	Ἀναφοραί		Ὡριαῖοι χρόνοι	Λεπτά
α	ξς	με	ιθ	γ	ργ	θ	ιη	κα	ρμβ	κβ	ις	μζ
β	ξζ	νγ	ιθ	γ	ρδ	κς	ιη	ιθ	ρμγ	μ	ις	μδ
γ	ξθ	α	ιθ	γ	ρε	μδ	ιη	ις	ρμδ	νη	ις	μ
δ	ο	θ	ιθ	β	ρζ	β	ιη	ιγ	ρμς	ις	ις	λζ
ε	οα	ιη	ιθ	β	ρη	κ	ιη	ια	ρμζ	ιδ	ις	λγ
ς	οβ	κς	ιθ	α	ρθ	λη	ιη	η	ρμη	νγ	ις	λ
ζ	ογ	λε	ιθ	ō	ρι	νς	ιη	ε	ρν	ια	ις	κς
η	οδ	μγ	ιθ	α	ριβ	ιδ	ιη	β	ρνα	κθ	ις	κβ
θ	οε	νβ	ιθ	ō	ριγ	λβ	ιη	ō	ρνβ	μζ	ις	ιθ
ι	οζ	α	ιη	νθ	ριδ	να	ιζ	νζ	ρνδ	ε	ις	ιε
ια	οη	ιγ	ιη	νη	ρις	ι	ιζ	νδ	ρνε	κγ	ις	ια
ιβ	οθ	κς	ιη	νς	ριζ	κη	ιζ	να	ρνς	μα	ις	η
ιγ	π	λη	ιη	νε	ριη	μζ	ιζ	μη	ρνζ	νθ	ις	δ
ιδ	πα	να	ιη	νδ	ρκ	ς	ιζ	μδ	ρνθ	ιζ	ις	ō
ιε	πγ	γ	ιη	νβ	ρκα	κδ	ιζ	μα	ρξ	λε	ιε	νς
ις	πδ	ις	ιη	να	ρκβ	μγ	ιζ	λη	ρξα	νγ	ιε	νγ
ιζ	πε	κθ	ιη	ν	ρκδ	β	ιζ	λε	ρξγ	ια	ιε	μθ
ιη	πς	μβ	ιη	μθ	ρκε	κα	ιζ	λβ	ρξδ	κθ	ιε	με
ιθ	πζ	νς	ιη	μζ	ρκς	λθ	ιζ	κθ	ρξε	μζ	ιε	μα
κ	πθ	θ	ιη	με	ρκζ	νη	ιζ	κε	ρξζ	ε	ιε	λη
κα	ϟ	κε	ιη	μγ	ρκθ	ιζ	ιζ	κβ	ρξη	κς	ιε	λδ
κβ	ϟα	μα	ιη	μα	ρλ	λε	ιζ	ιθ	ρξθ	μ	ιε	λ
κγ	ϟβ	νζ	ιη	λθ	ρλα	νδ	ιζ	ιε	ρο	νζ	ιε	κς
κδ	ϟδ	ιγ	ιη	λζ	ρλγ	ιβ	ιζ	ιβ	ροβ	ιε	ιε	κγ
κε	ϟε	κθ	ιη	λε	ρλδ	λα	ιζ	η	ρογ	λβ	ιε	ιθ
κς	ϟς	με	ιη	λγ	ρλε	ν	ιζ	ε	ροδ	ν	ιε	ιε
κζ	ϟη	α	ιη	λα	ρλζ	θ	ιζ	α	ρος	ζ	ιε	ια
κη	ϟθ	ιη	ιη	κθ	ρλη	κζ	ις	νη	ροζ	κε	ιε	η
κθ	ρ	λε	ιη	κς	ρλθ	με	ις	νδ	ροη	μβ	ιε	δ
λ	ρα	να	ιη	κδ	ρμα	δ	ις	να	ρπ	ō	ιε	ō
		δ		δ		δ		δ		δ		δ
	λς	ιδ			λθ	ιγ			λη	νς		

Table of the ascensions on the parallel of Byzantium,
15 ¼ equinoctial hours, 43 degrees, 5 minutes

Sun	Cancer				Leo				Virgo			
D	Ascensions		H-t	M	Ascensions		H-t	M	Ascensions		H-t	M
1	66	45	19	3	103	9	18	21	142	22	16	47
2	67	53	19	3	104	26	18	19	143	40	16	44
3	69	1	19	3	105	44	18	16	144	58	16	40
4	70	9	19	2	107	2	18	13	146	16	16	37
5	71	18	19	2	108	20	18	11	147	14	16	33
6	72	26	19	1	109	38	18	8	148	53	16	30
7	73	35	19	0	110	56	18	5	150	11	16	26
8	74	43	19	1	112	14	18	2	151	29	16	22
9	75	52	19	0	113	32	18	0	152	47	16	19
10	77	1	18	59	114	51	17	57	154	5	16	15
11	78	13	18	58	116	10	17	54	155	23	16	11
12	79	26	18	56	117	28	17	51	156	41	16	8
13	80	38	18	55	118	47	17	48	157	59	16	4
14	81	51	18	54	120	6	17	44	159	17	16	0
15	83	3	18	52	121	24	17	41	160	35	15	56
16	84	16	18	51	122	43	17	38	161	53	15	53
17	85	29	18	50	124	2	17	35	163	11	15	49
18	86	42	18	49	125	21	17	32	164	29	15	45
19	87	56	18	47	126	39	17	29	165	47	15	41
20	89	9	18	45	127	58	17	25	167	5	15	38
21	90	25	18	43	129	17	17	22	168	26	15	34
22	91	41	18	41	130	35	17	19	169	40	15	30
23	92	57	18	39	131	54	17	15	170	57	15	26
24	94	13	18	37	133	12	17	12	172	15	15	23
25	95	29	18	35	134	31	17	8	173	32	15	19
26	96	45	18	33	135	50	17	5	174	50	15	15
27	98	1	18	31	137	9	17	1	176	7	15	11
28	99	18	18	29	138	27	16	58	177	25	15	8
29	100	35	18	26	139	45	16	54	178	42	15	4
30	101	51	18	24	141	4	16	51	180	0	15	0
		4		4		4		4		4		4
	36	14			39	13			38	56		

Κανὼν ἀναφορῶν τοῦ διὰ Βυζαντίου κλίματος, ὡρῶν ἰσημερινῶν ιε δ, μοιρῶν μγ, λεπτῶν ε΄

Ἡλίου	Ζυγοῦ				Σκορπίου				Τοξότου			
Μοῖραι	Ἀναφοραί		Ὡριαῖοι χρόνοι	Λεπτά	Ἀναφοραί		Ὡριαῖοι χρόνοι	Λεπτά	Ἀναφοραί		Ὡριαῖοι χρόνοι	Λεπτά
α	ρπα	ιη	ιδ	νς	σκ	ιδ	ιγ	ς	σνθ	κε	ια	λδ
β	ρπβ	λε	ιδ	νβ	σκα	ιγ	ιγ	β	σξ	μβ	ια	λα
γ	ρπγ	νγ	ιδ	μθ	σκβ	να	ιβ	νθ	σξα	νθ	ια	κθ
δ	ρπε	ι	ιδ	με	σκδ	ι	ιβ	νε	σξγ	ιε	ια	κζ
ε	ρπς	κη	ιδ	μα	σκε	κθ	ιβ	νβ	σξδ	λα	ια	κε
ς	ρπζ	με	ιδ	λζ	σκς	μη	ιβ	μη	σξε	μζ	ια	κγ
ζ	ρπθ	γ	ιδ	λδ	σκη	ς	ιβ	με	σξζ	γ	ια	κα
η	ρϟ	κ	ιδ	λ	σκθ	κε	ιβ	μα	σξη	ιθ	ια	ιθ
θ	ρϟα	λη	ιδ	κς	σλ	μγ	ιβ	λη	σξθ	λε	ια	ιζ
ι	ρϟβ	νε	ιδ	κβ	σλβ	β	ιβ	λε	σο	να	ια	ιε
ια	ρϟδ	ιγ	ιδ	ιθ	σλγ	κα	ιβ	λα	σοβ	δ	ια	ιγ
ιβ	ρϟε	λα	ιδ	ιε	σλδ	λθ	ιβ	κη	σογ	ιη	ια	ια
ιγ	ρϟς	μθ	ιδ	ια	σλε	νη	ιβ	κε	σοδ	λα	ια	ι
ιδ	ρϟη	ζ	ιδ	ζ	σλζ	ιζ	ιβ	κβ	σοε	μδ	ια	θ
ιε	ρϟθ	κε	ιδ	δ	σλη	λς	ιβ	ιθ	σος	νζ	ια	η
ις	σ	μγ	ιδ	ō	σλθ	νδ	ιβ	ις	σοη	θ	ια	ς
ιζ	σβ	λ	ιγ	νς	σμα	ιγ	ιβ	ιβ	σοθ	κβ	ια	ε
ιη	σγ	ιθ	ιγ	νβ	σμβ	λβ	ιβ	θ	σπ	λδ	ια	δ
ιθ	σδ	λζ	ιγ	μθ	σμγ	ν	ιβ	ς	σπα	μζ	ια	β
κ	σε	νε	ιγ	με	σμε	θ	ιβ	γ	σπβ	νθ	ια	α
κα	σζ	ιγ	ιγ	μα	σμς	κη	ιβ	ō	σπδ	η	ια	ō
κβ	ση	λα	ιγ	λη	σμζ	μς	ια	νη	σπε	ιζ	ια	ō
κγ	σθ	μθ	ιγ	λδ	σμθ	δ	ια	νε	σπς	κε	ι	νθ
κδ	σια	ζ	ιγ	λ	σν	κβ	ια	νβ	σπζ	λδ	ι	νθ
κε	σιβ	κς	ιγ	κζ	σνα	μ	ια	μθ	σπη	μβ	ι	νη
κς	σιγ	μδ	ιγ	κγ	σνβ	νη	ια	μζ	σπθ	να	ι	νη
κζ	σιε	β	ιγ	κ	σνδ	ις	ια	μδ	σϟ	νθ	ι	νζ
κη	σις	κ	ιγ	ις	σνε	λδ	ια	μα	σϟβ	ζ	ι	νζ
κθ	σιζ	λη	ιγ	ιγ	σνς	να	ια	λθ	σϟγ	ιε	ι	νζ
λ	σιη	λς	ιγ	θ	σνη	θ	ια	λς	σϟδ	κγ	ι	νς

Table of the ascensions on the parallel of Byzantium,
15 ¼ equinoctial hours, 43 degrees, 5 minutes

Sun	Libra				Scorpio				Sagittarius			
D	Ascensions		H-t	M	Ascensions		H-t	M	Ascensions		H-t	M
1	181	18	14	56	220	14	13	6	259	25	11	34
2	182	35	14	52	221	13	13	2	260	42	11	31
3	183	53	14	49	222	51	12	59	261	59	11	29
4	185	10	14	45	224	10	12	55	263	15	11	27
5	186	28	14	41	225	29	12	52	264	31	11	25
6	187	45	14	37	226	48	12	48	265	47	11	23
7	189	3	14	34	228	6	12	45	267	3	11	21
8	190	20	14	30	229	25	12	41	268	19	11	19
9	191	38	14	26	230	43	12	38	269	35	11	17
10	192	55	14	22	232	2	12	35	270	51	11	15
11	194	13	14	19	233	21	12	31	272	4	11	13
12	195	31	14	15	234	39	12	28	273	18	11	11
13	196	49	14	11	235	58	12	25	274	31	11	10
14	198	7	14	7	237	17	12	22	275	44	11	9
15	199	25	14	4	238	36	12	19	276	57	11	8
16	200	43	14	0	239	54	12	16	278	9	11	6
17	202	30	13	56	241	13	12	12	279	22	11	5
18	203	19	13	52	242	32	12	9	280	34	11	4
19	204	37	13	49	243	50	12	6	281	47	11	2
20	205	55	13	45	245	9	12	3	282	59	11	1
21	207	13	13	41	246	28	12	0	284	8	11	0
22	208	31	13	38	247	46	11	58	285	17	11	0
23	209	49	13	34	249	4	11	55	286	25	10	59
24	211	7	13	30	250	22	11	52	287	34	10	59
25	212	26	13	27	251	40	11	49	288	42	10	58
26	213	44	13	23	252	58	11	47	289	51	10	58
27	215	2	13	20	254	16	11	44	290	59	10	57
28	216	20	13	16	255	34	11	41	292	7	10	57
29	217	38	13	13	256	51	11	39	293	15	10	57
30	218	36	13	9	258	9	11	36	294	23	10	56

Κανὼν ἀναφορῶν τοῦ διὰ Βυζαντίου κλίματος, ὡρῶν ἰσημερινῶν ιε δ, μοιρῶν μγ, λεπτῶν ε΄

Ἡλίου	Αἰγοκέρωτος				Ὑδροχόου				Ἰχθύων			
Μοῖραι	Ἀναφοραί		Ὡριαῖοι χρόνοι	Λεπτά	Ἀναφοραί		Ὡριαῖοι χρόνοι	Λεπτά	Ἀναφοραί		Ὡριαῖοι χρόνοι	Λεπτά
α	σϚε	κϛ	ι	νζ	τκγ	κζ	ια	λθ	τμγ	νβ	ιγ	ιγ
β	σϚϛ	κθ	ι	νζ	τκδ	ιβ	ια	μα	τμδ	κη	ιγ	ιϛ
γ	σϚζ	λβ	ι	νζ	τκδ	νη	ια	μδ	τμε	γ	ιγ	κ
δ	σϚη	λε	ι	νη	τκε	μγ	ια	μζ	τμε	λη	ιγ	κγ
ε	σϚθ	λζ	ι	νη	τκϛ	κθ	ια	μθ	τμϛ	ιγ	ιγ	κζ
ϛ	τ	μ	ι	νθ	τκζ	ιδ	ια	νβ	τμϛ	μη	ιγ	λ
ζ	τα	μβ	ι	νθ	τκζ	νθ	ια	νε	τμζ	κβ	ιγ	λδ
η	τβ	με	ια	ō	τκη	μδ	ια	νη	τμζ	νζ	ιγ	λη
θ	τγ	μζ	ια	ō	τκθ	κθ	ιβ	ō	τμη	λα	ιγ	μα
ι	τδ	μθ	ια	α	τλ	ιδ	ιβ	γ	τμθ	ϛ	ιγ	μδ
ια	τε	μϛ	ια	β	τλ	νϛ	ιβ	ϛ	τμθ	λθ	ιγ	μη
ιβ	τϛ	μγ	ια	δ	τλα	λζ	ιβ	θ	τν	ιγ	ιγ	να
ιγ	τζ	μ	ια	ε	τλβ	ιθ	ιβ	ιβ	τν	μϛ	ιγ	νδ
ιδ	τη	λϛ	ια	ϛ	τλγ	ō	ιβ	ιϛ	τνα	κ	ιδ	ō
ιε	τθ	λγ	ια	η	τλγ	μα	ιβ	ιθ	τνα	νγ	ιδ	δ
ιϛ	τι	κθ	ια	θ	τλδ	κα	ιβ	κβ	τνβ	κϛ	ιδ	ζ
ιζ	τια	κϛ	ια	ι	τλε	β	ιβ	κε	τνβ	νθ	ιδ	ια
ιη	τιβ	κβ	ια	ια	τλε	μδ	ιβ	κη	τνγ	λα	ιδ	ιε
ιθ	τιγ	ιθ	ια	ιγ	τλϛ	κγ	ιβ	λα	τνδ	δ	ιδ	ιθ
κ	τιδ	ιε	ια	ιε	τλζ	γ	ιβ	λε	τνδ	λϛ	ιδ	κβ
κα	τιε	ϛ	ια	ιζ	τλζ	μα	ιβ	λη	τνε	θ	ιδ	κϛ
κβ	τιε	νζ	ια	ιθ	τλη	ιη	ιβ	μα	τνε	μα	ιδ	λ
κγ	τιϛ	μη	ια	κα	τλη	νϛ	ιβ	με	τνϛ	ιδ	ιδ	λδ
κδ	τιζ	λθ	ια	κγ	τλθ	λγ	ιβ	μη	τνϛ	μϛ	ιδ	λζ
κε	τιη	λ	ια	κε	τμ	ια	ιβ	νβ	τνζ	ιη	ιδ	μα
κϛ	τιθ	κ	ια	κζ	τμ	μη	ιβ	νε	τνζ	να	ιδ	με
κζ	τκ	ια	ια	κθ	τμα	κϛ	ιβ	νθ	τνη	κδ	ιδ	μθ
κη	τκα	α	ια	λα	τμβ	γ	ιγ	β	τνη	νϛ	ιδ	νβ
κθ	τκα	να	ια	λδ	τμβ	μ	ιγ	ϛ	τνθ	κη	ιδ	νϛ
λ	τκβ	μα	ια	λϛ	τμγ	ιζ	ιγ	θ	τξ	ō	ιε	ō

Table of the ascensions on the parallel of Byzantium,
15 ¼ equinoctial hours, 43 degrees, 5 minutes

Sun	Capricorn				Aquarius				Pisces			
D	Ascensions		H-t	M	Ascensions		H-t	M	Ascensions		H-t	M
1	295	26	10	57	323	27	11	39	343	52	13	13
2	296	29	10	57	324	12	11	41	344	28	13	16
3	297	32	10	57	324	58	11	44	345	3	13	20
4	298	35	10	58	325	43	11	47	345	38	13	23
5	299	37	10	58	326	29	11	49	346	13	13	27
6	300	40	10	59	327	14	11	52	346	48	13	30
7	301	42	10	59	327	59	11	55	347	22	13	34
8	302	45	11	0	328	44	11	58	347	57	13	38
9	303	47	11	0	329	29	12	0	348	31	13	41
10	304	49	11	1	330	14	12	3	349	6	13	44
11	305	46	11	2	330	56	12	6	349	39	13	48
12	306	43	11	4	331	37	12	9	350	13	13	51
13	307	40	11	5	332	19	12	12	350	46	13	54
14	308	36	11	6	333	0	12	16	351	20	14	0
15	309	33	11	8	333	41	12	19	351	53	14	4
16	310	29	11	9	334	21	12	22	352	26	14	7
17	311	26	11	10	335	2	12	25	352	59	14	11
18	312	22	11	11	335	44	12	28	353	31	14	15
19	313	19	11	13	336	23	12	31	354	4	14	19
20	314	15	11	15	337	3	12	35	354	36	14	22
21	315	6	11	17	337	41	12	38	355	9	14	26
22	315	57	11	19	338	18	12	41	355	41	14	30
23	316	48	11	21	338	56	12	45	356	14	14	34
24	317	39	11	23	339	33	12	48	356	46	14	37
25	318	30	11	25	340	11	12	52	357	18	14	41
26	319	20	11	27	340	48	12	55	357	51	14	45
27	320	11	11	29	341	26	12	59	358	24	14	49
28	321	1	11	31	342	3	13	2	358	56	14	52
29	321	51	11	34	342	40	13	6	359	28	14	56
30	322	41	11	36	343	17	13	9	360	0	15	0

Οἱ μὲν οὖν κανόνες τῶν ἀναφορῶν τῆς τε ὀρθῆς σφαίρας καὶ ἑκάστου τῶν ἑπτὰ κλιμάτων εἰσὶν οὗτοι. ῥάδιον δέ ἐστιν, ὡς προείρηται, καὶ τὰς ἐπὶ τῶν ἐν μέσῳ ὡντινωνοῦν κλιμάτων ἀναφορὰς καὶ τοὺς ὡριαίους χρόνους κατασυλλογίζεσθαι ἐν τῇ παρεμπιπτούσῃ ἑκάστοτε χρήσει, κατὰ ἀναλογίαν ἐκ τῶν ἑκατέρωθεν αὐτῆς τῆς οἰκήσεως κλιμάτων καὶ τῶν κατ᾽ αὐτὰ ἐκτεθειμένων κανόνων. ἡμεῖς δ᾽, ἐπεὶ ἡ παροῦσα σπουδὴ πρόθεσιν ἔχει περὶ τῶν ἑκάστοτε γενησομένων ψηφοφοριῶν ἐν τῇ καθ᾽ ἡμᾶς τῇδε μεγαλοπόλει βασιλίδι, πρὸς τοῖς ἄλλοις ἐξεθέμεθα καὶ κανόνα ἀναφορῶν ἐπὶ τῆς κατ᾽ αὐτὴν οἰκήσεως τὸ εὐμεταχείριστον καὶ ἕτοιμον τῆς χρήσεως πάντα τρόπον πραγματευόμενοι καὶ οἰκονομοῦντες.

These are the tables for *sphaera recta* and for each of the seven zones. For specific locations between the zones it is easy to compute the ascensions and hourly times by interpolating the distances between the adjacent zones and then using the tables. Furthermore, since in the present study we intend to refer computations [and results] for our Queen City, we presented in addition a table of ascensions for its parallel, in order to have it ready and easily available to use in any application.

29 περὶ τῆς τοῦ ἡλίου λοξώσεως ἑκάστοτε ἀπὸ τοῦ ἰσημερινοῦ, καὶ κανὼν τῆς αὐτοῦ λοξώσεως καὶ διασάφησις τοῦ αὐτοῦ κανόνος

Ἐπεὶ δὲ χρήσιμον εἰς τοὺς περὶ ἡλίου λόγους πρὸς τοῖς ἄλλοις κατά τινα διδόμενον ἑκάστοτε χρόνον ἔχοντας ἀπὸ τῆς κατὰ μῆκος αὐτοῦ ψηφοφορίας τὴν ἐποχὴν αὐτοῦ καὶ τὸν ἐπὶ τοῦ ζωδιακοῦ δηλονότι τόπον εἰδέναι καὶ ὅσον ἀφίσταται τηνικαῦτα δὴ τοῦ μεγίστου ἐν τῇ οὐρανίᾳ σφαίρᾳ κύκλου τοῦ ἰσημερινοῦ—ταὐτὸν δὲ εἰπεῖν, ὅσον ἕκαστον τῶν τοῦ ζωδιακοῦ καὶ λοξοῦ κύκλου τμημάτων, ἤτοι μοιρῶν, ἐφ᾽ οὗ φέρεται ὁ ἥλιος διέστηκε τοῦ ἰσημερινοῦ—ἐκτιθέμεθα καὶ τὸν περὶ τούτου κανόνα ἐν τῷ Προχείρῳ καταγεγραμμένον, ὡς ἂν εὐχερῶς ἑκάστοτε ἐξ αὐτοῦ τὴν περὶ τούτου εὕρεσιν ἔχοιμεν.

Ἐφοδεύεται μὲν οὖν Πτολεμαίῳ ἐν τῇ Συντάξει θαυμασιώτατα διὰ γεωμετρικῶν καὶ σφαιρικῶν ἀποδείξεων ὅση περιφέρειά ἐστιν ἀφ᾽ ἑκάστης μοίρας τοῦ ζωδιακοῦ πρὸς τὸν ἰσημερινόν, τοῦ διὰ τῶν πόλων καὶ ἀμφοτέρων τῶν κύκλων, τοῦ λοξοῦ δηλονότι καὶ τοῦ ἰσημερινοῦ, διϊόντος μεγίστου κύκλου καὶ τέμνοντος αὐτούς τε ἀμφοτέρους καὶ τὴν ὅλην σφαῖραν εἰς δύο ἶσα, ὃν δὴ καὶ τὸν αὐτόν ἐστι νοεῖν τῷ μεσημβρινῷ. καὶ δῆτα ἀνευρὼν οὕτω τὰς ποσότητας τῶν διαστάσεων τὰς καθ᾽ ἑκάστην μοῖραν, ἐκτίθεται ταύτας ἔπειτα καὶ κανονικῶς ἐφ᾽ ἑνὸς τεταρτημορίου τοῦ κύκλου, ἤτοι ϙ μοιρῶν. εὖ γὰρ δῆλον, ὡς ἶσαί εἰσιν αἱ διαστάσεις ἐπὶ τῶν τεττάρων τεταρτημορίων τοῦ ζωδιακοῦ ἑκάστης τῶν τοῦ λοξοῦ μοιρῶν πρὸς τὸν ἰσημερινὸν ἀρχομένων ἀπ᾽ αὐτῶν τῶν ἰσημερινῶν σημείων, ἤτοι τῆς ἀρχῆς τοῦ Κριοῦ καὶ τῆς ἀρχῆς τοῦ Ζυγοῦ, καὶ λοξουμένων ἐφ᾽ ἑκάτερα, βόρεια δηλαδὴ καὶ νότια, κατ᾽ ὀλίγον ἑξῆς μέχρις ἐνενηκονταμοιρίας, οἷον πρὸς τὰ βόρεια μέρη τρία δωδεκατημόρια ἀπὸ τοῦ ἰσημερινοῦ σημείου τοῦ κατὰ τὴν ἀρχὴν τοῦ Κριοῦ, αὐτὸ τὸ τοῦ Κριοῦ, τὸ τοῦ Ταύρου καὶ τὸ τῶν Διδύμων.

8 Cf. Ptol. *Proch. Kan.* vol. 1 p. 144 Halma; Ptol. *Alm.* 1.15 **10** Cf. Ptol. *Alm.* 1.14

25 τὸ[1] om. V

29 The obliquity of the Sun with a Table and its explanation

In discussions and calculations concerning the Sun it is useful to have in addition to the longitude at any given time, i.e. its position on the ecliptic, also how much [the Sun] deviates at that time from the great circle of the equator. Or, to say it another way, how many degrees each section of the inclined is away from the equator. We present this topic in a table written simply so that we can find the answer for any occasion.

Ptolemy in the *Syntaxis* explores methodically and admirably, with proofs from plane and spherical geometry, how large is the arc for each degree between the ecliptic and the equator, [measured] along a meridian. Once he finds the arcs for each degree [of the equator], he presents them in an orderly fashion for one quadrant of the circle, that is for 90°. It is evident that the arcs between the inclined and the equator recorded for each degree are the same for the four quadrants, which start at the beginning of Aries and the beginning of Libra. Then he records the obliquity, advancing slowly on the inclined in either direction to the north and to the south, in each case up to 90°. To the north of the node, at the beginning of Aries, there are three signs: Aries, Taurus, and Gemini.

Καὶ πάλιν ἕτερα τρία δωδεκατημόρια ἀπὸ τοῦ ἑτέρου ἰσημερινοῦ σημείου, ἤτοι τῆς ἀρχῆς τοῦ Ζυγοῦ ὡς πρὸς τὰ βόρεια μέρη, τὸ τῆς Παρθένου, τοῦ Λέοντος καὶ τοῦ Καρκίνου. καὶ αὖθις ἀπὸ τῶν αὐτῶν ἰσημερινῶν σημείων ἐπὶ θάτερα τὰ νότια δηλονότι μέρη ἕτερα ἓξ δωδεκατημόρια, τρία ἀπὸ τῆς ἀρχῆς τοῦ Ζυγοῦ, αὐτὸ τὸ τοῦ Ζυγοῦ, τὸ τοῦ Σκορπίου καὶ τὸ τοῦ Τοξότου καὶ ἀπὸ τῆς ἀρχῆς τοῦ Κριοῦ τὸ τῶν Ἰχθύων, τὸ τοῦ Ὑδροχόου καὶ τὸ τοῦ Αἰγοκέρωτος. ἔστι τοίνυν τὸ βορειότερον τοῦ λοξοῦ σημεῖον αὕτη ἡ ἀρχὴ τοῦ Καρκίνου ἢ τὸ τέλος τῶν Διδύμων· καὶ τὸ νοτιώτερον τὸ κατὰ διάμετρον, ἤτοι ἡ ἀρχὴ τοῦ Αἰγοκέρωτος ἢ τὸ τέλος τοῦ Τοξότου. καὶ ἴση οὖσα ἡ διάστασις τῶν αὐτῶν σημείων ἀπὸ τοῦ ἰσημερινοῦ, εὕρηται διὰ τῆς ἐποπτείας τοῦ ἀστρολαβικοῦ κρίκου περιφέρεια οὖσα μοιρῶν κγ καὶ λεπτῶν να´ τοῦ προειρημένου μεγίστου κύκλου τοῦ διὰ τῶν πόλων ἀμφοτέρων τοῦ λοξοῦ καὶ τοῦ ἰσημερινοῦ. αἱ δὲ ἑξῆς μοῖραι τῶν τοιούτων ἄκρων σημείων τοῦ βορείου τὲ καὶ νοτίου ἐφ᾽ ἑκάτερα προχωροῦσαι εἰς αὐτὰ τὰ εἰρημένα ἰσημερινὰ σημεῖα κατ᾽ ὀλίγον ἐλαττοῦσι τὴν διάστασιν τῆς λοξώσεως.

Τὸν μὲν οὖν περὶ τούτων κανόνα, ὡς ἔφημεν, ἐν ἑνὶ τεταρτημορίῳ ὁ Πτολεμαῖος ἐκτίθεται, ἐν δὲ τοῖς Προχείροις Κανόσι τοῦ Θέωνος ὁ περὶ τούτων κανὼν ὅλα τὰ τεταρτημόρια ἐκτίθεται ἀπὸ τῆς ἀρχῆς τοῦ Καρκίνου, ἤτοι τοῦ βορειοτάτου σημείου. καὶ ἔστι καὶ καλεῖται τὸ ἓν τεταρτημόριον ἀπὸ τῆς ἀρχῆς τοῦ Καρκίνου μέχρι καὶ ϛ μοιρῶν ἀπὸ βορρᾶ κατάβασις. ἐντεῦθεν πάλιν ἀπὸ τοῦ ἰσημερινοῦ σημείου τοῦ κατὰ τὸν Ζυγὸν μέχρι καὶ ϛ μοιρῶν, ἤτοι μέχρι καὶ ρπ ἀπὸ τῆς ἀρχῆς τοῦ Καρκίνου μέχρι καὶ τῆς ἀρχῆς τοῦ Αἰγοκέρωτος, ἔστι τὲ καὶ καλεῖται νότου κατάβασις. κἀκεῖθεν μέχρι καὶ ϛ μοιρῶν αὖθις, ἤτοι μέχρι καὶ σο ἀπὸ τῆς ἀρχῆς τοῦ Καρκίνου εἰς αὐτὸ τὸ κατὰ τὴν ἀρχὴν τοῦ Κριοῦ ἰσημερινὸν σημεῖον, ἔστι τὲ καὶ καλεῖται νότου ἀνάβασις. κἀκεῖθεν αὖθις μέχρι καὶ ϛ μοιρῶν, ἤτοι μέχρι τξ ἀπὸ τῆς ἀρχῆς τοῦ Κριοῦ μέχρις εἰς αὐτὴν πάλιν τὴν ἀρχὴν τοῦ Καρκίνου, ἔστι τὲ καὶ καλεῖται βορρᾶ

44 Cf. Ptol. *Alm.* 1.15

31 τὸ τοῦ Τοξότου καὶ τὸ Σκορπίου V C^{ac} : corr. C^{pc} **45** τεταρμόρια (sic) C

There are again three other signs from the other node of the equator that is beginning from Libra toward the north: Virgo, Leo and Cancer. Then again, from the same points of the equator toward the other places to the south there are six additional signs; three are from the beginning of Libra, those being Libra, Scorpio and Sagittarius and from the beginning of Aries those being Pisces, Aquarius and Capricorn. The point of the inclined furthest to the north is at the beginning of Cancer, or the end of Gemini. The southernmost [point] is diametrically opposite, that is at the beginning of Capricorn or the end of Sagittarius. The distance of these points from the equator is equal, and was observed — using the ring of the astrolabe — to be 23° 51′ of the meridian passing through the poles of the inclined and of the equator. For all other points of the ecliptic, the declination decreases gradually, as we move from the extreme points toward the nodes of the equator.

As we mentioned, Ptolemy presents a table covering these points for one quadrant. Theon, on the other hand, presents in the *Handy Tables* all quadrants, starting at the beginning of Cancer, that is from the northernmost point. The quadrant from the beginning of Cancer up to 90° is called northern descending. From the point on the equator at the sign of Libra and for 90° to the beginning of Capricorn, that is 180° from the beginning of Cancer, it is called the southern descent. From then on and for 90° to the beginning of Aries, that is 270° from the beginning of Cancer, it is called the southern ascent. From the beginning of Aries and for 90° to the beginning of Cancer — to which we arrive after a complete revolution — it is called the northern

ἀνάβασις. τοῦτον οὖν τὸν ἐν τῷ Προχείρῳ περὶ τῶν τοιούτων κανόνα ἀεὶ τοῦ εὐμεταχειριστοτέρου πρὸς τὴν χρῆσιν φροντίδα ποιούμενοι καὶ νῦν ἐν τῇ παρούσῃ πραγματείᾳ ἐκτιθέμεθα. καὶ ἔστιν ὁ κανὼν οὗτος:

Κανὼν ἡλίου λοξώσεως

Ἀριθμοί κοινοί		Ἡλίου λόξωσις			Ἀριθμοί κοινοί		Ἡλίου λόξωσις	
ō	τξ	κγ	να		ϙγ	σξζ	α	ιγ
γ	τνζ	κγ	με		ϙϛ	σξδ	β	κε
ϛ	τνδ	κγ	μγ					
θ	τνα	κγ	λϛ		ϙθ	σξα	γ	λη
ιβ	τμη	κγ	ιη		ρβ	σνη	δ	μθ
ιε	τμε	κγ	ō		ρε	σνε	ϛ	α
ιη	τμβ	κβ	λζ		ρη	σνβ	ζ	ια
κα	τλθ	κβ	ια		ρια	σμθ	η	κ
κδ	τλϛ	κα	μα		ριδ	σμϛ	θ	κη
κζ	τλγ	κα	ζ		ριζ	σμγ	ι	λε
λ	τλ	κ	λ		ρκ	σμ	ια	μ
λγ	τκζ	ιθ	ν		ρκγ	σλζ	ιβ	μγ
λϛ	τκδ	ιθ	ϛ		ρκϛ	σλδ	ιγ	με
λθ	τκα	ιη	ιθ		ρκθ	σλα[7]	ιδ	με
μβ	τιη	ιζ	κθ		ρλβ	σκη	ιε	μβ
με	τιε	ιϛ	λζ		ρλε	σκε	ιϛ	λζ
μη	τιβ	ιε	μβ		ρλη	σκβ	ιζ	κθ
να	τθ	ιδ	μδ		ρμα	σιθ	ιη	ιθ
νδ	τϛ	ιγ	με		ρμδ	σιϛ	ιθ	ϛ
νζ	τγ	ιβ	μγ		ρμζ	σιγ	ιθ	ν
ξ	τ	ια	μ		ρν	σι	κ	λ
ξγ	σϙζ	ι	λε		ρνγ	σζ	κα	ζ
ξϛ	σϙδ	θ	κη		ρνϛ	σδ	κα	μα
ξθ	σϙα	η	κ		ρνθ	σα	κβ	ια
οβ	σπη	ζ	ια		ρξβ	ρϙη	κβ	λζ
οε	σπε	ϛ	α		ρξε	ρϙε	κγ	ō
οη	σπβ	δ	μθ		ρξη	ρϙβ	κγ	ιη
πα	σοθ	γ	λη		ροα	ρπθ	κγ	λβ
πδ	σοϛ	β	κε		ροδ	ρπϛ	κγ	μγ
πζ	σογ	α	ιγ		ροζ	ρπγ	κγ	μθ
ϙ	σο	ō	ō		ρπ	ρπ	κγ	να

ascent. In the present treatise we present here the table from the *Handy Tables*, which is prepared with care for easy use. The table is as follows:

Table of Sun's Obliquity

Common Numbers		Obliquity			Common Numbers		Obliquity	
0	360	23	51		93	267	1	13
3	357	23	45		96	264	2	25
6	354	23	43					
9	351	23	36		99	261	3	38
12	348	23	18		102	258	4	49
15	345	23	0		105	255	6	1
18	342	22	37		108	252	7	11
21	339	22	11		111	248	8	20
24	336	21	41		114	246	9	28
27	333	21	7		117	243	10	35
30	330	20	30		120	240	11	40
33	327	19	50		123	237	12	43
36	324	19	6		126	234	13	45
39	321	18	19		129	231	14	45
42	318	17	29		132	228	15	42
45	315	16	37		135	225	16	37
48	312	15	42		138	222	17	29
51	309	14	44		141	218	18	19
54	306	13	45		144	216	19	6
57	303	12	43		147	213	19	50
60	300	11	40		150	210	20	30
63	297	10	35		153	207	21	7
66	294	9	28		156	204	21	41
69	291	8	20		159	201	22	11
72	288	7	11		162	198	22	37
75	285	6	1		165	195	23	0
78	282	4	49		168	192	23	18
81	279	3	38		171	189	23	32
84	276	2	25		174	186	23	43
87	273	1	13		177	183	23	49
90	270	0	0		180	180	23	51

Τὰ μὲν οὖν πρῶτα δύο σελίδια τοῦ τοιούτου κανόνος περιέχουσι τοὺς ἀριθμοὺς τῶν ἀπὸ τοῦ Καρκίνου μοιρῶν, τὸ δὲ ἑξῆς τρίτον καταγράφει τὴν ποσότητα τῆς διαστάσεως ἀπὸ τοῦ ἰσημερινοῦ ἑκάστης μοίρας. καὶ ἔστιν αὐτόθεν δῆλον ὡς, ὅταν βουλώμεθα καταλαμβάνειν ὁπόσον ἀφίσταται ὁ ἥλιος ἀπὸ τοῦ ἰσημερινοῦ, ἀνευρίσκομεν τὴν μοῖραν, ἣν ἐπέχει τηνικαῦτα τοῦ ζωδιακοῦ, ὅση ἐστὶν ἀπὸ τῆς ἀρχῆς τοῦ Καρκίνου, καὶ ταύτην θεωροῦντες ἐπὶ τοῦ εἰρημένου κανόνος εὑρίσκομεν εὐχερῶς πόσος ἀριθμὸς διαστάσεως ἀπὸ τοῦ ἰσημερινοῦ παράκειται αὐτῇ ἐν τῷ τρίτῳ σελιδίῳ τοῦ αὐτοῦ κανόνος μοιρῶν ἢ μοιρῶν καὶ ἑξηκοστῶν. δῆλον δὲ καὶ τοῦτο, ὡς, ὅταν ὁ ἀριθμὸς τῶν εἰσαγομένων μοιρῶν ἀπὸ μιᾶς μέχρις ϛ ἢ ἀπὸ σο μέχρι τξ ὑπάρχῃ, ἐν τοῖς βορειοτέροις μέρεσι πρὸς τὸν ἰσημερινὸν ἔστιν ὁ ἥλιος· ὅταν δὲ ἀπὸ ϛ μέχρι σο ὑπάρχῃ, ἐν τοῖς νοτίοις μέρεσιν. ἀλλ’ οὕτω μὲν καὶ τὰ περὶ τούτων ἱκανῶς διώρισται.

The first two columns of the present table contain the number of degrees starting at Cancer. The other column, the third, records for each degree the declination from the equator. It is evident from this that when we want to determine how far the Sun is from the equator, we find the degree where the Sun is located on the ecliptic, i.e. the distance from Cancer. Then, searching the table, we find easily the distance from the equator recorded in the third column in degrees or degrees and minutes. It is also clear, when the number of degrees is from 1° to 90° or from 270° to 360°, that the Sun is located to the north of the equator; when it is from 90° to 270°, [the Sun] is located to the south. In this manner we described the subject thoroughly.

30 περὶ τῆς εὑρέσεως τοῦ ὡροσκοποῦντος καὶ μεσουρανοῦντος ἑκάστοτε τμήματος τοῦ ζῳδιακοῦ

Ἑπόμενον δ᾿ ἂν εἴη προσθεῖναι νῦν ἐνταῦθα καὶ περὶ τοῦ ὡροσκοποῦντος τμήματος τοῦ ζῳδιακοῦ, ἤτοι τοῦ ἐν τῷ ἀνατολικῷ ὁρίζοντι εὑρισκομένου ἑκάστοτε καὶ ἐφ᾿ ἡστινοσοῦν καιρικῆς ὥρας δεδομένης ἢ ἡμέρας ἢ νυκτός. ἔτι δὲ περὶ τοῦ τηνικαῦτα μεσουρανοῦντος ὑπὲρ γῆν διαλαβεῖν καὶ διατάξασθαι πῶς ἂν ἡ τούτων εὕρεσις ἐκ τοῦ προχείρου καταλαμβάνηται, ὡς ἐξανάγκης πάντως συναναφαινομένου, καὶ τὰ ἐκ διαμέτρου τούτοις, τὸ μὲν ἐπὶ τοῦ δυτικοῦ ὁρίζοντος εἶναι τηνικαῦτα, τὸ δὲ ἐπὶ τοῦ ὑπὸ γῆν μεσουρανήματος.

Ἀνευρίσκομεν οὖν τὸ ὡροσκοποῦν, ὡς εἴρηται, τμῆμα ἐφ᾿ ἑκάστης δεδομένης ὥρας καιρικῆς ἢ καθ᾿ ἡμέραν ἢ κατὰ νύκτα τὸν τρόπον τοῦτον. καὶ ἔστω πρῶτον ἐπὶ τῆς δεδομένης καιρικῆς ὥρας καθ᾿ ἡντιναοῦν ἡμέραν—εἰσάγοντες γὰρ ἣν ἐπέχει τηνικαῦτα μοῖραν τοῦ ζῳδιακοῦ ὁ ἥλιος ἐν τῷ κανόνι τῶν ἀναφορῶν τοῦ κλίματος ἐκείνου, καθ᾿ ὃ ψηφοφοροῦμεν, καὶ ἐπὶ τοῦ δωδεκατημορίου ἐκείνου, οὗ ἐστιν ἡ τοιαύτη μοῖρα, ἀνευρίσκομεν ὅσοι παράκεινται τῇ τοιαύτῃ ἡλιακῇ μοίρᾳ καὶ ἐποχῇ ὡριαῖοι χρόνοι κατὰ τὸ τρίτον σελίδιον τοῦ κανόνος. τοὺς τοιούτους οὖν ὡριαίους χρόνους πολλαπλασιάσαντες ἐπὶ τὰς δεδομένας τηνικαῦτα καιρικὰς ἡμῖν ἡμερινὰς ὥρας, τὸν συναγόμενον ἐντεῦθεν ἀριθμὸν ἐπιλογιζόμεθα καὶ τοῦτον προστιθέαμεν τῷ ἀριθμῷ τῶν ἀναφορῶν τῷ παρακειμένῳ ἐν τῷ δευτέρῳ σελιδίῳ τοῦ κανόνος τῇ ἡλιακῇ ἐποχῇ, ἤτοι μοίρᾳ, καθ᾿ ἣν εὑρίσκεται τηνικαῦτα ὁ ἥλιος. εἶτα τὸν συναχθέντα ὅλον ἀριθμὸν τῶν ἀναφορῶν σκοποῦμεν τίνι μοίρᾳ τίνος δωδεκατημορίου παράκειται ἐν τῷ αὐτῷ κανόνι τῶν ἀναφορῶν τοῦ αὐτοῦ κλίματος, καὶ τὴν μοῖραν ἐκείνην λέγομεν τηνικαῦτα ὡροσκοπεῖν, ἤτοι ἐπὶ τοῦ ἀνατολικοῦ ὁρίζοντος εἶναι.

Συμβαίνει δὲ πολλάκις τὸν συναχθέντα ὅλον ἀριθμὸν ὑπερβαίνειν τὸν τξ ἀριθμόν, ἤτοι κύκλον ὁλόκληρον. καὶ τότε δεῖ ἀφαιρεῖν ἀπὸ τοῦ ὅλου ἀριθμοῦ τὸν ἀριθμὸν τὸν τξ, καὶ τὸν καταλειπόμενον ἀριθμὸν ὡς

3 διδομένης C **15** ὅσοι παράκεινται bis exhib. C[ac] **18** τηνικαῦτα bis exhib. C[ac]

30 Determination of the horoscope and the culmination for each section of the zodiac

The next topic to add here is the determination of the horoscope, that is finding the section of the ecliptic located on the eastern horizon of a location [on Earth] at a certain seasonal hour of a day or night; in addition, to determine and record the point of the ecliptic that culminates above the Earth and how they are determined easily. A by-product is the determination of the diametrically opposite points: the sunset in the western horizon and the culmination under the Earth.

For a seasonal hour of a day or night, we use the following method in order to find the horoscope. Let us first consider a given seasonal hour of a specific day. In the table of ascensions for the zone we find the degrees within a sign where the Sun is located [on that day]. Then we find in the third column of the table the hourly-times in degrees corresponding to that longitude. We multiply the hourly-times with the seasonal hours of the given day, and the product is added to the number of rising times appearing in the second column of the table for the longitude under consideration. Finally, we search in the table of the same zone and find which degree of a constellation corresponds to the computed rising time. The degree [of the constellation that we find] is the so-called horoscope of ascent that is [the specific degree] lies on the eastern horizon.

Many times it happens that the resulting number [of the multiplication] exceeds 360° of a complete circle. In that case, we subtract from the resulting number 360°, and the remainder is considered as a new starting point to be

ἐξ ἄλλης ἀρχῆς εἰσάγειν, ὡς εἴρηται, εἰς τὸν κανόνα καὶ ἀνευρίσκειν τὴν μοῖραν, ᾗ παράκειται, ὡς ἔφημεν, καὶ ταύτην ὡσαύτως ὡροσκοποῦσαν ἀποφαίνεσθαι. ἰστέον μέντοι καὶ τοῦτο, ὡς οἱ συναγόμενοι ἀριθμοὶ ἔστιν ὅτε οὐχ εὑρίσκονται καταγεγραμμένοι ἐν τοῖς κανόσιν, ἀλλ᾽ ἑκατέρωθεν αὐτῶν, ὀλίγῳ τινὶ πλείους καὶ ὀλίγῳ τινὶ ἐλάττους. εὑρίσκεται δὲ γεωμετρικῶς ἐξ ἀναλόγου διὰ τῶν ἑκατέρωθεν τοιούτων ἀριθμῶν, ὡς ἔφημεν, τῶν καταγραφομένων ἐν τῷ κανόνι τῶν ὑπερβαλλόντων καὶ ἐλλειπόντων τὸ ἐπιβάλλον τμῆμα τοῦ ζῳδιακοῦ. ὅντινα δὲ τρόπον τὸ ἐξ ἀναλόγου τοῖς ἐπιστήμοσι μεθοδεύεται καὶ κατὰ τίνα λόγον ἐν τοῖς ἑξῆς ἐροῦμεν, ἐγκαίρως ἐμοὶ δοκεῖ, νυνὶ δὲ τοσοῦτο προστίθημι, ὅτι κἂν μὴ τὸ ἐπιβάλλον ἠκριβωμένως ἀνευρίσκῃ καὶ ἐπιλογίζηταί τις καὶ ἀποφαίνηται τμῆμα τοῦ ζῳδιακοῦ πρὸς τὴν ἑκάστοτε χρῆσιν, ἀλλά τινα τιθείη τῶν ἐν τῷ κανόνι ἔγγιστα ἀριθμῶν, οὐ πολλή τις ἐντεῦθεν οὐδ᾽ ἀξιόλογος ἡ διαμαρτία ἔσται τῷ λέγοντι.

Ἀλλ᾽ οὕτω μὲν ἐν ταῖς ἑκάστοτε δεδομέναις καιρικαῖς ἡμεριναῖς ὥραις μεθοδεύεται ἡ εὕρεσις τοῦ ὡροσκοποῦντος τμήματος τοῦ ζῳδιακοῦ, ἐὰν δὲ ὦσιν αἱ δεδομέναι καιρικαὶ ὧραι ἐν νυκτί, οὕτω πάλιν μεθοδεύεται ἡ εὕρεσις τοῦ ὡροσκοποῦντος τηνικαῦτα τμήματος τοῦ ζῳδιακοῦ· ἐπισκοποῦμεν πόσοι παράκεινται ὡριαῖοι χρόνοι τῇ διαμετρούσῃ μοίρᾳ τὴν ἡλιακὴν ἐποχήν, ἤτοι τὴν μοῖραν, καθ᾽ ἣν εὑρίσκεται ὁ ἥλιος τηνικαῦτα, καὶ τοὺς τοιούτους ὡριαίους χρόνους, εἰ δὲ βούλεταί τις (ταὐτὸν γάρ ἐστι), τοὺς λείποντας εἰς τοὺς λ´ μετὰ τοὺς παρακειμένους ὡριαίους χρόνους τῇ ἡλιακῇ ἐποχῇ, πολλαπλασιάζομεν ἐπὶ τὰς ἀναδοθείσας καιρικὰς ἀπὸ τῆς ἡλίου δύσεως ὥρας. εἶτα ὡσαύτως, ὡς ἀναγέγραπται, ποιοῦντες, ἀνευρίσκομεν τὸ ὡροσκοποῦν τμῆμα τοῦ ζῳδιακοῦ τηνικαῦτα.

Ἐὰν δὲ καὶ τὸ μεσουρανοῦν ἑκάστοτε τμῆμα ὑπὲρ γῆν θελήσομεν εὑρεῖν, ποιήσομεν οὕτως. λαμβάνοντες τὰς ἀπὸ τῆς παρελθούσης μεσημβρίας καιρικὰς ὥρας μέχρι τῆς δεδομένης ἡμῖν ὥρας, πολλαπλασιάζομεν αὐτὰς ἐπὶ τοὺς εἰρημένους ὡριαίους χρόνους τοὺς οἰκείους, ταῖς τε καιρικαῖς ὥραις τῆς ἡμέρας καὶ ταῖς καιρικαῖς ὥραις τῆς νυκτός, εἰ καὶ νυκτεριναί εἰσι· καὶ τὸν συναγόμενον ἐκ τοῦ πολλαπλασιασμοῦ ἀριθμὸν ἐκβάλλομεν ὡς πρὸς τὰ ἑπόμενα τῶν ζῳδίων ἀπὸ τῆς ἡλιακῆς ἐποχῆς κατὰ τὸν κανόνα τῶν ἀναφορῶν

introduced in the table in order to find the degree [within a constellation], this being the corresponding horoscope. Let it also be emphasized that sometimes the resulting number is not found in the entries of the table; then [we find] the adjacent numbers, which are a little larger or smaller, and we interpolate between the numbers on each side in order to find the appropriate segment of the ecliptic. This method of interpolation is used by scientists, and we shall discuss the reason at the appropriate place.[4] For now [it suffices] to add this: if someone in whatever application does not obtain, compute and display the precise number but inserts the nearby number from the table, the resulting error is not large or worth mentioning.

With this method we find the horoscope for a section of the ecliptic, when we are given seasonal hours of a day. Whenever the seasonal hours we consider are for a night, we find the horoscope for the section of the ecliptic by following another method. We inspect and find how many hourly-times are next to the longitude diametrically opposite to the Sun's position at that day – or alternatively one computes the difference of 30° minus the hourly-times that we found next to the Sun's longitude (this gives the same result). Next, we multiply the degrees of hourly-times obtained with the seasonal hours accumulated after sunset. Thereupon, following the same steps described earlier, we find the horoscope for that specific time.

If we wish to find the point of the ecliptic at upper culmination, we proceed in the following way. We take the intervening seasonal hours from the previous midday and multiply them with the hourly-times corresponding to that day and with the hourly-times corresponding to that of the night, if it happens to involve [seasonal hours] of the night. We add the product to the rising time we find for the Sun's longitude in the table

[4] The interpolation is discussed in section 11 of the second book.

τῶν ἐπὶ τῆς ὀρθῆς σφαίρας. εἴ γε δὴ πλεονάζει ἐνίοτε, ἀφαιροῦντες αὐτοῦ κατὰ τὴν μικρῷ πρόσθεν εἰρημένην μέθοδον ἑνὸς κύκλου τμήματα τξ, καὶ ἔνθα ἂν καταντήσῃ καὶ εἰς οἵαν μοῖραν εὑρίσκηται παρακείμενος ἐπὶ τοῦ κανόνος τῶν ἀναφορῶν τῆς ὀρθῆς σφαίρας, ἐκείνην τότε φήσομεν μεσουρανεῖν. πῶς δὲ τοῦτό φαμεν τὸ ἐκβάλλειν τὸν συναγόμενον ἀριθμὸν τῶν ἀναφορῶν ἐπὶ τοῦ κανόνος τῆς ὀρθῆς σφαίρας, οὐκ ἐπὶ τοῦ κανόνος τοῦ οἰκείου κλίματος, τοῖς ἐπιμελῶς τὸν νοῦν προσέχουσι καὶ ἐπισκεπτομένοις τὰ λεγόμενα ὁ λόγος δηλώσει, ὅτι τὰς διὰ τοῦ μεσημβρινοῦ παντὸς κλίματος καὶ ἀναφορὰς τοῦ ἰσημερινοῦ ἰσοδυναμεῖν προέφημεν ταῖς παρόδοις καὶ ἀναφοραῖς τοῦ ἐπὶ τῆς ὀρθῆς σφαίρας ὁρίζοντος. καὶ οὕτω μὲν καὶ ἡ εὕρεσις τοῦ μεσουρανοῦντος ὑπὲρ γῆν τμήματος τοῦ ζωδιακοῦ μεθοδεύεται.

Ἔστι δ᾽ ὅτε καὶ δοθέντος καὶ εὑρεθέντος τοῦ ἀνατέλλοντος ζωδιακοῦ τμήματος ἐν ὥραις ἑκάσταις καιρικαῖς ἀνευρεῖν ἐντεῦθεν καὶ τὸ μεσουρανοῦν τμῆμα τοῦ ζωδιακοῦ τρόπον ἄλλον. ἀφελόντες γὰρ ἀπὸ τοῦ παρακειμένου αὐτῇ, ἤγουν τῇ ὡροσκοπούσῃ, ἀριθμοῦ τῶν ἀναφορῶν κατὰ τὸ οἰκεῖον κλίμα τεταρτημορίου τοῦ κύκλου ἀριθμόν, ἤτοι ϛ ἀναφοράς, προσθέντες τούτῳ δηλονότι κατὰ τὸν ἀναγεγραμμένον τρόπον ἑνὸς κύκλου τμήματα τξ, ὅταν ἐλλιπὴς ᾖ πρὸς τὴν ἀφαίρεσιν ὁ πρῶτος ἀριθμός, εἰς ἣν ἂν παράκειται ὁ καταληφθεὶς ἀριθμὸς μοῖραν τοῦ ζωδιακοῦ ἐπὶ τῆς ὀρθῆς σφαίρας, ταύτην φήσομεν τηνικαῦτα μεσουρανεῖν ὑπὲρ γῆν. καὶ ἀνάπαλιν ἔστιν ὅτε τοῦ μεσουρανοῦντος τμήματος τοῦ ζωδιακοῦ δοθέντος, εἴτουν εὑρεθέντος, ἀνευρίσκεται καὶ τὸ ὡροσκοποῦν, ἤτοι τὸ ἀνατέλλον τοῦ ζωδιακοῦ τμῆμα, τὸν τρόπον τοῦτον. λαβόντες γὰρ τὰς τῷ τοιούτῳ μεσουρανοῦντι τμήματι τοῦ ζωδιακοῦ παρακειμένας ἀναφορὰς ἐν τῷ κανόνι τῆς ὀρθῆς σφαίρας καὶ προσθέντες αὐτῷ πάλιν ἑνὸς τεταρτημορίου κύκλου χρόνους, εἴτουν ἀναφορὰς ϛ, ἀπὸ τοῦ ἐπισυναχθέντος ἀριθμοῦ ἀφαιροῦντες

77 εἴτουν μοῖραν post τμῆμα add. V^{sl} **77 sqq.** ἰστέον δὲ ὅτι ὁ τρόπος οὗτος παραδίδοται τῷ Πτολεμαίῳ ἐν τῷ δευτέρῳ τῆς Συντάξεως ὡς τοῦ κανόνος τῶν ἐπὶ τῆς ὀρθῆς σφαίρας ἀναφορῶν ἀπὸ τοῦ Κριοῦ ἀρχομένου, ἀλλ᾽ οὐ καθὼς ἔκκειται ἐν τῷ Προχείρῳ ἀπὸ τοῦ Αἰγοκέρωτος καὶ ἐν τῷ παρόντι βιβλίῳ. καὶ χρὴ ἐπισκέπτεσθαι τοῦτο ἐν τῇ ψηφοφορίᾳ ἑκάστοτε ὡς ἂν ἀσφαλῶς γίνηται sch. in mg. V (Greg.) C G **78** ἤγουν τῇ ὡροσκοπούσῃ G^{sl} : τῇ ὡροσκοπούσῃ V^{sl} C^{sl} **86** τοῦ ζωδιακοῦ post τμῆμα transp. C

of *sphaera recta*. In case the result exceeds 360°, we subtract a complete cycle, in the usual method described a little earlier, and the number to which one arrives and the degree found in the table of ascensions of *sphaera recta* is the point that culminates above the Earth. Why did we say we add the product to the rising time found in the table of *sphaera recta* and not that found in the table of the zone? For those who think carefully and study the text, it is clear that crossings of the meridian of any zone and risings on the equator are equivalent to crossings and risings on the horizon of *sphaera recta*. With this method we find the upper culmination for a section of the zodiac.

Sometimes we are given, or we have found, the section of the ecliptic that ascends at some specific seasonal hour and wish to find from this the point of the ecliptic that culminates [at that moment]. For this we use an alternative method. [In the table] of the zone we subtract from the ascension a quadrant, in other words 90 ascensions; if the first number is small, so that the subtraction is impossible,[5] we add 360° in the standard method. We carry the result to the table of *sphaera recta* and read off the corresponding degree within a constellation — this is the point that culminates. Conversely, we are given, or we have found the culmination of a point of the ecliptic, and we wish to find the horoscope that is the point that ascends. In the table for *sphaera recta* we find, next the degree of the ecliptic that culminates, the ascension. Then, adding to it a quadrant, we arrive at a new rising time

77ff. (note by the hand of Gregoras in the margin of V, copied also in C and G): Notice that the Table for Ascensions of *Sphaera Recta* presented by Ptolemy in the second book of the *Syntaxis* begins at Aries, which is not the same in the *Handy Tables*, and in this book where the Table begins at Capricorn. In calculations it is necessary to correct for the difference.

[5] In modern terminology it means after subtraction the resulting number is negative.

ἐνίοτε δηλονότι πάλιν ἑνὸς κύκλου τμήματα τξ, ὅταν ὑπερβαίνῃ τὸν τῶν τξ ἀριθμόν, τὸν ἐναπολειφθέντα ἀριθμόν, εἰ δὲ μὴ ὑπερβαίνει τὸν τῶν τξ ἀριθμόν, αὐτὸν ἐκεῖνον τὸν συναχθέντα ἀριθμὸν εἰσάξομεν εἰς τὸν κανόνα τῶν ἀναφορῶν τοῦ οἰκείου κλίματος, ἐφ᾽ οὗ δηλονότι ψηφοφοροῦμεν. καὶ εἰς οἷον ἂν τμῆμα τοῦ ζωδιακοῦ ὁ τοιοῦτος ἀριθμὸς παράκειται, ἐκεῖνο τηνικαῦτα φήσομεν ὡροσκοπεῖν, ἤτοι ἐπὶ τοῦ ἀνατολικοῦ ὁρίζοντος εὑρίσκεσθαι.

(whenever the result exceeds 360° we subtract a complete circle). We carry over the new rising time to the table of the zone and read the corresponding degree within a constellation; this is the horoscope, i.e. the point located on the eastern horizon.

Appendix

The 24 tables between the four on Meroë and the four on Byzantium (ch. 28)

Abbreviations
M = Μοῖραι
Λ = Λεπτά
Ὡρ χρ = Ὡριαῖοι χρόνοι

Κανὼν ἀναφορῶν τοῦ διὰ **Συήνης** δευτέρου κλίματος, ὡρῶν ιγ, μοιρῶν κγ, λεπτῶν να΄

Ἡλίου	Κριοῦ				Ταύρου				Διδύμων			
Μ	Ἀναφοραί		Ὡρ χρ	Λ	Ἀναφοραί		Ὡρ χρ	Λ	Ἀναφοραί		Ὡρ χρ	Λ
α	ō	μδ	ιε	β	κγ	κε	ιε	νδ	μθ	ιβ	ις	λς
β	α	κη	ιε	δ	κδ	ιγ	ιε	νε	ν	θ	ις	λζ
γ	β	ιβ	ιε	ε	κε	α	ιε	νζ	να	ς	ις	λη
δ	β	νς	ιε	ζ	κε	μθ	ιε	νθ	νβ	γ	ις	λθ
ε	γ	μ	ιε	θ	κς	λζ	ις	ō	νγ	α	ις	μ
ς	δ	κδ	ιε	ια	κζ	κε	ις	β	νγ	νθ	ις	μα
ζ	ε	η	ιε	ιβ	κη	ιδ	ις	γ	νδ	νζ	ις	μβ
η	ε	νγ	ιε	ιδ	κθ	γ	ις	ε	νε	νε	ις	μγ
θ	ς	λη	ιε	ις	κθ	νβ	ις	ζ	νς	νγ	ις	μδ
ι	ζ	κγ	ιε	ιη	λ	μα	ις	η	νζ	να	ις	με
ια	η	η	ιε	ιθ	λα	λβ	ις	θ	νη	νβ	ις	με
ιβ	η	νγ	ιε	κα	λβ	κγ	ις	ια	νθ	νγ	ις	μς
ιγ	θ	λη	ιε	κγ	λγ	ιδ	ις	ιγ	ξ	νδ	ις	μς
ιδ	ι	κγ	ιε	κε	λδ	ε	ις	ιδ	ξα	νε	ις	μζ
ιε	ια	η	ιε	κζ	λδ	νς	ις	ιε	ξβ	νς	ις	μζ
ις	ια	νγ	ιε	κη	λε	μζ	ις	ιζ	ξγ	νζ	ις	μη
ιζ	ιβ	λη	ιε	λ	λς	λη	ις	ιη	ξδ	νη	ις	μθ
ιη	ιγ	κγ	ιε	κβ	λζ	κθ	ις	κ	ξε	νθ	ις	μθ
ιθ	ιδ	η	ιε	λδ	λη	κ	ις	κα	ξζ	ō	ις	ν
κ	ιδ	νγ	ιε	λε	λθ	ια	ις	κγ	ξη	β	ις	ν
κα	ιε	λθ	ιε	λζ	μ	ς	ις	κδ	ξθ	ς	ις	να
κβ	ις	κε	ιε	λθ	μα	ō	ις	κε	ο	ι	ις	να
κγ	ιζ	ια	ιε	μ	μα	νδ	ις	κς	οα	ιδ	ις	να
κδ	ιζ	νζ	ιε	μβ	μβ	μη	ις	κη	οβ	ιη	ις	νβ
κε	ιη	μγ	ιε	μδ	μγ	μβ	ις	κθ	ογ	κβ	ις	νβ
κς	ιθ	κθ	ιε	με	μδ	λς	ις	λ	οδ	κς	ις	νβ
κζ	κ	ις	ιε	μζ	με	λ	ις	λα	οε	λ	ις	νβ
κη	κα	γ	ιε	μθ	μς	κε	ις	λβ	ος	λθ	ις	νβ
κθ	κα	ν	ιε	ν	μζ	κ	ις	λδ	οζ	μ	ις	νβ
λ	κβ	λζ	ιε	νβ	μη	ιε	ις	λε	οη	με	ις	νβ
	κβ	λζ			κε	λ			λ	λ		

Κανὼν ἀναφορῶν τοῦ διὰ **Συήνης** δευτέρου κλίματος, ὡρῶν ιγ, μοιρῶν κγ, λεπτῶν να΄

Ἡλίου	Καρκίνου				Λέοντος				Παρθένου			
Μ	Ἀναφοραί		Ὥρ χρ	Λ	Ἀναφοραί		Ὥρ χρ	Λ	Ἀναφοραί		Ὥρ χρ	Λ
α	οθ	να	ις	νβ	ριγ	νς	ις	λδ	ρμη	δ	ιε	ν
β	π	νζ	ις	νβ	ριε	ε	ις	λβ	ρμθ	ια	ιε	μθ
γ	πβ	γ	ις	νβ	ρις	ιδ	ις	λα	ρν	ιη	ιε	μζ
δ	πγ	ι	ις	νβ	ριζ	κγ	ις	λ	ρνα	κε	ιε	με
ε	πδ	ιζ	ις	νβ	ριη	λβ	ις	κθ	ρνβ	λβ	ιε	μδ
ς	πε	κδ	ις	να	ριθ	μα	ις	κη	ρνγ	λη	ιε	μβ
ζ	πς	λα	ις	να	ρκ	ν	ις	κς	ρνδ	μδ	ιε	μ
η	πζ	λη	ις	να	ρκα	νθ	ις	κε	ρνε	ν	ιε	λθ
θ	πη	με	ις	να	ρκγ	η	ις	κδ	ρνς	νς	ιε	λζ
ι	πθ	νβ	ις	ν	ρκδ	ιζ	ις	κγ	ρνη	β	ιε	λε
ια	Ϙα	ō	ις	ν	ρκε	κε	ις	κα	ρνθ	η	ιε	λδ
ιβ	Ϙβ	η	ις	μθ	ρκς	λε	ις	κ	ρξ	ιδ	ιε	λβ
ιγ	Ϙγ	ις	ις	μθ	ρκζ	μδ	ις	ιη	ρξα	κ	ιε	λ
ιδ	Ϙδ	κδ	ις	μη	ρκη	νγ	ις	ιζ	ρξβ	κς	ιε	κη
ιε	Ϙε	λβ	ις	μζ	ρλ	α	ις	ιε	ρξγ	λβ	ιε	κζ
ις	Ϙς	μ	ις	μζ	ρλα	θ	ις	ιδ	ρξδ	λη	ιε	κε
ιζ	Ϙζ	μη	ις	μς	ρλβ	ιζ	ις	ιγ	ρξε	μδ	ιε	κγ
ιη	Ϙη	νζ	ις	μς	ρλγ	κε	ις	ια	ρξς	ν	ιε	κα
ιθ	ρ	ς	ις	με	ρλδ	λγ	ις	θ	ρξζ	νς	ιε	ιθ
κ	ρα	ιε	ις	μδ	ρλε	μα	ις	η	ρξθ	β	ιε	ιη
κα	ρβ	κδ	ις	μδ	ρλς	μθ	ις	ζ	ρο	η	ιε	ις
κβ	ργ	λγ	ις	μγ	ρλζ	νζ	ις	ε	ροα	ιδ	ιε	ιδ
κγ	ρδ	μβ	ις	μβ	ρλθ	ε	ις	γ	ροβ	κ	ιε	ιβ
κδ	ρε	να	ις	μα	ρμ	ιγ	ις	β	ρογ	κς	ιε	ια
κε	ρζ	α	ις	μ	ρμα	κα	ις	ō	ροδ	λβ	ιε	θ
κς	ρη	ια	ις	λθ	ρμβ	κθ	ιε	νθ	ροε	λη	ιε	ζ
κζ	ρθ	κ	ις	λη	ρμγ	λς	ιε	νζ	ρος	μδ	ιε	ε
κη	ρι	κθ	ις	λζ	ρμδ	μγ	ιε	νε	ροζ	ν	ιε	δ
κθ	ρια	λη	ις	λς	ρμε	ν	ιε	νδ	ροη	νε	ιε	β
λ	ριβ	μζ	ις	λε	ρμς	νζ	ιε	νβ	ρπ	ō	ιε	ō
	λδ	β			λδ	ν			λγ	γ		

Κανὼν ἀναφορῶν τοῦ διὰ **Συήνης** δευτέρου κλίματος, ὡρῶν ιγ, μοιρῶν κγ, λεπτῶν να΄

Ἡλίου	Ζυγοῦ				Σκορπίου				Τοξότου			
Μ	Ἀναφοραί		Ὥρ. χρ.	Λ	Ἀναφοραί		Ὥρ. χρ.	Λ	Ἀναφοραί		Ὥρ. χρ.	Λ
α	ρπα	ε	ιδ	νη	σιδ	ι	ιδ	ϛ	σμη	κβ	ιγ	κδ
β	ρπβ	ι	ιδ	νϛ	σιε	ιζ	ιδ	ε	σμθ	λα	ιγ	κγ
γ	ρπγ	ιϛ	ιδ	νε	σιϛ	κδ	ιδ	γ	σν	μ	ιγ	κβ
δ	ρπδ	κβ	ιδ	νγ	σιζ	λα	ιδ	α	σνα	μθ	ιγ	κα
ε	ρπε	κη	ιδ	να	σιη	λθ	ιδ	ō	σνβ	νθ	ιγ	κ
ϛ	ρπϛ	λδ	ιδ	μθ	σιθ	μζ	ιδ	νη	σνδ	θ	ιγ	ιθ
ζ	ρπζ	μ	ιδ	μη	σκ	νε	ιδ	νζ	σνε	ιη	ιγ	ιη
η	ρπη	μϛ	ιδ	μϛ	σκβ	γ	ιδ	νε	σνϛ	κζ	ιγ	ιζ
θ	ρπθ	νβ	ιδ	μδ	σκγ	ια	ιδ	νγ	σνζ	λϛ	ιγ	ιϛ
ι	ρϙ	νη	ιδ	μβ	σκδ	ιθ	ιδ	νβ	σνη	με	ιγ	ιε
ια	ρϙβ	δ	ιδ	μα	σκε	κζ	ιδ	να	σνθ	νδ	ιγ	ιε
ιβ	ρϙγ	ι	ιδ	λθ	σκϛ	λε	ιδ	μθ	σξα	ν	ιγ	ιδ
ιγ	ρϙδ	ιϛ	ιδ	λζ	σκζ	μγ	ιδ	μζ	σξβ	ιβ	ιγ	ιδ
ιδ	ρϙε	κβ	ιδ	λε	σκη	να	ιδ	μϛ	σξγ	κ	ιγ	ιγ
ιε	ρϙϛ	κη	ιδ	λγ	σκθ	νθ	ιδ	με	σξδ	κη	ιγ	ιγ
ιϛ	ρϙζ	λδ	ιδ	λβ	σλα	ζ	ιδ	μγ	σξε	λϛ	ιγ	ιβ
ιζ	ρϙη	μ	ιδ	λ	σλβ	ιϛ	ιδ	μβ	σξϛ	μδ	ιγ	ια
ιη	ρϙθ	μϛ	ιδ	κη	σλγ	κε	ιδ	μ	σξζ	νβ	ιγ	ια
ιθ	σ	νβ	ιδ	κϛ	σλδ	λδ	ιδ	λθ	σξθ	ō	ιγ	ι
κ	σα	νη	ιδ	κε	σλε	μγ	ιδ	λζ	σο	η	ιγ	ι
κα	σγ	δ	ιδ	κγ	σλϛ	νβ	ιδ	λϛ	σοα	ιε	ιγ	θ
κβ	σδ	ι	ιδ	κα	σλη	α	ιδ	λε	σοβ	κβ	ιγ	θ
κγ	σε	ιϛ	ιδ	κ	σλθ	ι	ιδ	λδ	σογ	κθ	ιγ	θ
κδ	σϛ	κβ	ιδ	ιη	σμ	ιθ	ιδ	λβ	σοδ	λϛ	ιγ	θ
κε	σζ	κη	ιδ	ιϛ	σμα	κη	ιδ	λα	σοε	μγ	ιγ	η
κϛ	ση	λε	ιδ	ιε	σμβ	λζ	ιδ	λ	σοϛ	ν	ιγ	η
κζ	σθ	μβ	ιδ	ιγ	σμγ	μϛ	ιδ	κθ	σοζ	νζ	ιγ	η
κη	σι	μθ	ιδ	ια	σμδ	νε	ιδ	κη	σοθ	γ	ιγ	η
κθ	σια	νϛ	ιδ	ι	σμϛ	δ	ιδ	κϛ	σπ	θ	ιγ	η
λ	σιγ	γ	ιδ	η	σμζ	ιγ	ιδ	κε	σπα	ιε	ιγ	η
	λγ	γ			λδ	ι			λγ	β		

Κανὼν ἀναφορῶν τοῦ διὰ **Συήνης** δευτέρου κλίματος, ὡρῶν ιγ, μοιρῶν κγ, λεπτῶν να΄

Ἡλίου	Αἰγοκέρωτος				Ὑδροχόου				Ἰχθύων			
Μ	Ἀναφοραί		Ὡρ. χρ.	Λ	Ἀναφοραί		Ὡρ. χρ.	Λ	Ἀναφοραί		Ὡρ. χρ.	Λ
α	σπβ	κ	ιγ	η	τιβ	μ	ιγ	κς	τλη	ι	ιδ	ι
β	σπγ	κε	ιγ	η	τιγ	λε	ιγ	κη	τλη	νζ	ιδ	ια
γ	σπδ	λ	ιγ	η	τιδ	λ	ιγ	κθ	τλθ	μδ	ιδ	ιγ
δ	σπε	λδ	ιγ	η	τιε	κδ	ιγ	λ	τμ	λα	ιδ	ιε
ε	σπς	λη	ιγ	η	τις	ιη	ιγ	λα	τμα	ιζ	ιδ	ις
ς	σπζ	μβ	ιγ	θ	τιζ	ιβ	ιγ	λβ	τμβ	γ	ιδ	ιη
ζ	σπη	μς	ιγ	θ	τιη	ς	ιγ	λδ	τμβ	μθ	ιδ	κ
η	σπθ	ν	ιγ	θ	τιθ	ō	ιγ	λε	τμγ	λε	ιδ	κα
θ	σϟ	νδ	ιγ	θ	τιθ	νδ	ιγ	λς	τμδ	κα	ιδ	κγ
ι	σϟα	νη	ιγ	ι	τκ	μη	ιγ	λζ	τμε	ζ	ιδ	κε
ια	σϟγ	ō	ιγ	ι	τκα	μ	ιγ	λθ	τμε	νβ	ιδ	κς
ιβ	σϟδ	α	ιγ	ια	τκβ	λα	ιγ	μ	τμς	λζ	ιδ	κη
ιγ	σϟε	β	ιγ	ια	τκγ	κβ	ιγ	μβ	τμζ	κβ	ιδ	λ
ιδ	σϟς	γ	ιγ	ιβ	τκδ	ιγ	ιγ	μγ	τμη	ζ	ιδ	λβ
ιε	σϟζ	δ	ιγ	ιγ	τκε	δ	ιγ	με	τμη	νβ	ιδ	λγ
ις	σϟη	ε	ιγ	ιγ	τκε	νε	ιγ	μς	τμθ	λζ	ιδ	λε
ιζ	σϟθ	ς	ιγ	ιδ	τκς	μς	ιγ	μζ	τν	κβ	ιδ	λζ
ιη	τ	ζ	ιγ	ιδ	τκζ	λζ	ιγ	μθ	τνα	ζ	ιδ	λθ
ιθ	τα	η	ιγ	ιε	τκη	κη	ιγ	να	τνα	νβ	ιδ	μα
κ	τβ	θ	ιγ	ις	τκθ	ιθ	ιγ	νβ	τνβ	λζ	ιδ	μβ
κα	τγ	ζ	ιγ	ις	τλ	η	ιγ	νγ	τνγ	κβ	ιδ	μδ
κβ	τδ	ε	ιγ	ιζ	τλ	νζ	ιγ	νε	τνδ	ζ	ιδ	μς
κγ	τε	γ	ιγ	ιη	τλα	μς	ιγ	νζ	τνδ	νβ	ιδ	μη
κδ	τς	α	ιγ	ιθ	τλβ	λε	ιγ	νη	τνε	λς	ιδ	μθ
κε	τς	νθ	ιγ	κ	τλγ	κγ	ιδ	ō	τνς	κ	ιδ	να
κς	τζ	νζ	ιγ	κα	τλδ	ια	ιδ	α	τνζ	δ	ιδ	νγ
κζ	τη	νδ	ιγ	κβ	τλδ	νθ	ιδ	γ	τνζ	μη	ιδ	νε
κη	τθ	να	ιγ	κγ	τλε	μη	ιδ	ε	τνη	λβ	ιδ	νς
κθ	τι	μη	ιγ	κδ	τλς	λς	ιδ	ς	τνθ	ις	ιδ	νη
λ	τια	με	ιγ	κε	τλζ	κγ	ιδ	η	τξ	ō	ιε	ō
	λ	λ			κε	λη			κβ	λζ		

Κανὼν ἀναφορῶν τοῦ διὰ τῆς **Κάτω Χώρας** τρίτου κλίματος, ὡρῶν ιδ, μοιρῶν λ, λεπτῶν κβ΄

Ἡλίου	Κριοῦ				Ταύρου				Διδύμων			
Μ	Ἀναφοραί		Ὡρ. χρ.	Λ	Ἀναφοραί		Ὡρ. χρ.	Λ	Ἀναφοραί		Ὡρ. χρ.	Λ
α	ō	μ	ιε	β	κα	λη	ις	ιβ	μς	ō	ιζ	η
β	α	κ	ιε	ε	κβ	κγ	ις	ιδ	μς	νε	ιζ	θ
γ	β	α	ιε	ζ	κγ	η	ις	ις	μζ	ν	ιζ	ι
δ	β	μβ	ιε	θ	κγ	νγ	ις	ιη	μη	μς	ιζ	ιβ
ε	γ	κγ	ιε	ιβ	κδ	λη	ις	κ	μθ	μβ	ιζ	ιγ
ς	δ	δ	ιε	ιδ	κε	κγ	ις	κβ	ν	λη	ιζ	ιδ
ζ	δ	με	ιε	ιζ	κς	η	ις	κδ	να	λδ	ιζ	ιε
η	ε	κς	ιε	ιθ	κς	νδ	ις	κς	νβ	λ	ιζ	ιζ
θ	ς	ζ	ιε	κα	κζ	μ	ις	κθ	νγ	κς	ιζ	ιη
ι	ς	μη	ιε	κδ	κη	κζ	ις	λα	νδ	κβ	ιζ	ιθ
ια	ζ	κθ	ιε	κς	κθ	ιδ	ις	λγ	νε	κβ	ιζ	κ
ιβ	η	ι	ιε	κη	λ	β	ις	λε	νς	κβ	ιζ	κα
ιγ	η	να	ιε	λα	λ	ν	ις	λς	νζ	κβ	ιζ	κβ
ιδ	θ	λβ	ιε	λγ	λα	λη	ις	λη	νη	κβ	ιζ	κβ
ιε	ι	ιγ	ιε	λε	λβ	κς	ις	μ	νθ	κβ	ιζ	κγ
ις	ι	νε	ιε	λη	λγ	ιδ	ις	μβ	ξ	κβ	ιζ	κδ
ιζ	ια	λζ	ιε	μ	λδ	β	ις	μδ	ξα	κβ	ιζ	κε
ιη	ιβ	ιθ	ιε	μβ	λδ	ν	ις	μς	ξβ	κβ	ιζ	κς
ιθ	ιγ	α	ιε	με	λε	λθ	ις	μη	ξγ	κβ	ιζ	κζ
κ	ιγ	μγ	ιε	μζ	λς	κη	ις	ν	ξδ	κβ	ιζ	κζ
κα	ιδ	κς	ιε	μθ	λζ	ιθ	ις	νβ	ξε	κε	ιζ	κη
κβ	ιε	θ	ιε	να	λη	ι	ις	νγ	ξς	κη	ιζ	κη
κγ	ιε	νβ	ιε	νδ	λθ	α	ις	νε	ξζ	λβ	ιζ	κη
κδ	ις	λε	ιε	νς	λθ	νγ	ις	νζ	ξη	λς	ιζ	κη
κε	ιζ	ιη	ιε	νη	μ	με	ις	νη	ξθ	μ	ιζ	κθ
κς	ιη	α	ις	ō	μα	λζ	ιζ	ō	ο	μδ	ιζ	κθ
κζ	ιη	μδ	ις	γ	μβ	κθ	ιζ	β	οα	μη	ιζ	κθ
κη	ιθ	κζ	ις	ε	μγ	κα	ιζ	γ	οβ	νβ	ιζ	κθ
κθ	κ	ι	ις	ζ	μδ	ιγ	Ιζ	ε	ογ	νε	ιζ	λ
λ	κ	νγ	ις	θ	με	ε	ιζ	ς	οε	ō	ιζ	λ
	κ	νγ			κδ	ιβ						

Κανὼν ἀναφορῶν τοῦ διὰ τῆς **Κάτω Χώρας** τρίτου κλίματος, ὡρῶν ιδ, μοιρῶν λ, λεπτῶν κβ΄

Ἡλίου	Καρκίνου				Λέοντος				Παρθένου			
Μ	Ἀναφοραί		Ὡρ. χρ.	Λ	Ἀναφοραί		Ὡρ. χρ.	Λ	Ἀναφοραί		Ὡρ. χρ.	Λ
α	ος	ζ	ιζ	λ	ρι	ν	Ιζ	ε	ρμς	κγ	ις	ζ
β	οζ	ιδ	ζ	κθ	ριβ	β	Ιζ	γ	ρμζ	λγ	ις	ε
γ	οη	κα	ιζ	κθ	ριγ	ιδ	Ιζ	β	ρμη	μγ	ις	γ
δ	οθ	κη	ιζ	κθ	ριδ	κς	Ιζ	ō	ρμθ	νγ	ις	ō
ε	π	λε	ιζ	κθ	ριε	λη	ις	νη	ρνα	γ	ιε	νη
ς	πα	μβ	ιζ	κη	ρις	ν	ις	νζ	ρνβ	ιγ	ιε	νς
ζ	πβ	μθ	ιζ	κη	ριη	β	ις	νε	ρνγ	κγ	ιε	νδ
η	πγ	νς	ιζ	κη	ριθ	ιδ	ις	νγ	ρνδ	λγ	ιε	να
θ	πε	γ	ιζ	κη	ρκ	κε	ις	νβ	ρνε	μγ	ιε	μθ
ι	πς	ιβ	ιζ	κζ	ρκα	λς	ις	ν	ρνς	νγ	ιε	μζ
ια	πζ	κα	ιζ	κζ	ρκβ	μζ	ις	μη	ρνη	γ	ιε	με
ιβ	πη	λ	ιζ	κς	ρκγ	νη	ις	μς	ρνθ	ιγ	ιε	μβ
ιγ	πθ	λθ	ιζ	κε	ρκε	θ	ις	μδ	ρξ	κγ	ιε	μ
ιδ	Ϙ	μη	ιζ	κδ	ρκς	κ	ις	μβ	ρξα	λγ	ιε	λη
ιε	Ϙ α	νζ	ιζ	κγ	ρκζ	λα	ις	μ	ρξβ	μγ	ιε	λε
ις	Ϙ γ	ς	ιζ	κβ	ρκη	μβ	ις	λη	ρξγ	νγ	ιε	λγ
ιζ	Ϙ δ	ις	ιζ	κβ	ρκθ	νγ	ις	λς	ρξε	γ	ιε	λα
ιη	Ϙ ε	κς	ιζ	κα	ρλα	δ	ις	λε	ρξς	ιβ	ιε	κη
ιθ	Ϙ ς	λς	ιζ	κ	ρλβ	ιε	ις	λγ	ρξζ	κα	ιε	κς
κ	Ϙ ζ	μς	ιζ	ιθ	ρλγ	κς	ις	λα	ρξη	λ	ιε	κδ
κα	Ϙ η	νζ	ιζ	ιη	ρλδ	λζ	ις	κθ	ρξθ	λθ	ιε	κα
κβ	ρ	η	ιζ	ιζ	ρλε	μη	ις	κς	ρο	μη	ιε	ιθ
κγ	ρα	ιθ	ιζ	ιε	ρλς	νθ	ις	κδ	ροα	νζ	ιε	ιζ
κδ	ρβ	λ	ιζ	ιδ	ρλη	ι	ις	κβ	ρογ	ς	ιε	ιδ
κε	ργ	μα	ιζ	ιγ	ρλθ	κα	ις	κ	ροδ	ιε	ιε	ιβ
κς	ρδ	νβ	ιζ	ιβ	ρμ	λβ	ις	ιη	ροε	κδ	ιε	θ
κζ	ρς	γ	ιζ	ι	ρμα	μγ	ις	ις	ρος	λγ	ιε	ζ
κη	ρζ	ιδ	ιζ	θ	ρμβ	νγ	ις	ιδ	ροζ	μβ	ιε	ε
κθ	ρη	κς	ιζ	η	ρμδ	γ	ις	ιβ	ροη	να	ιε	β
λ	ρθ	λζ	ιζ	ς	ρμε	ιγ	ις	θ	ρπ	ō	ιε	ō
	λδ	λζ			λε	λς			λδ	μζ		

Κανὼν ἀναφορῶν τοῦ διὰ τῆς **Κάτω Χώρας** τρίτου κλίματος, ὡρῶν ιδ, μοιρῶν λ, λεπτῶν κβ΄

Ἡλίου	Ζυγοῦ				Σκορπίου				Τοξότου			
Μ	Ἀναφοραί		Ὡρ. χρ.	Λ	Ἀναφοραί		Ὡρ. χρ.	Λ	Ἀναφοραί		Ὡρ. χρ.	Λ
α	ρπα	θ	ιδ	νη	σιε	νζ	ιγ	μη	σνα	λδ	ιβ	νβ
β	ρπβ	ιη	ιδ	νε	σιζ	ζ	ιγ	μϛ	σνβ	μϛ	ιβ	να
γ	ρπγ	κζ	ιδ	νγ	σιη	ιζ	ιγ	μδ	σνγ	νζ	ιβ	ν
δ	ρπδ	λϛ	ιδ	να	σιθ	κη	ιγ	μβ	σνε	η	ιβ	μη
ε	ρπε	με	ιδ	μη	σκ	λθ	ιγ	μ	σνϛ	ιθ	ιβ	μζ
ϛ	ρπϛ	νδ	ιδ	μϛ	σκα	ν	ιγ	λη	σνζ	λ	ιβ	μϛ
ζ	ρπη	γ	ιδ	μγ	σκγ	α	ιγ	λϛ	σνη	μα	ιβ	με
η	ρπθ	ιβ	ιδ	μα	σκδ	ιβ	ιγ	λδ	σνθ	νβ	ιβ	μγ
θ	ρϙ	κα	ιδ	λθ	σκε	κγ	ιγ	λα	σξα	γ	ιβ	μβ
ι	ρϙα	λ	ιδ	λϛ	σκϛ	λδ	ιγ	κθ	σξβ	ιδ	ιβ	μα
ια	ρϙβ	λθ	ιδ	λδ	σκζ	με	ιγ	κζ	σξγ	κδ	ιβ	μ
ιβ	ρϙγ	μη	ιδ	λβ	σκη	νϛ	ιγ	κε	σξδ	λδ	ιβ	λθ
ιγ	ρϙδ	νζ	ιδ	κθ	σλ	ζ	ιγ	κδ	σξε	μδ	ιβ	λη
ιδ	ρϙϛ	ζ	ιδ	κζ	σλα	ιη	ιγ	κβ	σξϛ	νδ	ιβ	λη
ιε	ρϙζ	ιζ	ιδ	κε	σλβ	κθ	ιγ	κ	σξη	γ	ιβ	λζ
ιϛ	ρϙη	κζ	ιδ	κβ	σλγ	μ	ιγ	ιη	σξθ	ιβ	ιβ	λϛ
ιζ	ρϙθ	λζ	ιδ	κ	σλδ	να	ιγ	ιϛ	σο	κα	ιβ	λε
ιη	σ	μζ	ιδ	ιη	σλϛ	β	ιγ	ιδ	σοα	λ	ιβ	λδ
ιθ	σα	νζ	ιδ	ιε	σλζ	ιγ	ιγ	ιβ	σοβ	λθ	ιβ	λγ
κ	σγ	ζ	ιδ	ιγ	σλη	κδ	ιγ	ι	σογ	μη	ιβ	λγ
κα	σδ	ιζ	ιδ	ια	σλθ	λε	ιγ	η	σοδ	νϛ	ιβ	λβ
κβ	σε	κζ	ιδ	θ	σμ	μϛ	ιγ	ζ	σοϛ	δ	ιβ	λβ
κγ	σϛ	λζ	ιδ	ϛ	σμα	νη	ιγ	ε	σοζ	ια	ιβ	λβ
κδ	σζ	μζ	ιδ	δ	σμγ	ι	ιγ	γ	σοη	ιη	ιβ	λβ
κε	ση	νζ	ιδ	β	σμδ	κβ	ιγ	β	σου	κε	ιβ	λα
κϛ	σι	ζ	ιδ	ō	σμε	λδ	ιγ	ō	σπ	λβ	ιβ	λα
κζ	σια	ιζ	ιγ	νζ	σμϛ	μϛ	ιβ	νη	σπα	λθ	ιβ	λα
κη	σιβ	κζ	ιγ	νε	σμζ	νη	ιβ	νζ	σπβ	μϛ	ιβ	λα
κθ	σιγ	λζ	ιγ	νγ	σμθ	ι	ιβ	νε	σπγ	νγ	ιβ	λ
λ	σιδ	μζ	ιγ	να	σν	κγ	ιβ	νδ	σπε	ō	ιβ	λ
	λδ	μζ			λε	λϛ			λδ	λζ		

Κανὼν ἀναφορῶν τοῦ διὰ τῆς **Κάτω Χώρας** τρίτου κλίματος, ὡρῶν ιδ, μοιρῶν λ, λεπτῶν κβ΄

Ἡλίου	Αἰγοκέρωτος				Ὑδροχόου				Ἰχθύων			
Μ	Ἀναφοραί		Ὡρ. χρ.	Λ	Ἀναφοραί		Ὡρ. χρ.	Λ	Ἀναφοραί		Ὡρ. χρ.	Λ
α	σπς	δ	ιβ	λ	τιε	μζ	ιβ	νε	τλθ	ν	ιγ	νγ
β	σπζ	η	ιβ	λα	τις	λθ	ιβ	νζ	τμ	λγ	ιγ	νε
γ	σπη	ιβ	ιβ	λα	τιζ	λα	ιβ	νη	τμα	ις	ιγ	νζ
δ	σπθ	ις	ιβ	λα	τιη	κγ	ιγ	ō	τμα	νθ	ιδ	ō
ε	σϟ	κ	ιβ	λα	τιθ	ιε	ιγ	β	τμβ	μβ	ιδ	β
ς	σϟα	κδ	ιβ	λβ	τκ	ζ	ιγ	γ	τμγ	κε	ιδ	δ
ζ	σϟβ	κη	ιβ	λβ	τκ	νθ	ιγ	ε	τμδ	η	ιδ	ς
η	σϟγ	λβ	ιβ	λβ	τκα	ν	ιγ	ζ	τμδ	να	ιδ	θ
θ	σϟδ	λε	ιβ	λβ	τκβ	μα	ιγ	η	τμε	λδ	ιδ	ια
ι	σϟε	λη	ιβ	λγ	τκγ	λβ	ιγ	ι	τμς	ιζ	ιδ	ιγ
ια	σϟς	λη	ιβ	λγ	τκδ	κγ	ιγ	ιβ	τμς	νθ	ιδ	ιε
ιβ	σϟζ	λη	ιβ	λδ	τκε	ι	ιγ	ιδ	τμζ	μα	ιδ	ιη
ιγ	σϟη	λη	ιβ	λε	τκε	νη	ιγ	ις	τμη	κγ	ιδ	κ
ιδ	σϟθ	λη	ιβ	λς	τκς	μς	ιγ	ιη	τμθ	ε	ιδ	κβ
ιε	τ	λη	ιβ	λζ	τκζ	λδ	ιγ	κ	τμθ	μζ	ιδ	κε
ις	τα	λη	ιβ	λη	τκη	κβ	ιγ	κβ	τν	κη	ιδ	κζ
ιζ	τβ	λη	ιβ	λη	τκθ	ι	ιγ	κδ	τνα	θ	ιδ	κθ
ιη	τγ	λη	ιβ	λθ	τκθ	νη	ιγ	κε	τνα	ν	ιδ	λβ
ιθ	τδ	λη	ιβ	μ	τλ	μς	ιγ	κζ	τνβ	λα	ιδ	λδ
κ	τε	λη	ιβ	μα	τλα	λδ	ιγ	κθ	τνγ	ιβ	ιδ	λς
κα	τς	λδ	ιβ	μβ	τλβ	κβ	ιγ	λα	τνγ	νγ	ιδ	λθ
κβ	τζ	λ	ιβ	μγ	τλγ	ς	ιγ	λδ	τνδ	λδ	ιδ	μα
κγ	τη	κς	ιβ	με	τλγ	νβ	ιγ	λς	τνε	ιε	ιδ	μγ
κδ	τθ	κβ	ιβ	μς	τλδ	λζ	ιγ	λη	τνε	νς	ιδ	μς
κε	τι	ιη	ιβ	μζ	τλε	κβ	ιγ	μ	τνς	λζ	ιδ	μη
κς	τια	ιδ	ιβ	μη	τλς	ζ	ιγ	μβ	τνζ	ιη	ιδ	να
κζ	τιβ	ι	ιβ	ν	τλς	νβ	ιγ	μδ	τνζ	νθ	ιδ	νγ
κη	τιγ	ε	ιβ	να	τλζ	λζ	ιγ	μς	τνη	λε	ιδ	νε
κθ	τιδ	ō	ιβ	νβ	τλη	κβ	ιγ	μη	τνθ	κ	ιδ	νη
λ	τιδ	νε	ιβ	νδ	τλθ	ζ	ιγ	να	τξ	ō	ιε	ō
	κθ	νε			κδ	ιβ			κ	νγ		

Κανὼν ἀναφορῶν τοῦ διὰ **Ῥόδου** τετάρτου κλίματος, ὡρῶν ιδ, μοιρῶν λς´

Ἡλίου	Κριοῦ				Ταύρου				Διδύμων			
Μ	Ἀναφοραί		Ὥρ. χρ.	Λ	Ἀναφοραί		Ὥρ. χρ.	Λ	Ἀναφοραί		Ὥρ. χρ.	Λ
α	ō	λζ	ιε	γ	ιθ	νδ	ις	κθ	μβ	να	ιζ	λθ
β	α	ιδ	ιε	ς	κ	λς	ις	λβ	μγ	μδ	ιζ	μα
γ	α	να	ιε	θ	κα	ιη	ις	λδ	μδ	λζ	ιζ	μγ
δ	β	κη	ιε	ιβ	κβ	ō	ις	λζ	με	λ	ιζ	μδ
ε	γ	ε	ιε	ιε	κβ	μβ	ις	μ	μς	κδ	ιζ	μς
ς	γ	μβ	ιε	ιη	κγ	κδ	ις	μβ	μζ	ιη	ιζ	μζ
ζ	δ	κ	ιε	κ	κδ	ς	ις	με	μη	ιβ	ιζ	μθ
η	δ	νη	ιε	κγ	κδ	μη	ις	μη	μθ	ς	ιζ	να
θ	ε	λς	ιε	κς	κε	λ	ις	ν	ν	ō	ιζ	νβ
ι	ς	ιδ	ιε	κθ	κς	ιγ	ις	νγ	να	νδ	ιζ	νδ
ια	ς	νβ	ιε	λβ	κς	νη	ις	νε	να	νβ	ιζ	νε
ιβ	ζ	λ	ιε	λε	κζ	μγ	ις	νη	νβ	ν	ιζ	νς
ιγ	η	η	ιε	λη	κη	κη	ιζ	ō	νγ	μζ	ιζ	νζ
ιδ	η	μς	ιε	μα	κθ	ιγ	ιζ	γ	νδ	μς	ιζ	νη
ιε	θ	κδ	ιε	μδ	κθ	νη	ιζ	ε	νε	μς	ιζ	νθ
ις	ι	β	ιε	μζ	λ	μγ	ιζ	ζ	νς	με	ιη	ō
ιζ	ι	μ	ιε	ν	λα	κη	ιζ	ι	νζ	μδ	ιη	α
ιη	ια	ιη	ιε	νβ	λβ	ιγ	ιζ	ιβ	νη	μγ	ιη	β
ιθ	ια	νς	ιε	νε	λγ	ō	ιζ	ιε	νθ	μβ	ιη	γ
κ	ιβ	λε	ιε	νη	λγ	μς	ιζ	ιζ	ξ	μα	ιη	δ
κα	ιγ	ιδ	ις	α	λδ	λε	ιζ	ιθ	ξα	μδ	ιη	δ
κβ	ιγ	νγ	ις	δ	λε	κδ	ιζ	κα	ξβ	μζ	ιη	ε
κγ	ιδ	λβ	ις	ζ	λς	ιγ	ιζ	κγ	ξγ	ν	ιη	ε
κδ	ιε	ιβ	ις	ι	λζ	β	ιζ	κε	ξδ	νγ	ιη	ε
κε	ιε	νβ	ις	ιβ	λζ	να	ιζ	κζ	ξε	νς	ιη	ς
κς	ις	λβ	ις	ιε	λη	μ	ιζ	λ	ξς	νθ	ιη	ς
κζ	ιζ	ιβ	ις	ιη	λθ	κθ	ιζ	λβ	ξη	γ	ιη	ς
κη	ιζ	νβ	ις	κα	μ	ιη	ιζ	λδ	ξθ	ζ	ιη	ζ
κθ	ιη	λβ	ις	κδ	μα	η	ιζ	λς	ο	ια	ιη	ζ
λ	ιθ	ιβ	ις	κς	μα	νη	ιζ	λη	οα	ιε	ιη	ζ
	ιθ	ιβ			κβ	μς			κθ	ιζ		

Κανὼν ἀναφορῶν τοῦ διὰ **Ῥόδου** τετάρτου κλίματος, ὡρῶν ιδ, μοιρῶν λς´

Ἡλίου	Καρκίνου				Λέοντος				Παρθένου			
Μ	Ἀναφοραί		Ὡρ. χρ.	Λ	Ἀναφοραί		Ὡρ. χρ.	Λ	Ἀναφοραί		Ὡρ. χρ.	Λ
α	οβ	κβ	ιη	ζ	ρζ	μδ	ιζ	λς	ρμδ	μζ	ις	κδ
β	ογ	κθ	ιη	ζ	ρη	νη	ιζ	λδ	ρμς	ō	ις	κα
γ	οδ	λς	ιη	ς	ρι	ιβ	ιζ	λβ	ρμζ	ιγ	ις	ιη
δ	οε	μγ	ιη	ς	ρια	κς	ιζ	λ	ρμη	κς	ις	ιε
ε	ος	να	ιη	ς	ριβ	μ	ιζ	κζ	ρμθ	λθ	ις	ιβ
ς	οζ	νθ	ιη	ε	ριγ	νδ	ιζ	κε	ρν	νβ	ις	ι
ζ	οθ	ζ	ιη	ε	ριε	η	ιζ	κγ	ρνβ	ε	ις	ζ
η	π	ιε	ιη	ε	ρις	κβ	ιζ	κα	ρνγ	ιη	ις	δ
θ	πα	κγ	ιη	δ	ριζ	λς	ιζ	ιθ	ρνδ	λα	ις	α
ι	πβ	λα	ιη	δ	ριη	ν	ιζ	ιζ	ρνε	μδ	ιε	νη
ια	πγ	μα	ιη	δ	ρκ	ε	ιζ	ιε	ρνς	νζ	ιε	νε
ιβ	πδ	να	ιη	β	ρκα	κ	ιζ	ιβ	ρνη	ι	ιε	νβ
ιγ	πς	α	ιη	α	ρκβ	λε	ιζ	ι	ρνθ	κγ	ιε	ν
ιδ	πζ	ιβ	ιη	ō	ρκγ	μθ	ιζ	ζ	ρξ	λς	ιε	μζ
ιε	πη	κγ	ιζ	νθ	ρκε	γ	ιζ	ε	ρξα	μθ	ιε	μδ
ις	πθ	λδ	ιζ	νη	ρκς	ιζ	ιζ	γ	ρξγ	β	ιε	μα
ιζ	ϙ	με	ιζ	νζ	ρκζ	λα	ιζ	ō	ρξδ	ιε	ιε	λη
ιη	ϙα	νς	ιζ	νς	ρκη	με	ις	νη	ρξε	κη	ιε	λε
ιθ	ϙγ	ζ	ιζ	νε	ρκθ	νθ	ις	νε	ρξς	μα	ιε	λβ
κ	ϙδ	ιη	ιζ	νδ	ρλα	ιγ	ις	νγ	ρξζ	νδ	ιε	κθ
κα	ϙε	λα	ιζ	νβ	ρλβ	κζ	ις	ν	ρξθ	ζ	ιε	κς
κβ	ϙς	μδ	ιζ	να	ρλγ	μα	ις	μη	ρο	κ	ιε	κγ
κγ	ϙζ	νζ	ιζ	μθ	ρλδ	νε	ις	με	ροα	λγ	ιε	κ
κδ	ϙθ	ι	ιζ	μζ	ρλς	θ	ις	μβ	ροβ	μς	ιε	ιη
κε	ρ	κγ	ιζ	μς	ρλζ	κγ	ις	μ	ρογ	νθ	ιε	ιε
κς	ρα	λς	ιζ	μδ	ρλη	λζ	ις	λζ	ροε	ιβ	ιε	ιβ
κζ	ρβ	μθ	ιζ	μγ	ρλθ	να	ις	λδ	ρος	κδ	ιε	θ
κη	ρδ	β	ιζ	μα	ρμα	ε	ις	λβ	ροζ	λς	ιε	ς
κθ	ρε	ις	ιζ	λθ	ρμβ	ιθ	ις	κθ	ροη	μη	ιε	γ
λ	ρς	λ	ιζ	λη	ρμγ	λγ	ις	κς	ρπ	ō	ιε	ō
	λε	ιε			λζ	γ			λς	κζ		

Κανὼν ἀναφορῶν τοῦ διὰ **Ῥόδου** τετάρτου κλίματος, ὡρῶν ιδ, μοιρῶν λς΄

Ἡλίου	Ζυγοῦ				Σκορπίου				Τοξότου			
Μ	Ἀναφοραί		Ὡρ. χρ.	Λ	Ἀναφοραί		Ὡρ. χρ.	Λ	Ἀναφοραί		Ὡρ. χρ.	Λ
α	ρπα	ιβ	ιδ	νζ	σιζ	μα	ιγ	λα	σνδ	μδ	ιβ	κα
β	ρπβ	κδ	ιδ	νδ	σιη	νε	ιγ	κη	σνε	νη	ιβ	ιθ
γ	ρπγ	λς	ιδ	να	σκ	θ	ιγ	κς	σνζ	ιβ	ιβ	ιζ
δ	ρπδ	μη	ιδ	μη	σκα	κγ	ιγ	κγ	σνη	κδ	ιβ	ις
ε	ρπς	α	ιδ	με	σκβ	λζ	ιγ	κ	σνθ	λζ	ιβ	ιδ
ς	ρπζ	ιδ	ιδ	μβ	σκγ	να	ιγ	ιη	σξ	ν	ιβ	ιγ
ζ	ρπη	κζ	ιδ	μ	σκε	ε	ιγ	ιε	σξβ	γ	ιβ	ια
η	ρπθ	μ	ιδ	λζ	σκς	ιθ	ιγ	ιβ	σξγ	ις	ιβ	θ
θ	ρϟ	νγ	ιδ	λδ	σκζ	λγ	ιγ	ι	σξδ	κθ	ιβ	η
ι	ρϟβ	ς	ιδ	λα	σκη	μζ	ιγ	ζ	σξε	μβ	ιβ	ς
ια	ρϟγ	ιθ	ιδ	κη	σλ	α	ιγ	ε	σξς	νγ	ιβ	ε
ιβ	ρϟδ	λβ	ιδ	κε	σλα	ιε	ιγ	β	σξη	δ	ιβ	δ
ιγ	ρϟε	με	ιδ	κβ	σλβ	κθ	ιγ	ō	σξθ	ιε	ιβ	γ
ιδ	ρϟς	νη	ιδ	ιθ	σλγ	μγ	ιβ	νζ	σο	κς	ιβ	β
ιε	ρϟη	ια	ιδ	ις	σλδ	νζ	ιβ	νε	σοα	λζ	ιβ	α
ις	ρϟθ	κδ	ιδ	ιγ	σλς	ια	ιβ	νγ	σοβ	μη	ιβ	ō
ιζ	σ	λζ	ιδ	ι	σλζ	κε	ιβ	ν	σογ	νθ	ια	νθ
ιη	σα	ν	ιδ	η	σλη	μ	ιβ	μη	σοε	θ	ια	νη
ιθ	σγ	γ	ιδ	ε	σλθ	νε	ιβ	με	σος	ιθ	ια	νζ
κ	σδ	ις	ιδ	β	σμα	ι	ιβ	μγ	σοζ	κθ	ια	νς
κα	σε	κθ	ιγ	νθ	σμβ	κδ	ιβ	μα	σοη	λθ	ια	νς
κβ	σς	μβ	ιγ	νς	σμγ	λη	ιβ	λθ	σοθ	μζ	ια	νε
κγ	σζ	νε	ιγ	νγ	σμδ	νβ	ιβ	λζ	σπ	νδ	ια	νε
κδ	σθ	η	ιγ	ν	σμς	ς	ιβ	λε	σπβ	α	ια	νε
κε	σι	κα	ιγ	μη	σμζ	κ	ιβ	λγ	σπγ	θ	ια	νδ
κς	σια	λδ	ιγ	με	σμη	λδ	ιβ	λ	σπδ	ιζ	ια	νδ
κζ	σιβ	μζ	ιγ	μβ	σμθ	μη	ιβ	κη	σπε	κδ	ια	νδ
κη	σιδ	ō	ιγ	λθ	σνα	β	ιβ	κς	σπς	λα	ια	νγ
κθ	σιε	ιγ	ιγ	λς	σνβ	ις	ιβ	κδ	σπζ	λη	ια	νγ
λ	σις	κζ	ιγ	λδ	σνγ	λ	ιβ	κβ	σπη	με	ια	νγ
	λς	κζ			λζ	γ			λε	ιε		

Κανὼν ἀναφορῶν τοῦ διὰ **Ῥόδου** τετάρτου κλίματος, ὡρῶν ιδ, μοιρῶν λς΄

Ἡλίου	Αἰγοκέρωτος				Ὑδροχόου				Ἰχθύων			
Μ	Ἀναφοραί		Ὡρ. χρ.	Λ	Ἀναφοραί		Ὡρ. χρ.	Λ	Ἀναφοραί		Ὡρ. χρ.	Λ
α	σπθ	μθ	ια	νγ	τιη	νβ	ιβ	κδ	τμα	κη	ιγ	λς
β	σϙ	νγ	ια	νγ	τιθ	μβ	ιβ	κς	τμβ	η	ιγ	λθ
γ	σϙα	νζ	ια	νδ	τκ	λα	ιβ	κη	τμβ	μη	ιγ	μβ
δ	σϙγ	α	ια	νδ	τκα	κ	ιβ	λ	τμγ	κη	ιγ	με
ε	σϙδ	δ	ια	νδ	τκβ	θ	ιβ	λγ	τμδ	ζ	ιγ	μη
ς	σϙε	ζ	ια	νε	τκβ	νη	ιβ	λε	τμδ	μς	ιγ	ν
ζ	σϙς	ι	ια	νε	τκγ	μζ	ιβ	λζ	τμε	κς	ιγ	νγ
η	σϙζ	ιγ	ια	νε	τκδ	λς	ιβ	λθ	τμς	ς	ιγ	νς
θ	σϙη	ις	ια	νς	τκε	κε	ιβ	μα	τμς	μς	ιγ	νθ
ι	σϙθ	ιθ	ια	νς	τκς	ιδ	ιβ	μς	τμζ	κε	ιδ	β
ια	τ	ιη	ια	νς	τκζ	ō	ιβ	με	τμη	δ	ιδ	ε
ιβ	τα	ιζ	ια	νη	τκζ	μς	ιβ	μη	τμη	μβ	ιδ	η
ιγ	τβ	ις	ια	νθ	τκη	λβ	ιβ	ν	τμθ	κ	ιδ	ι
ιδ	τγ	ιε	ιβ	ō	τκθ	ιζ	ιβ	νγ	τμθ	νη	ιδ	ιγ
ιε	τδ	ιδ	ιβ	α	τλ	β	ιβ	νε	τν	λς	ιδ	ις
ις	τε	ιγ	ιβ	β	τλ	μζ	ιβ	νζ	τνα	ιδ	ιδ	ιθ
ιζ	τς	ιβ	ιβ	γ	τλα	λβ	ιγ	ō	τνα	νβ	ιδ	κβ
ιη	τζ	ια	ιβ	δ	τλβ	ιζ	ιγ	β	τνβ	λ	ιδ	κε
ιθ	τη	η	ιβ	ε	τλγ	β	ιγ	ε	τνγ	η	ιδ	κη
κ	τθ	ς	ιβ	ς	τλγ	μζ	ιγ	ζ	τνγ	μς	ιδ	λα
κα	τι	ō	ιβ	η	τλδ	λ	ιγ	ι	τνδ	κδ	ιδ	λδ
κβ	τι	νδ	ιβ	θ	τλε	ιβ	ιγ	ιβ	τνε	β	ιδ	λζ
κγ	τια	μη	ιβ	ια	τλε	νδ	ιγ	ιε	τνε	μ	ιδ	μ
κδ	τιβ	μβ	ιβ	ιγ	τλς	λς	ιγ	ιη	τνς	ιη	ιδ	μβ
κε	τιγ	λς	ιβ	ιδ	τλζ	ιη	ιγ	κ	τνς	νε	ιδ	με
κς	τιδ	λ	ιβ	ις	τλη	ō	ιγ	κγ	τνζ	λβ	ιδ	μη
κζ	τιε	κγ	ιβ	ιζ	τλη	μβ	ιγ	κς	τνη	θ	ιδ	να
κη	τις	ις	ιβ	ιθ	τλθ	κδ	ιγ	κη	τνη	μς	ιδ	νδ
κθ	τιζ	θ	ιβ	κα	τμ	ς	ιγ	λα	τνθ	κγ	ιδ	νζ
λ	τιη	β	ιβ	κβ	τμ	μη	ιγ	λδ	τξ	ō	ιδ	ō

Κανὼν ἀναφορῶν τοῦ διὰ **Ἑλλησπόντου** πέμπτου κλίματος, ὡρῶν ιε, μοιρῶν μ, λεπτῶν νς′

Ἡλίου	Κριοῦ				Ταύρου				Διδύμων			
Μ	Ἀναφοραί		Ὡρ. χρ.	Λ	Ἀναφοραί		Ὡρ. χρ.	Λ	Ἀναφοραί		Ὡρ. χρ.	Λ
α	ō	λδ	ιε	γ	ιη	ι	ις	μς	λθ	μα	ιη	ια
β	α	η	ιε	ζ	ιη	μθ	ις	μθ	μ	λγ	ιη	ιγ
γ	α	μβ	ιε	ι	ιθ	κη	ις	νγ	μα	κε	ιη	ιε
δ	β	ις	ιε	ιδ	κ	ζ	ις	νς	μβ	ιζ	ιη	ιζ
ε	β	ν	ιε	ιζ	κ	μς	ις	νθ	μγ	θ	ιη	ιη
ς	γ	κδ	ιε	κα	κα	κε	ιζ	β	μδ	α	ιη	κ
ζ	γ	νη	ιε	κδ	κβ	δ	ιζ	ε	μδ	νγ	ιη	κβ
η	δ	λβ	ιε	κζ	κβ	μγ	ιζ	η	με	με	ιη	κδ
θ	ε	ς	ιε	λα	κγ	κβ	ιζ	ιβ	μς	λζ	ιη	κς
ι	ε	μ	ιε	λε	κδ	α	ιζ	ιε	μζ	κθ	ιη	κη
ια	ς	ιδ	ιε	λη	κδ	μγ	ιζ	ιη	μη	κς	ιη	κθ
ιβ	ς	μη	ιε	μβ	κε	κε	ιζ	κα	μθ	κγ	ιη	λα
ιγ	ζ	κβ	ιε	με	κς	ζ	ιζ	κδ	ν	κ	ιη	λβ
ιδ	ζ	νζ	ιε	μθ	κς	μθ	ιζ	κς	να	ιζ	ιη	λγ
ιε	η	λβ	ιε	νβ	κζ	λα	ιζ	κθ	νβ	ιδ	ιη	λδ
ις	θ	ζ	ιε	νς	κη	ιγ	ιζ	λβ	νγ	ια	ιη	λς
ιζ	θ	μβ	ιε	νθ	κη	νς	ιζ	λε	νδ	η	ιη	λζ
ιη	ι	ιζ	ις	γ	κθ	λθ	ιζ	λη	νε	ε	ιη	λη
ιθ	ι	νβ	ις	ς	λ	κβ	ιζ	μα	νς	γ	ιη	λθ
κ	ια	κζ	ις	θ	λα	ε	ιζ	μδ	νζ	α	ιη	μ
κα	ιβ	γ	ις	ιγ	λα	να	ιζ	μς	νη	γ	ιη	μα
κβ	ιβ	λθ	ις	ις	λβ	λζ	ιζ	μθ	νθ	ς	ιη	μβ
κγ	ιγ	ιε	ις	κ	λγ	κγ	ιζ	να	ξ	θ	ιη	μγ
κδ	ιγ	να	ις	κγ	λδ	θ	ιζ	νδ	ξα	ιβ	ιη	μγ
κε	ιδ	κζ	ις	κς	λδ	νς	ιζ	νς	ξβ	ιε	ιη	μδ
κς	ιε	δ	ις	λ	λε	μγ	ιζ	νθ	ξγ	ιη	ιη	μδ
κζ	ιε	μα	ις	λγ	λς	λ	ιη	α	ξδ	κα	ιη	με
κη	ις	ιη	ις	λς	λζ	ιζ	ιη	δ	ξε	κδ	ιη	με
κθ	ις	νε	ις	μ	λη	δ	ιη	ς	ξς	κζ	ιη	μς
λ	ις	λβ	ις	μγ	λη	να	ιη	θ	ξζ	λ	ιη	μς
	ιζ	λβ			κα	ιθ			κη	λθ		

Κανὼν ἀναφορῶν τοῦ διὰ **Ἑλλησπόντου** πέμπτου κλίματος, ὡρῶν ιε, μοιρῶν μ, λεπτῶν νϛʹ

Ἡλίου	Καρκίνου				Λέοντος				Παρθένου			
Μ	Ἀναφοραί		Ὡρ. χρ.	Λ	Ἀναφοραί		Ὡρ. χρ.	Λ	Ἀναφοραί		Ὡρ. χρ.	Λ
α	ξη	λη	ιη	με	ρδ	λθ	ιη	ϛ	ρμγ	ια	ιϛ	μ
β	ξθ	μϛ	ιη	μδ	ρε	νε	ιη	δ	ρμδ	κη	ιϛ	λϛ
γ	ο	νδ	ιη	μδ	ρζ	ια	ιη	α	ρμε	με	ιϛ	λγ
δ	οβ	β	ιη	μγ	ρη	κζ	ιζ	νθ	ρμϛ	β	ιϛ	λ
ε	ογ	ι	ιη	μγ	ρθ	μδ	ιζ	νϛ	ρμη	ιθ	ιϛ	κϛ
ϛ	οδ	ιη	ιη	μβ	ρια	α	ιζ	νδ	ρμθ	λϛ	ιϛ	κγ
ζ	οε	κϛ	ιη	μβ	ριβ	ιη	ιζ	να	ρν	νβ	ιϛ	κ
η	οϛ	λδ	ιη	μβ	ριγ	λθ	ιζ	μθ	ρνβ	η	ιϛ	ιϛ
θ	οζ	μβ	ιη	μα	ριδ	νβ	ιζ	μϛ	ρνγ	κδ	ιϛ	ιγ
ι	οη	να	ιη	μ	ριϛ	θ	ιζ	μδ	ρνδ	μ	ιϛ	θ
ια	π	γ	ιη	λθ	ριζ	κϛ	ιζ	μα	ρνε	νϛ	ιϛ	ϛ
ιβ	πα	ιε	ιη	λη	ριη	μγ	ιζ	λη	ρνζ	ιβ	ιϛ	γ
ιγ	πβ	κζ	ιη	λζ	ρκ	ō	ιζ	λε	ρνη	κη	ιε	νθ
ιδ	πγ	λθ	ιη	λζ	ρκα	ιζ	ιζ	λβ	ρνθ	μδ	ιε	νϛ
ιε	πδ	να	ιη	λδ	ρκβ	λδ	ιζ	κθ	ρξα	ō	ιε	νβ
ιϛ	πϛ	γ	ιη	λγ	ρκγ	να	ιζ	κϛ	ρξβ	ιϛ	ιε	μθ
ιζ	πζ	ιε	ιη	λβ	ρκε	η	ιζ	κδ	ρξγ	λβ	ιε	με
ιη	πη	κζ	ιη	λα	ρκϛ	κϛ	ιζ	κα	ρξδ	μη	ιε	μβ
ιθ	πθ	λθ	ιη	κθ	ρκζ	μδ	ιζ	ιη	ρξϛ	δ	ιε	λη
κ	ϟ	να	ιη	κη	ρκθ	β	ιζ	ιε	ρξζ	κ	ιε	λε
κα	ϟβ	η	ιη	κϛ	ρλ	κ	ιζ	ιβ	ρξη	λϛ	ιε	λα
κβ	ϟγ	κγ	ιη	κδ	ρλα	λη	ιζ	η	ρξθ	νβ	ιε	κη
κγ	ϟδ	λη	ιη	κβ	ρλβ	νε	ιζ	ε	ροα	η	ιε	κδ
κδ	ϟε	νγ	ιη	κ	ρλδ	ιβ	ιζ	β	ροβ	κδ	ιε	κα
κε	ϟζ	η	ιη	ιη	ρλε	κθ	ιϛ	νθ	ρογ	μ	ιε	ιζ
κϛ	ϟη	κγ	ιη	ιζ	ρλϛ	μϛ	ιϛ	νϛ	ροδ	νϛ	ιε	ιδ
κζ	ϟθ	λη	ιη	ιε	ρλη	γ	ιϛ	νγ	ροϛ	ιβ	ιε	ι
κη	ρ	νγ	ιη	ιγ	ρλθ	κ	ιϛ	μθ	ροζ	κη	ιε	ζ
κθ	ρβ	η	ιη	ια	ρμ	λζ	ιϛ	μϛ	ροη	μδ	ιε	δ
λ	ργ	κγ	ιη	ιθ	ρμα	νδ	ιϛ	μγ	ρπ	ō	ιε	ō
	λε	νγ			λη	λα			λη	ϛ		

Κανὼν ἀναφορῶν τοῦ διὰ **Ἑλλησπόντου** πέμπτου κλίματος, ὡρῶν ιε, μοιρῶν μ, λεπτῶν νς′

Ἡλίου	Ζυγοῦ				Σκορπίου				Τοξότου			
Μ	Ἀναφοραί		Ὡρ. χρ.	Λ	Ἀναφοραί		Ὡρ. χρ.	Λ	Ἀναφοραί		Ὡρ. χρ.	Λ
α	ρπα	ις	ιδ	νζ	σιθ	κγ	ιγ	ιδ	σνζ	νβ	ια	μθ
β	ρπβ	λβ	ιδ	νγ	σκ	μ	ιγ	ια	σνθ	ζ	ια	μζ
γ	ρπγ	μη	ιδ	ν	σκα	νζ	ιγ	ζ	σξ	κβ	ια	με
δ	ρπε	δ	ιδ	μς	σκγ	ιδ	ιγ	δ	σξα	λζ	ια	μγ
ε	ρπς	κ	ιδ	μγ	σκδ	λα	ιγ	α	σξβ	νβ	ια	μβ
ς	ρπζ	λς	ιδ	λθ	σκε	μη	ιβ	νη	σξδ	ζ	ια	μ
ζ	ρπη	νβ	ιδ	λς	σκζ	ε	ιβ	νε	σξε	κβ	ια	λη
η	ρϟ	η	ιδ	λγ	σκη	κβ	ιβ	νβ	σξς	λζ	ια	λς
θ	ρϟα	κδ	ιδ	κθ	σκθ	μ	ιβ	μη	σξζ	νβ	ια	λδ
ι	ρϟβ	μ	ιδ	κε	σλ	νη	ιβ	με	σξθ	ζ	ια	λβ
ια	ρϟγ	νς	ιδ	κβ	σλβ	ις	ιβ	μβ	σο	κβ	ια	λα
ιβ	ρϟε	ιβ	ιδ	ιη	σλγ	λδ	ιβ	λθ	σοα	λγ	ια	κθ
ιγ	ρϟς	κη	ιδ	ιε	σλδ	νβ	ιβ	λς	σοβ	με	ια	κη
ιδ	ρϟζ	μδ	ιδ	ια	σλς	θ	ιβ	λδ	σογ	νζ	ια	κζ
ιε	ρϟθ	ō	ιδ	η	σλζ	κς	ιβ	λα	σοε	θ	ια	κς
ις	σ	ις	ιδ	δ	σλη	μγ	ιβ	κη	σος	κα	ια	κδ
ιζ	σα	λβ	ιδ	α	σμ	ō	ιβ	κε	σοζ	λγ	ια	κγ
ιη	σβ	μη	ιγ	νζ	σμα	ιζ	ιβ	κβ	σοη	με	ια	κβ
ιθ	σδ	δ	ιγ	νδ	σμβ	λδ	ιβ	ιθ	σοθ	νζ	ια	κα
κ	σε	κ	ιγ	να	σμγ	να	ιβ	ις	σπα	θ	ια	κ
κα	σς	λς	ιγ	μζ	σμε	η	ιβ	ιδ	σπβ	ιη	ια	ιθ
κβ	σζ	νβ	ιγ	μδ	σμς	κε	ιβ	ια	σπγ	κς	ια	ιη
κγ	σθ	η	ιγ	μ	σμζ	μβ	ιβ	θ	σπδ	λδ	ια	ιζ
κδ	σι	κδ	ιγ	λζ	σμη	νθ	ιβ	ς	σπε	μβ	ια	ιζ
κε	σια	μα	ιγ	λδ	σν	ις	ιβ	δ	σπς	ν	ια	ις
κς	σιβ	νη	ιγ	λ	σνα	λγ	ιβ	α	σπζ	νη	ια	ις
κζ	σιδ	ιε	ιγ	κζ	σνβ	μθ	ια	νθ	σπθ	ς	ια	ιε
κη	σιε	λβ	ιγ	κδ	σνδ	ε	ια	νς	σϟ	ιδ	ια	ιε
κθ	σις	μθ	ιγ	κ	σνε	κα	ια	νδ	σϟα	κβ	ια	ιδ
λ	σιη	ς	ιγ	ιζ	σνς	λζ	ια	να	σϟβ	λ	ια	ιδ
	λη	ς			λη	λα			λε	νγ		

Κανὼν ἀναφορῶν τοῦ διὰ **Ἑλλησπόντου** πέμπτου κλίματος, ὡρῶν ιε, μοιρῶν μ, λεπτῶν νς΄

Ἡλίου	Αἰγοκέρωτος				Ὑδροχόου				Ἰχθύων			
Μ	Ἀναφοραί		Ὡρ. χρ.	Λ	Ἀναφοραί		Ὡρ. χρ.	Λ	Ἀναφοραί		Ὡρ. χρ.	Λ
α	σϟγ	λγ	ια	ιε	τκα	ν	ια	νδ	τμγ	ε	ιγ	κ
β	σϟδ	λς	ια	ις	τκβ	μγ	ια	νς	τμγ	μβ	ιγ	κδ
γ	σϟε	λθ	ια	ις	τκγ	λ	ια	νθ	τμδ	ιθ	ιγ	κζ
δ	σϟς	μβ	ια	ιζ	τκδ	ιζ	ιβ	α	τμδ	νς	ιγ	λ
ε	σϟζ	με	ια	ιζ	τκε	δ	ιβ	δ	τμε	λγ	ιγ	λδ
ς	σϟη	μη	ια	ιη	τκε	να	ιβ	ς	τμς	θ	ιγ	λζ
ζ	σϟθ	να	ια	ιη	τκς	λζ	ιβ	θ	τμς	με	ιγ	μ
η	τ	νδ	ια	ιθ	τκζ	κγ	ιβ	ια	τμζ	κα	ιγ	μδ
θ	τα	νζ	ια	ιθ	τκη	θ	ιβ	ιδ	τμζ	νζ	ιγ	μζ
ι	τβ	νθ	ια	κ	τκη	νε	ιβ	ις	τμη	λγ	ιγ	να
ια	τγ	νζ	ια	κα	τκθ	λη	ιβ	ιθ	τμθ	η	ιγ	νδ
ιβ	τδ	νε	ια	κβ	τλ	κα	ιβ	κβ	τμθ	μγ	ιγ	νζ
ιγ	τε	νβ	ια	κγ	τλα	δ	ιβ	κε	τν	ιη	ιδ	α
ιδ	τς	μθ	ια	κδ	τλα	μζ	ιβ	κη	τν	νγ	ιδ	δ
ιε	τζ	μς	ια	κς	τλβ	κθ	ιβ	λα	τνα	κη	ιδ	η
ις	τη	μγ	ια	κζ	τλγ	ια	ιβ	λδ	τνβ	γ	ιδ	ια
ιζ	τθ	μ	ια	κη	τλγ	νγ	ιβ	λς	τνβ	λη	ιδ	ιε
ιη	τι	λζ	ια	κθ	τλδ	λε	ιβ	λθ	τνγ	ιβ	ιδ	ιη
ιθ	τια	λδ	ια	λα	τλε	ιζ	ιβ	μβ	τνγ	μς	ιδ	κβ
κ	τιβ	λα	ια	λβ	τλε	νθ	ιβ	με	τνδ	κ	ιδ	κε
κα	τιγ	κγ	ια	λδ	τλς	λη	ιβ	μη	τνδ	νδ	ιδ	κθ
κβ	τιδ	ιε	ια	λς	τλζ	ιζ	ιβ	νβ	τνε	κη	ιδ	λβ
κγ	τιε	νδ	ια	λη	τλζ	νς	ιβ	νε	τνς	β	ιδ	λς
κδ	τιε	νθ	ια	μ	τλη	λε	ιβ	νη	τνς	λς	ιδ	λθ
κε	τις	να	ια	μβ	τλθ	ιδ	ιγ	α	τνζ	ι	ιδ	μγ
κς	τιζ	μγ	ια	μγ	τλθ	νγ	ιγ	δ	τνζ	μδ	ιδ	μς
κζ	τιη	λε	ια	με	τμ	λβ	ιγ	ζ	τνη	ιη	ιδ	ν
κη	τιθ	κζ	ια	μζ	τμα	ια	ιγ	ια	τνη	νβ	ιδ	νγ
κθ	τκ	ιη	ια	μθ	τμα	ν	ιγ	ιδ	τνθ	κ	ιδ	νς
λ	τκ	θ	ια	να	τμβ	κη	ιγ	ιζ	τξ	ō	ιε	ō

Κανὼν ἀναφορῶν τοῦ διὰ **Μέσου Πόντου** ἕκτου κλίματος, ὡρῶν ιε, μοιρῶν με, λεπτῶν λ΄

Ἡλίου	Κριοῦ				Ταύρου				Διδύμων			
Μ	Ἀναφοραί		Ὡρ. χρ.	Λ	Ἀναφοραί		Ὡρ. χρ.	Λ	Ἀναφοραί		Ὡρ. χρ.	Λ
α	ō	λ	ιε	δ	ις	λ	ιζ	γ	λς	λς	ιη	μβ
β	α	ō	ιε	η	ιζ	ε	ιζ	ζ	λζ	κε	ιη	μδ
γ	α	λα	ιε	ιβ	ιζ	μα	ιζ	ι	λη	ιδ	ιη	μς
δ	β	β	ιε	ις	ιη	ιζ	ιζ	ιδ	λθ	γ	ιη	μθ
ε	β	λγ	ιε	κ	ιη	νγ	ιζ	ιη	λθ	νβ	ιη	να
ς	γ	δ	ιε	κδ	ιθ	κθ	ιζ	κα	μ	μβ	ιη	νγ
ζ	γ	λε	ιε	κη	κ	ε	ιζ	κε	μα	λβ	ιη	νς
η	δ	ς	ιε	λβ	κ	μα	ιζ	κθ	μβ	κβ	ιη	νη
θ	δ	λζ	ιε	λς	κα	ιζ	ιζ	λβ	μγ	ιβ	ιθ	ō
ι	ε	η	ιε	μ	κα	νγ	ιζ	λς	μδ	β	ιθ	β
ια	ε	λθ	ιε	μδ	κβ	λβ	ιζ	λθ	μδ	νζ	ιθ	δ
ιβ	ς	ι	ιε	μη	κγ	ια	ιζ	μγ	με	νβ	ιθ	ε
ιγ	ς	μα	ιε	νβ	κγ	ν	ιζ	μς	μς	μζ	ιθ	ζ
ιδ	ζ	ιβ	ιε	νς	κδ	κθ	ιζ	ν	μζ	με	ιθ	η
ιε	ζ	μγ	ις	ō	κε	η	ιζ	νγ	μη	μα	ιθ	ι
ις	η	ιδ	ις	δ	κε	μζ	ιζ	νζ	μθ	λζ	ιθ	ια
ιζ	η	μς	ις	η	κς	κζ	ιη	ō	ν	λγ	ιθ	ιγ
ιη	θ	ιη	ις	ιβ	κζ	ζ	ιη	γ	να	κθ	ιθ	ιδ
ιθ	θ	ν	ις	ις	κζ	μζ	ιη	ζ	νβ	κε	ιθ	ις
κ	ι	κβ	ις	κ	κη	κζ	ιη	ι	νγ	κα	ιθ	ιζ
κα	ι	νε	ις	κδ	κθ	ια	ιη	ιγ	νδ	κγ	ιθ	ιη
κβ	ια	κη	ις	κη	κθ	νε	ιη	ις	νε	κε	ιθ	ιη
κγ	ιβ	α	ις	λβ	λ	λθ	ιη	ιθ	νς	κζ	ιθ	ιθ
κδ	ιβ	λδ	ις	λς	λα	κγ	ιη	κβ	νζ	κθ	ιθ	ιθ
κε	ιγ	ζ	ις	μ	λβ	ζ	ιη	κε	νη	λα	ιθ	κ
κς	ιγ	μ	ις	μδ	λβ	να	ιη	κη	νθ	λγ	ιθ	κ
κζ	ιδ	ιγ	ις	μη	λγ	λε	ιη	λα	ξ	λς	ιθ	κα
κη	ιδ	μζ	ις	να	λδ	ιθ	ιη	λδ	ξα	λθ	ιθ	κα
κθ	ιε	κα	ις	νε	λε	γ	ιη	λζ	ξβ	μβ	ιθ	κβ
λ	ιε	νε	ις	νθ	λε	μζ	ιη	λθ	ξγ	με	ιθ	κβ

Κανὼν ἀναφορῶν τοῦ διὰ **Μέσου Πόντου** ἕκτου κλίματος, ὡρῶν ιε, μοιρῶν με, λεπτῶν λ΄

Ἡλίου	Καρκίνου				Λέοντος				Παρθένου			
Μ	Ἀναφοραί		Ὥρ. χρ.	Λ	Ἀναφοραί		Ὥρ. χρ.	Λ	Ἀναφοραί		Ὥρ. χρ.	Λ
α	ξδ	νγ	ιθ	κβ	ρα	λη	ιη	λζ	ρμα	λς	ις	νε
β	ξς	α	ιθ	κα	ρβ	νζ	ιη	λδ	ρμβ	νς	ις	να
γ	ξζ	θ	ιθ	κα	ρδ	ις	ιη	λα	ρμδ	ις	ις	μη
δ	ξη	ιζ	ιθ	κ	ρε	λε	ιη	κη	ρμε	λς	ις	μδ
ε	ξθ	κς	ιθ	κ	ρς	νδ	ιη	κε	ρμς	νς	ις	μ
ς	ο	λε	ιθ	ιθ	ρη	ιγ	ιη	κβ	ρμη	ις	ις	λς
ζ	οα	μδ	ιθ	ιθ	ρθ	λβ	ιη	ιθ	ρμθ	λς	ις	λβ
η	οβ	νγ	ιθ	ιη	ρι	να	ιη	ις	ρν	νς	ις	κη
θ	οδ	β	ιθ	ιη	ριβ	ια	ιη	ιγ	ρνβ	ις	ις	κδ
ι	οε	ια	ιθ	ιζ	ριγ	λα	ιη	ι	ρνγ	λς	ις	κ
ια	ος	κδ	ιθ	ις	ριδ	να	ιη	ζ	ρνδ	νς	ις	ις
ιβ	οζ	λζ	ιθ	ιδ	ρις	ια	ιη	γ	ρνς	ις	ις	ιβ
ιγ	οη	ν	ιθ	ιγ	ριζ	λα	ιη	ō	ρνζ	λς	ις	η
ιδ	π	γ	ιθ	ια	ριη	να	ιζ	νζ	ρνη	νς	ις	δ
ιε	πα	ις	ιθ	ι	ρκ	ια	ιζ	νγ	ρξ	ιε	ις	ō
ις	πβ	λ	ιθ	η	ρκα	λα	ιζ	ν	ρξα	λδ	ιε	νς
ιζ	πγ	μδ	ιθ	ζ	ρκβ	να	ιζ	μς	ρξβ	νγ	ιε	νβ
ιη	πδ	νη	ιθ	ε	ρκδ	ια	ιζ	μγ	ρξδ	ιβ	ιε	μη
ιθ	πς	ιβ	ιθ	δ	ρκε	λβ	ιζ	μ	ρξε	λα	ιε	μδ
κ	πζ	κς	ιθ	β	ρκς	νγ	ιζ	λς	ρξς	ν	ιε	μ
κα	πη	μγ	ιθ	ō	ρκη	ιε	ιζ	λβ	ρξη	θ	ιε	λς
κβ	ϙ	ō	ιη	νη	ρκθ	λε	ιζ	κθ	ρξθ	κη	ιε	λβ
κγ	ϙα	ιζ	ιη	νς	ρλ	νς	ιζ	κε	ρο	κζ	ιε	κη
κδ	ϙβ	λδ	ιη	νγ	ρλβ	ις	ιζ	κα	ροβ	ς	ιε	κδ
κε	ϙγ	να	ιη	να	ρλγ	λς	ιζ	ιη	ρογ	κε	ιε	κ
κς	ϙε	η	ιη	μθ	ρλδ	νς	ιζ	ιδ	ροδ	μδ	ιε	ις
κζ	ϙς	κε	ιη	μς	ρλς	ις	ιζ	ι	ρος	γ	ιε	ιβ
κη	ϙζ	μγ	ιη	μδ	ρλζ	λς	ιζ	ζ	ροζ	κβ	ιε	η
κθ	ϙθ	α	ιη	μβ	ρλη	νς	ιζ	γ	ροη	μα	ιε	δ
λ	ρ	ιθ	ιη	λθ	ρμ	ις	ιζ	νθ	ρπ	ō	ιε	ō

Κανὼν ἀναφορῶν τοῦ διὰ **Μέσου Πόντου** ἕκτου κλίματος, ὡρῶν ιε, μοιρῶν με, λεπτῶν λʹ

Ἡλίου	Ζυγοῦ				Σκορπίου				Τοξότου			
Μ	Ἀναφοραί		Ὡρ. χρ.	Λ	Ἀναφοραί		Ὡρ. χρ.	Λ	Ἀναφοραί		Ὡρ. χρ.	Λ
α	ρπα	ιθ	ιδ	νς	σκα	δ	ιβ	νζ	σξ	νθ	ια	ιη
β	ρπβ	λη	ιδ	νβ	σκβ	κδ	ιβ	νγ	σξβ	ιζ	ια	ις
γ	ρπγ	νζ	ιδ	μη	σκγ	μδ	ιβ	ν	σξγ	λε	ια	ιδ
δ	ρπε	ις	ιδ	μδ	σκε	δ	ιβ	μς	σξδ	νβ	ια	ια
ε	ρπς	λε	ιδ	μ	σκς	κδ	ιβ	μβ	σξς	θ	ια	θ
ς	ρπζ	νδ	ιδ	λς	σκζ	μδ	ιβ	λθ	σξζ	κς	ια	ζ
ζ	ρπθ	ιγ	ιδ	λβ	σκθ	δ	ιβ	λε	σξη	μγ	ια	δ
η	ρϟ	λβ	ιδ	κη	σλ	κε	ιβ	λα	σο	ō	ια	β
θ	ρϟα	να	ιδ	κδ	σλα	μς	ιβ	κη	σοα	ιζ	ια	ō
ι	ρϟγ	ι	ιδ	κ	σλγ	ζ	ιβ	κδ	σοβ	λδ	ι	νη
ια	ρϟδ	κθ	ιδ	ις	σλδ	κη	ιβ	κα	σογ	μη	ι	νς
ιβ	ρϟε	μη	ιδ	ιβ	σλε	μθ	ιβ	ιζ	σοε	β	ι	νε
ιγ	ρϟζ	ζ	ιδ	η	σλζ	θ	ιβ	ιδ	σος	ις	ι	νγ
ιδ	ρϟη	κς	ιδ	δ	σλη	κθ	ιβ	ι	σοζ	λ	ι	νβ
ιε	ρϟθ	με	ιδ	ō	σλθ	μθ	ιβ	ζ	σοη	μδ	ι	ν
ις	σα	δ	ιγ	νς	σμα	θ	ιβ	γ	σοθ	νζ	ι	μθ
ιζ	σβ	κδ	ιγ	νβ	σμβ	κθ	ιβ	ō	σπα	ι	ι	μζ
ιη	σγ	μδ	ιγ	μη	σμγ	μθ	ια	νζ	σπβ	κγ	ι	μς
ιθ	σε	δ	ιγ	μδ	σμε	θ	ια	νγ	σπγ	λς	ι	μδ
κ	σς	κδ	ιγ	μ	σμς	κθ	ια	ν	σπδ	μθ	ι	μγ
κα	σζ	μδ	ιγ	λς	σμζ	μθ	ια	μζ	σπε	νη	ι	μβ
κβ	σθ	δ	ιγ	λβ	σμθ	θ	ια	μδ	σπζ	ζ	ι	μβ
κγ	σι	κδ	ιγ	κη	σν	κη	ια	μα	σπη	ις	ι	μα
κδ	σια	μδ	ιγ	κδ	σνα	μζ	ια	λη	σπθ	κε	ι	μα
κε	σιγ	δ	ιγ	κ	σνγ	ς	ια	λε	σϟ	λδ	ι	μ
κς	σιδ	κδ	ιγ	ις	σνδ	κε	ια	λβ	σϟα	μγ	ι	μ
κζ	σιε	μδ	ιγ	ιβ	σνε	μδ	ια	κθ	σϟβ	να	ι	λθ
κη	σιζ	δ	ιγ	θ	σνζ	γ	ια	κς	σϟγ	νθ	ι	λθ
κθ	σιη	κδ	ιγ	ε	σνη	κβ	ια	κγ	σϟε	ζ	ι	λη
λ	σιθ	μδ	ιγ	α	σνθ	μα	ια	κα	σϟς	ιε	ι	λη

Κανὼν ἀναφορῶν τοῦ διὰ **Μέσου Πόντου** ἕκτου κλίματος, ὡρῶν ιε, μοιρῶν με, λεπτῶν λ΄

Ἡλίου	Αἰγοκέρωτος				Ὑδροχόου				Ἰχθύων			
Μ	Ἀναφοραί		Ὡρ. χρ.	Λ	Ἀναφοραί		Ὡρ. χρ.	Λ	Ἀναφοραί		Ὡρ. χρ.	Λ
α	σϚς	ιη	ι	λη	τκδ	νζ	ια	κς	τμδ	λθ	ιγ	ε
β	σϚη	κα	ι	λθ	τκε	μα	ια	κς	τμε	ιγ	ιγ	θ
γ	σϚθ	κδ	ι	λθ	τκς	κε	ια	κθ	τμε	μζ	ιγ	ιβ
δ	τ	κζ	ι	μ	τκζ	θ	ια	λβ	τμς	κ	ιγ	ις
ε	τα	κθ	ι	μ	τκζ	νγ	ια	λε	τμς	νγ	ιγ	κ
ς	τβ	λα	ι	μα	τκη	λζ	ια	λη	τμζ	κς	ιγ	κδ
ζ	τγ	λγ	ι	μα	τκθ	κα	ια	μα	τμζ	νθ	ιγ	κη
η	τδ	λε	ι	μβ	τλ	ε	ια	μδ	τμη	λβ	ιγ	λβ
θ	τε	λζ	ι	μβ	τλ	μθ	ια	μζ	τμθ	ε	ιγ	λς
ι	τς	λθ	ι	μγ	τλα	λγ	ια	ν	τμθ	λη	ιγ	μ
ια	τζ	λε	ι	μδ	τλβ	ιγ	ια	νγ	τν	ι	ιγ	μδ
ιβ	τη	λα	ι	μς	τλβ	νγ	ια	νζ	τν	μβ	ιγ	μη
ιγ	τθ	κζ	ι	μζ	τλγ	λγ	ιβ	ō	τνα	ιδ	ιγ	νβ
ιδ	τι	κγ	ι	μθ	τλδ	ιγ	ιβ	γ	τνα	μς	ιγ	νς
ιε	τια	ιθ	ι	ν	τλδ	νβ	ιβ	ζ	τνβ	ιζ	ιδ	ō
ις	τιβ	ιε	ι	νβ	τλε	λα	ιβ	ι	τνβ	μη	ιδ	δ
ιζ	τιγ	ια	ι	νγ	τλς	ι	ιβ	ιδ	τνγ	ιθ	ιδ	η
ιη	τιδ	ζ	ι	νε	τλς	μθ	ιβ	ιζ	τνγ	ν	ιδ	ιβ
ιθ	τιε	γ	ι	νς	τλζ	κη	ιβ	κ	τνδ	κα	ιδ	ις
κ	τιε	νη	ι	νη	τλη	ζ	ιβ	κδ	τνδ	νβ	ιδ	κ
κα	τις	μη	ια	ō	τλη	μγ	ιβ	κη	τνε	κγ	ιδ	κδ
κβ	τιζ	λη	ια	β	τλθ	ιθ	ιβ	λα	τνε	νδ	ιδ	κη
κγ	τιη	κη	ια	δ	τλθ	νε	ιβ	λε	τνς	κς	ιδ	λβ
κδ	τιθ	ιη	ια	ζ	τμ	λα	ιβ	λθ	τνς	νς	ιδ	λς
κε	τκ	η	ια	θ	τμα	ζ	ιβ	μβ	τνζ	κζ	ιδ	μ
κς	τκ	νζ	ια	ια	τμα	μγ	ιβ	μς	τνζ	νη	ιδ	μδ
κζ	τκα	μς	ια	ιδ	τμβ	ιθ	ιβ	ν	τνη	κθ	ιδ	μη
κη	τκβ	λε	ια	ις	τμβ	νε	ιβ	νγ	τνθ	ō	ιδ	νβ
κθ	τκγ	κδ	ια	ιη	τμγ	λα	ιβ	νζ	τνθ	λ	ιδ	νς
λ	τκδ	ιγ	ια	κα	τμδ	ε	ιβ	α	τξ	ō	ιε	ō

Κανὼν ἀναφορῶν τοῦ διὰ **Βορυσθένους** ἑβδόμου κλίματος, ὡρῶν ις, μοιρῶν μη, λεπτῶν λβ΄

Ἡλίου	Κριοῦ				Ταύρου				Διδύμων			
Μ	Ἀναφοραί		Ὡρ. χρ.	Λ	Ἀναφοραί		Ὡρ. χρ.	Λ	Ἀναφοραί		Ὡρ. χρ.	Λ
α	ō	κζ	ιε	ε	ιδ	νβ	ιζ	ιθ	λγ	λ	ιθ	ιγ
β	ō	νδ	ιε	θ	ιε	κδ	ιζ	κγ	λδ	ιζ	ιθ	ιε
γ	α	κα	ιε	ιδ	ιε	νς	ιζ	κη	λε	δ	ιθ	ιη
δ	α	μη	ιε	ιη	ις	κη	ιζ	λβ	λε	να	ιθ	κα
ε	β	ις	ιε	κγ	ιζ	α	ιζ	λς	λς	λη	ιθ	κγ
ς	β	μδ	ιε	κζ	ιζ	λδ	ιζ	μ	λζ	κε	ιθ	κς
ζ	γ	ιβ	ιε	λβ	ιη	ζ	ιζ	με	λη	ιβ	ιθ	κθ
η	γ	μ	ιε	λζ	ιη	μ	ιζ	μθ	λθ	ō	ιθ	λα
θ	δ	η	ιε	μα	ιθ	ιγ	ιζ	νγ	λθ	μη	ιθ	λδ
ι	δ	λς	ιε	μς	ιθ	μς	ιζ	νζ	μ	λς	ιθ	λζ
ια	ε	δ	ιε	ν	κ	κβ	ιη	α	μα	λ	ιθ	λθ
ιβ	ε	λβ	ιε	νε	κ	νη	ιη	ε	μβ	κδ	ιθ	μ
ιγ	ς	ō	ιε	νθ	κα	λδ	ιη	θ	μγ	ιη	ιθ	μβ
ιδ	ς	κη	ις	δ	κβ	ι	ιη	ιγ	μδ	ιβ	ιθ	μδ
ιε	ς	νς	ις	η	κβ	μς	ιη	ιζ	με	ς	ιθ	με
ις	ζ	κδ	ις	ιγ	κγ	κγ	ιη	κα	μς	α	ιθ	μζ
ιζ	ζ	νβ	ις	ιζ	κδ	ō	ιη	κε	μς	νς	ιθ	μθ
ιη	η	κα	ις	κβ	κδ	λζ	ιη	κη	μζ	να	ιθ	να
ιθ	η	ν	ις	κς	κε	ιδ	ιη	λβ	μη	μς	ιθ	νβ
κ	θ	ιθ	ις	λα	κε	να	ιη	λς	μθ	μα	ιθ	νδ
κα	θ	μθ	ις	λε	κς	λβ	ιη	μ	ν	μβ	ιθ	νε
κβ	ι	ιθ	ις	μ	κζ	ιγ	ιη	μγ	να	μδ	ιθ	νε
κγ	ι	μθ	ις	μδ	κζ	νδ	ιη	μς	νβ	μς	ιθ	νε
κδ	ια	ιθ	ις	μη	κη	λε	ιη	ν	νγ	μη	ιθ	νς
κε	ια	μθ	ις	νγ	κθ	ιε	ιη	νγ	νδ	ν	ιθ	νζ
κς	ιβ	ιθ	ις	νζ	κθ	νζ	ιη	νζ	νε	νβ	ιθ	νη
κζ	ιβ	μθ	ιζ	ō	λ	λη	ιθ	ō	νς	νδ	ιθ	νη
κη	ιγ	ιθ	ιζ	ς	λα	ιθ	ιθ	γ	νζ	νς	ιθ	νθ
κθ	ιγ	μθ	ιζ	ια	λβ	α	ιθ	ζ	νη	νη	ιθ	νθ
λ	ιδ	κ	ιζ	ιε	λβ	μγ	ιθ	ι	ξ	ō	κ	ō

Κανὼν ἀναφορῶν τοῦ διὰ **Βορυσθένους** ἑβδόμου κλίματος, ὡρῶν ις, μοιρῶν μη, λεπτῶν λβ΄

Ἡλίου	Καρκίνου				Λέοντος				Παρθένου			
Μ	Ἀναφοραί		Ὡρ. χρ.	Λ	Ἀναφοραί		Ὡρ. χρ.	Λ	Ἀναφοραί		Ὡρ. χρ.	Λ
α	ξα	θ	ιθ	νθ	ϟη	λζ	ιθ	ζ	ρμ	δ	ιζ	ια
β	ξβ	ιη	ιθ	νθ	ϟθ	νθ	ιθ	γ	ρμα	κη	ιζ	ς
γ	ξγ	κζ	ιθ	νη	ρα	κα	ιθ	ō	ρμβ	να	ιζ	β
δ	ξδ	λς	ιθ	νη	ρβ	μγ	ιη	νζ	ρμδ	ιδ	ις	νζ
ε	ξε	με	ιθ	νζ	ρδ	ε	ιη	νγ	ρμε	λζ	ις	νγ
ς	ξς	νδ	ιθ	νς	ρε	κζ	ιη	ν	ρμζ	ō	ις	μη
ζ	ξη	γ	ιθ	νς	ς	μθ	ιη	μς	ρμη	κγ	ις	μδ
η	ξθ	ιβ	ιθ	νε	ρη	ια	ιη	μγ	ρμθ	μς	ις	μ
θ	ο	κα	ιθ	νε	ρθ	λγ	ιη	μ	ρνα	θ	ις	λε
ι	οα	λα	ιθ	νδ	ρι	νε	ιη	λς	ρνβ	λβ	ις	λα
ια	οβ	με	ιθ	νβ	ριβ	ιη	ιη	λβ	ρνγ	νε	ις	κς
ιβ	οδ	ō	ιθ	να	ριγ	μβ	ιη	κη	ρνε	ιη	ις	κβ
ιγ	οε	ιε	ιθ	μθ	ριε	δ	ιη	κε	ρνς	μα	ις	ιζ
ιδ	ος	λ	ιθ	μζ	ρις	κζ	ιη	κα	ρνη	δ	ις	ιγ
ιε	οζ	με	ιθ	με	ριζ	ν	ιη	ιζ	ρνθ	κζ	ις	η
ις	οθ	ō	ιθ	μδ	ριθ	ιγ	ιη	ιγ	ρξ	ν	ις	δ
ιζ	π	ιε	ιθ	μβ	ρκ	λς	ιη	θ	ρξβ	ιγ	ιε	νθ
ιη	πα	λ	ιθ	μ	ρκα	νθ	ιη	ε	ρξγ	λς	ιε	νε
ιθ	πβ	με	ιθ	λθ	ρκγ	κβ	ιη	α	ρξδ	νη	ιε	ν
κ	πδ	ō	ιθ	λζ	ρκδ	με	ιζ	νζ	ρξς	κ	ιε	μς
κα	πε	ιθ	ιθ	λδ	ρκς	η	ιζ	νγ	ρξζ	μβ	ιε	μα
κβ	πς	λη	ιθ	λα	ρκζ	λα	ιζ	μθ	ρξθ	δ	ιε	λζ
κγ	πζ	νζ	ιθ	κθ	ρκη	νδ	ιζ	με	ρο	κς	ιε	λβ
κδ	πθ	ις	ιθ	κς	ρλ	ιζ	ιζ	μ	ροα	μη	ιε	κζ
κε	ϟ	λε	ιθ	κγ	ρλα	μ	ιζ	λς	ρογ	ι	ιε	κγ
κς	ϟα	νε	ιθ	κα	ρλγ	δ	ιζ	λβ	ροδ	λβ	ιε	ιη
κζ	ϟγ	ιε	ιθ	ιη	ρλδ	κη	ιζ	κη	ροε	νδ	ιε	ιδ
κη	ϟδ	λε	ιθ	ιε	ρλε	νβ	ιζ	κγ	ροζ	ις	ιε	θ
κθ	ϟε	νε	ιθ	ιγ	ρλζ	ις	ιζ	ιθ	ροη	λη	ιε	ε
λ	ϟζ	ιε	ιθ	ια	ρλη	μ	ιζ	ιε	ρπ	ō	ιε	ō

Κανὼν ἀναφορῶν τοῦ διὰ **Βορυσθένους** ἑβδόμου κλίματος, ὡρῶν ις, μοιρῶν μη, λεπτῶν λβ΄

Ἡλίου	Ζυγοῦ				Σκορπίου				Τοξότου			
Μ	Ἀναφοραί		Ὡρ. χρ.	Λ	Ἀναφοραί		Ὡρ. χρ.	Λ	Ἀναφοραί		Ὡρ. χρ.	Λ
α	ρπα	κβ	ιδ	νε	σκβ	μδ	ιβ	μα	σξδ	ε	ι	μζ
β	ρπβ	μδ	ιδ	να	σκδ	η	ιβ	λζ	σξε	κε	ι	με
γ	ρπδ	ς	ιδ	μς	σκε	λβ	ιβ	λβ	σξς	με	ι	μβ
δ	ρπε	κη	ιδ	μβ	σκς	νς	ιβ	κη	σξη	ε	ι	λθ
ε	ρπς	ν	ιδ	λζ	σκη	κ	ιβ	κδ	σξθ	κε	ι	λζ
ς	ρπη	ιβ	ιδ	λγ	σκθ	μγ	ιβ	κ	σο	μδ	ι	λδ
ζ	ρπθ	λδ	ιδ	κη	σλα	ς	ιβ	ιε	σοβ	γ	ι	λα
η	ρϟ	νς	ιδ	κγ	σλβ	κθ	ιβ	ια	σογ	κβ	ι	κθ
θ	ρϟβ	ιη	ιδ	ιθ	σλγ	νβ	ιβ	ζ	σοδ	μα	ι	κς
ι	ρϟγ	μ	ιδ	ιδ	σλε	ιε	ιβ	γ	σος	ō	ι	κγ
ια	ρϟε	β	ιδ	ι	σλς	λη	ια	νθ	σοζ	ιε	ι	κα
ιβ	ρϟς	κδ	ιδ	ε	σλη	α	ια	νε	σοη	λ	ι	κ
ιγ	ρϟζ	μζ	ιδ	α	σλθ	κδ	ια	να	σοθ	με	ι	ιη
ιδ	ρϟθ	ι	ιγ	νς	σμ	μζ	ια	μζ	σπα	ō	ι	ις
ιε	σ	λγ	ιγ	νβ	σμβ	ι	ια	μγ	σπβ	ιε	ι	ιε
ις	σα	νς	ιγ	μζ	σμγ	λγ	ια	λθ	σπγ	λ	ι	ιγ
ιζ	σγ	ιθ	ιγ	μγ	σμδ	νς	ια	λε	σπδ	με	ι	ια
ιη	σδ	μβ	ιγ	λη	σμς	ιθ	ια	λβ	σπς	ō	ι	θ
ιθ	σς	ε	ιγ	λδ	σμζ	μβ	ια	κη	σπζ	ιε	ι	η
κ	σζ	κη	ιγ	κθ	σμθ	ε	ια	κδ	σπη	κθ	ι	ς
κα	ση	να	ιγ	κε	σν	κζ	ια	κ	σπθ	λθ	ι	ε
κβ	σι	ιδ	ιγ	κ	σνα	μθ	ια	ιζ	σϟ	μη	ι	ε
κγ	σια	λζ	ιγ	ις	σνγ	ια	ια	ιδ	σϟα	νζ	ι	ε
κδ	σιγ	ō	ιγ	ιβ	σνδ	λγ	ια	ι	σϟγ	ς	ι	δ
κε	σιδ	κγ	ιγ	ζ	σνε	νε	ια	ζ	σϟδ	ιε	ι	γ
κς	σιε	μς	ιγ	γ	σνζ	ιζ	ια	γ	σϟε	κδ	ι	β
κζ	σιζ	θ	ιγ	ō	σνη	λθ	ια	ō	σϟς	λγ	ι	β
κη	σιη	λβ	ιβ	νδ	σξ	α	ι	νζ	σϟζ	μβ	ι	α
κθ	σιθ	νς	ιβ	μθ	σξα	κγ	ι	νγ	σϟη	να	ι	α
λ	σκα	κ	ιβ	με	σξβ	με	ι	ν	τ	ō	ι	ō

Κανὼν ἀναφορῶν τοῦ διὰ **Βορυσθένους** ἑβδόμου κλίματος, ὡρῶν ις, μοιρῶν μη, λεπτῶν λβ΄

Ἡλίου	Αἰγοκέρωτος				Ὑδροχόου				Ἰχθύων			
Μ	Ἀναφοραί		Ὡρ. χρ.	Λ	Ἀναφοραί		Ὡρ. χρ.	Λ	Ἀναφοραί		Ὡρ. χρ.	Λ
α	τα	β	ι	α	τκζ	νθ	ι	νγ	τμς	ια	ιβ	μθ
β	τβ	δ	ι	α	τκη	μα	ι	νζ	τμς	μα	ιβ	νδ
γ	τγ	ς	ι	β	τκθ	κβ	ια	ō	τμζ	ια	ιβ	νη
δ	τδ	η	ι	β	τλ	γ	ια	γ	τμζ	μα	ιγ	γ
ε	τε	ι	ι	γ	τλ	μδ	ια	ζ	τμη	ια	ιγ	ζ
ς	τς	ιβ	ι	δ	τλα	κε	ια	ι	τμη	μα	ιγ	ιβ
ζ	τζ	ιδ	ι	δ	τλβ	ς	ια	ιδ	τμθ	ια	ιγ	ις
η	τη	ις	ι	ε	τλβ	μζ	ια	ιζ	τμθ	μα	ιγ	κ
θ	τθ	ιη	ι	ε	τλγ	κη	ια	κ	τν	ια	ιγ	κε
ι	τι	ιθ	ι	ς	τλδ	θ	ια	κδ	τν	μα	ιγ	κθ
ια	τια	ιδ	ι	η	τλδ	μς	ια	κη	τνα	ι	ιγ	λδ
ιβ	τιβ	θ	ι	θ	τλε	κγ	ια	λβ	τνα	λθ	ιγ	λη
ιγ	τιγ	δ	ι	ια	τλς	ō	ια	λε	τνβ	η	ιγ	μγ
ιδ	τιγ	νθ	ι	ιγ	τλς	λζ	ια	λθ	τνβ	λς	ιγ	μζ
ιε	τιδ	νδ	ι	ιε	τλζ	ιδ	ια	μγ	τνγ	δ	ιγ	νβ
ις	τιε	μη	ι	ις	τλζ	ν	ια	μζ	τνγ	λβ	ιγ	νς
ιζ	τις	μβ	ι	ιη	τλη	κς	ια	να	τνδ	ō	ιδ	α
ιη	τιζ	λς	ι	κ	τλθ	β	ια	νδ	τνδ	κη	ιδ	ε
ιθ	τιη	λ	ι	κα	τλθ	λη	ια	νθ	τνδ	νς	ιδ	ι
κ	τιθ	κδ	ι	κγ	τμ	ιδ	ιβ	γ	τνε	κδ	ιδ	ιδ
κα	τκ	ιβ	ι	κς	τμ	μζ	ιβ	ζ	τνε	νβ	ιδ	ιθ
κβ	τκα	ō	ι	κθ	τμα	κ	ιβ	ια	τνε	κ	ιδ	κγ
κγ	τκα	μη	ι	λα	τμα	νγ	ιβ	ιε	τνς	μη	ιδ	κη
κδ	τκβ	λς	ι	λδ	τμβ	κς	ιβ	κ	τνζ	ις	ιδ	λγ
κε	τκγ	κβ	ι	λζ	τμβ	νθ	ιβ	κδ	τνζ	μδ	ιδ	λζ
κς	τκδ	θ	ι	λθ	τμγ	λβ	ιβ	κη	τνη	ιβ	ιδ	μβ
κζ	τκδ	νς	ι	μβ	τμδ	δ	ιβ	λβ	τνη	λθ	ιδ	μς
κη	τκε	μγ	ι	με	τμδ	λς	ιβ	λζ	τνθ	ς	ιδ	να
κθ	τκς	λ	ι	μζ	τμε	η	ιβ	μα	τνθ	λδ	ιδ	νε
λ	τκζ	ιζ	ι	μθ	τμε	μ	ιβ	με	τξ	ō	ιε	ō

Figure 1. An angel carries the stellar globe in the form of a snail shell. The picture is a detail from a fresco of The Last Judgement in the *parecclesion* (side chapel) of the *catholicon* (church) of the Chora Monastery in Constantinople, where Metochites resided during the last part of his life.

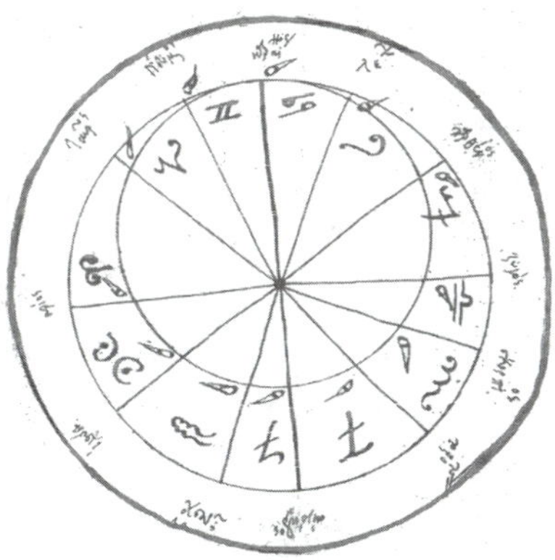

Figure 2. The model of the Sun with an eccentric trajectory and the signs of the zodiac. The figure is from Iveron Monastery, Mount Athos, manuscript 388, f. 116v. The models are described in chapter 17 of *Stoicheiosis*.

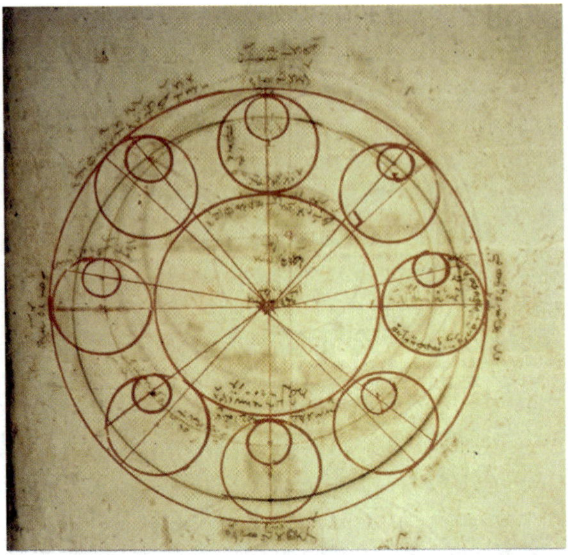

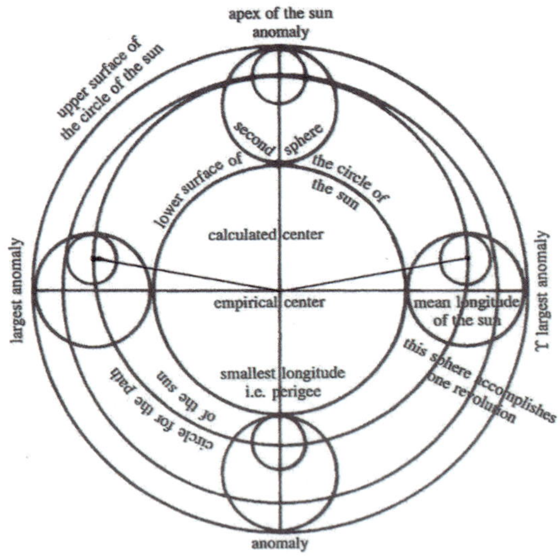

Figure 3. The model of the Sun with epicycles. The Sun corresponds to the smallest circle which rotates with the epicycle. The original diagram is from codex Vaticanus gr. 211, f. 116r; for the re-drawing of the diagram see Paschos - Sotiroudis (1998) 61.

Figure 4. The back side of an early modern French Astrolabe. It has multiple functions: for finding the position of the Sun on the zodiac for any day of the year, measuring the angle between two objects and converting equinoctial hours of a specific day to unequal (seasonal) hours.

6. Analysis

1. *Introduction* (Chapter 5)

In the fifth chapter Metochites describes the purpose of his book and his approach to the subject of Astronomy. He reserves high praise for those working in the field and asserts the old traditional opinion that Astronomy is part of Mathematics. The subject matter for astronomers is an abstract field that devises geometrical rotations in order to reproduce the observed rotations of the Sun, the Moon, the five planets and the stars. A second topic is the study of influences the stars may exert on objects within their orbits. In support of his opinions and for his great respect for astronomers he quotes *Epinomis*, which in the Byzantine tradition was incorrectly attributed to Plato but is now known to have been written after Plato's death.

Within his framework, Metochites makes several interesting statements.

1) Astronomy is one of the perfect sciences that motivated the creation of Mathematics. He emphasizes the concepts of numbers and continuity and says that numbers were invented from studies of the Heavens. Studies of the harmonious and continuous rotations of celestial bodies introduced the concept of number which permeates all of Mathematics and other branches depending on them. Numbers are applied in various branches in order to support or falsify statements; however, the theory of numbers stands logically by itself and does not depend on other branches.

2) There are also remarks that space and time are continuous quantities whose intervals (subdivisions) are described by numbers. It is stated that time is intimately related to motion, being a quantity that increases as motion advances forward. In general the notions of time and motion are closely related. This is evident because later in the work he will use the motion of the Sun to define the units of time: a year, a month, a day and hour and will motivate him to deal with the calendar.

The new development of replacing the static figures of geometry with points moving along the figures is typical of Astronomy. The combination of

several such motions produces small or large changes and can be traced back to the time of Proclus. In our time a typical example is the combination of two trigonometric functions whose superposition creates the Lissajous figures.

In the middle of the chapter Metochites includes a defence or apology for practicing Astronomy. This is included because it was probably necessary those days to refute criticisms that studies of Astronomy are contrary to the Christian faith i.e. to eliminate any possible accusation of practicing Astrology. The author makes a clear distinction. Studying the motions and relations of the planets and of the stars has no impact on Christian faith, because it involves the description of physical phenomena created by the Almighty.

But there may be the objection that studies of heavenly bodies lead to the predictions of future events — what was called *apotelesmatics*. Here he describes two methods borrowed from the introduction of *Tetrabiblos* of Ptolemy. The first method follows the *Almagest* and constructs the theoretical models which provide a precise description for the workings of the heavenly motions. This branch is established beyond doubt and makes predictions confirmed by observations (eclipses, conjunctions, oppositions etc.). The second method predicts events happening within the orbits of the Sun, the Moon and the five planets and is more difficult to describe and defend. Thus Metochites again divides the subject into two topics:

a) The claim that the stars have an influence on every aspect of human life is refuted as silly and deceitful. In addition such predictions are not reproducible. There is also a moral and legal issue involved, because if everything that happens is a consequence of the stars, then there is no cause for praise or blame…and there is nothing to strive for or be shunned from.[48] For these reasons he does not associate with persons professing these views.

b) The second option asserts that the stars influence large-scale phenomena within their orbits. As examples he mentions the changes with

[48] Cf. Basil, *On the Hexaemeron* 6.7 (ed. Giet 1968) αἱ δὲ μεγάλαι τῶν Χριστιανῶν ἐλπίδες φροῦδοι ἡμῖν οἰχήσονται, οὔτε δικαιοσύνης τιμωμένης, οὔτε κατακρινομένης τῆς ἁμαρτίας, διὰ τὸ μηδὲν κατὰ προαίρεσιν ὑπὸ τῶν ἀνθρώπων ἐπιτελεῖσθαι. Ὅπου γὰρ ἀνάγκη καὶ εἱμαρμένη κρατεῖ, οὐδεμίαν ἔχει χώραν τὸ πρὸς ἀξίαν, ὃ τῆς δικαιοκρισίας ὑπάρχει ἐξαίρετον ("We shall also see our great Christian hopes vanish, since from the moment that man does not act with freedom, there is neither reward for justice, nor punishment for sin. Under the reign of necessity and of fatality there is no place for merit, the first condition of all righteous judgment"). Cf. Bydén (2003) 351 and Hegedus (2007) 113-16.

seasons, the tides in rivers and the sea and the growth and decay of organic substances and organisms.

Recent authors have taken diverging views on these opinions. Bydén (2003: 353) states that the way Metochites presents the views of Ptolemy in *Tetrabiblos* is not completely honest because he is trying to soften the statements of Ptolemy. Magdalino[49] remarks that Metochites' ostensible rejection of astrology may simply be "a ritual disclaimer" repeating official statements and views of his time. After reading the many paragraphs on the chapter we obtained the impression that Metochites accepts that the Sun, the Moon and the five planets influence only large-scale events on the Earth. Obvious among them is the changing of seasons caused by the apparent rotation of the Sun. A recurring topic in the history of astronomy is the effect of the Moon in creating the tides. This problem attracted the attention of many scientists until it was solved much later using Newtonian mechanics.

A related question deals with the manner by which the stars influence objects that are at a great distance. The Aristotelian view was that the continuing motion of a body which is projected is sustained by a force through contact; in this case, by the air which has been set in motion behind it. This view is not mentioned by Metochites, instead he gives an example from heat transport. An object which is hot influences another object which is close or further away by warming it up. In a similar manner an action at a distance is attributed to the stars generating changes and influencing the motions on objects inside their orbits.

He describes the predictions of astrologers by giving examples from daily life which reveal his accute appreciation and assessment of human nature. Among the examples he includes their sayings: that one master is foreseen to be benevolent and another bad; or that neighbouring cities may be fighting each other and others terminating the conflicts by declaring peace; or finally that the Sun organizes the rotation of the planets like a king, who with precise laws, brings order to development. The interested reader will find additional examples of Metochites' opinions on matters of running the state and on ordinary life.

After the general introduction and declaration of his faith to the

[49] Magdalino (2002) 45 (n. 73).

canonical convictions on religious issues he turns to the description of Astronomy.

2. *The Nine Celestial Spheres* (Chapters 6-9)

The celestial model adopted by Metochites is the classical model of ancient astronomy which envisages a sequence of spherical shells one inside the other with the Earth at their centre. The shells are adjacent to each other and rotate around fixed axes with specific angular velocities; the stars are embedded on the shells and rotate with them. The combined effect of several rotations produces the changes we observe in the sky.

As we stand on the Earth and look at the stars they appear like points projected on a large sphere, the starry vault. When we observe them over several nights, we see that all of them appear to move from East to West along circles of various sizes and complete a revolution in 24 hours (the *nychthemeron*). This first revolution was attributed to the entire celestial body which carries along the other eight spheres with a great and unknown force. Since the depth of the celestial spheres is immense and all stars complete a revolution in 24 hours, the speed of the distant stars must be immense. By observing the diurnal rotation we can locate two points which do not move and they define the axis of rotation; these are the North and South poles. The major circle perpendicular to the axis of rotation is the equator, while major circles through its poles of rotation are the meridians. The parallels and the meridians define a celestial coordinate system with smaller parallel circles extending all the way to the north and south poles. This way we define the sphere called *sphaera recta* (ορθή σφαίρα). We show the celestial coordinates in Figure 1 (below). A star S is located in the northern hemisphere and the spring equinox at ♈. The coordinates are indicated with λ for the longitude, β for the latitude, δ for the declination, and ε for the obliquity.

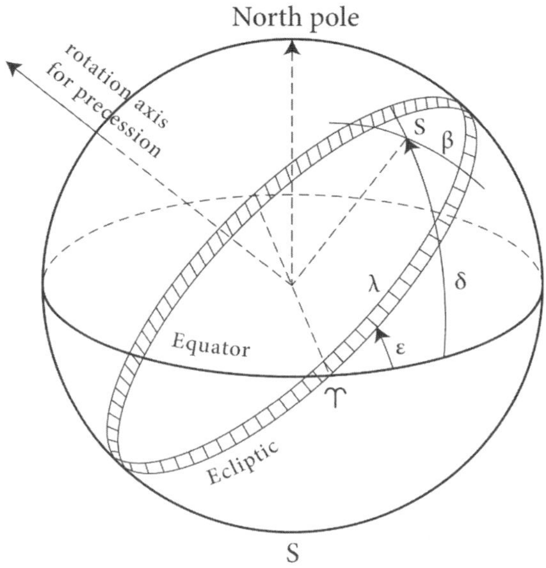

Figure 1: Celestial coordinate system

In addition, observations established that the Sun, the Moon and the five planets have additional rotations in the opposite direction close to another great circle inclined at an angle $\varepsilon \sim 24°$ relative to the equator. The Sun in a year completes a revolution on this inclined circle called the ecliptic. The two circles of the equator and the ecliptic intersect each other at two points called the nodes or the equinoxes. In a year the Sun crosses the equator twice: once traveling to the north at the beginning of Aries (vernal equinox) and again traveling to the south at the beginning of Libra (autumn equinox). For reasons that will soon become evident, we define the stellar longitude (λ) as the arc along the ecliptic beginning at the vernal equinox. We define the position of a star by its longitude and the parallel circle where it is located. The parallel is determined by giving the arc along the meridian measured from the equator. It is called the declination defined to be positive to the north and negative to the south of the equator. Finally we define stellar latitude to be the arc starting at the ecliptic and measured along a great circle perpendicular to it.

Observing the fixed stars over long periods of time it was established that their latitudes from the ecliptic did not change, but the declination changed very slowly. This requires a second rotation attributed to the eighth sphere having its own axis perpendicular to the ecliptic. The second rotation is very slow, so that many thought that it did not exist, but astronomers were already convinced of its existence. Hipparchos introduced this rotation and astronomers give him credit. Metochites does not mention Hipparchos in the first chapters of *Stoicheiosis* but later in chapter 89 (C, f. 236v) mentions him and also gives the value for the precession as 1° in a hundred years.

Chapters 6-9 outline the model adopted by Metochites for the revolutions of the celestial spheres. It is useful to present here a summary. Metochites describes the celestial body to be a three dimensional, transparent body, divided into spherical sections (κατανομές, shells) adjacent to each other. He explains the day-night rotation (diurnal) as a revolution of the entire celestial body. At a few places he mentions it as a sphere that completes a revolution in a day and a night carrying with it all the lower spheres. The stars are located in the lower eight spheres which rotate around the axis of the inclined toward the subsequent constellations of the zodiac. The first of these eight spheres contains all fixed stars and rotates very slowly from West to East, as we described earlier in this section. The remaining lower spheres carry the Sun, the Moon and the five planets. The description of the spheres is different in *Epinomis*, because at the time when *Epinomis* was written the precession of the equinoxes had not been discovered yet. In *Epinomis* all fixed stars are located on the highest sphere that produces the diurnal revolution from East to West. The Sun, the Moon and the five planets are located in the lower seven spheres. Astronomers developed models for the lower spheres with eccentric circles and/or with epicycles. Metochites expresses great respect for those who appreciate and understand the mechanics of the lower spheres, in contrast to those who, like Hesiod, limit their knowledge to a simple description of their trajectories. In his enthusiasm to praise astronomers who understand the theory of the lower spheres, he associates eight lower spheres with Plato and *Epinomis*, even though there are only seven lower spheres mentioned in *Epinomis*.[50]

[50] Ševčenko (1975) 43.

A second topic of contention is the value for the precession. Metochites adopts Hipparchos' value of 1° in 100 years, instead of the more accurate value of 1° in 66 years from Persian/Arabic sources.[51] In chapter 25 Metochites proposes a new calendar starting on the sixth of October of the year 1283. Furthermore, he explains at length (chapters 89 and 90) that he was going to adjust the locations of the fixed stars and bring their longitudes up to date. He will take the initial longitudes of the stars from *Megiste Syntaxis* and then correct them to the time of Andronikos. The *Syntaxis* was composed ca. 150 A. D. and the time interval up to 1283 is 1133 years. When approximated to 1100 years it implies an increase in longitude of ~ 11 degrees over this time period. We studied the tables in the two most important manuscripts. The oldest one is Vat. gr. 182 (V), written ca. 1317-1331, which we consider to be closest to Metochites' original work. The next complete and more popular manuscript is Vat. gr. 1365 (C), written ca. 1326-1332. In both manuscripts chapter 90 contains a description for latitudes and longitudes and chapter 91 tables for fixed stars of first and second magnitude. In the tables the names of the stars are a little different from those in *Syntaxis*, but even more interesting is the fact that the numerical values are different.

In the text Metochites states that the latitudes of the fixed stars remained the same, but their longitudes changed at the rate of 1° in 100 years. The correction of 11° appears in V. We found that the values of longitudes in V are Ptolemy's values +11°. The curious thing is that C, written later, has the longitudes from *Syntaxis*. C appears to be a copy of V and we are inclined to believe that V is closer to the original work of Metochites. Finally, in manuscript G (Vat. gr. 1087), which was copied by a number of scribes including Gregoras, most of the tables, including the relevant ones for our discussion, are missing. Perhaps Gregoras noticed that in the original the tables had been taken from the *Almagest* and he intended to review more manuscripts at a later date.

Differences in numerical values or in the conventions of the tables were noticed by careful readers who presented their comments in the margins of the manuscripts. In C, f. 75v, and in V, f.77v, there is a comment that the table for right ascensions starts at Capricorn and one must be careful when carrying over values to the table of oblique ascensions. In f. 233v of C a commentator

[51] Ševčenko (1962) 92 (n. 4); Fryde (2000) 348-49.

referring on the position of Venus relative to Sun wrote that it happens "according to hypotheses at the time of Ptolemy".

Since the second rotation is very slow and difficult to observe and establish it, Metochites considers it worthwhile to describe the methodology used by scientists. He declares from the beginning that the subject of Astronomy is based on observations which lead to syllogisms and hypotheses which eventually may develop into theories. The basic blocks in the development of science are observations and correlations among observations lead to the introduction of principles. He raises the question: what is the optimal method of selecting data? He asserts that, in selecting empirical facts, data closer to each other are more reliable. Among them one should not select those that are the most accurate but those which are closest to each other and fall within a logical framework, in other words taking an average. For the beginning of the 14th century this approach represents a very advanced point of view for which he quotes Ptolemy, Aristotle, and Porphyry. As a final conclusion he states that observations alone can not establish anything without logical thinking, nor is logic able to establish something that does not originate from observation. Thus the final arbitrator of a theory or a theorem is its verification or rejection by experiment.

By this method the astronomers established the inclined sphere whose great circle perpendicular to the axis of rotation is the zodiac. The zodiac is defined by the constellations: Aries, Taurus, Gemini, Cancer, Leo, Virgo, Libra, Scorpio, Sagittarius, Capricorn, Aquarius, and Pisces. The angle between the equator and the zodiac, as mentioned, is approximately 24° (later in chapter 29th he will give a more accurate value, of 23° 51′). Among the stars the Sun has the simplest trajectory moving through the middle of the zodiac. The five planets and the Moon also move within the zodiac but their trajectories deviate from that of the Sun sometimes being above and other times below.

3. *A Brief Survey of Geography* (Chapters 10-12)

The spheres and coordinates defined in the previous chapters refer to the celestial globe with the Earth located at the centre. The Earth is very small in comparison to the highest sphere but still spherical. Thus one can project

the celestial coordinates down into the surface of the Earth and define a terrestrial coordinate system. For each point on the Earth there is a one-to-one correspondence to the celestial coordinates.

At each point on the surface of the Earth we consider a tangent plane which defines the local horizon. Since the Earth is small relative to the celestial globe the extension of the horizon intersects the starry vault in a major circle. When the Sun is above the local horizon it is day and when below night. The horizon for every place in the equator passes through the North and South pole of the celestial sphere and the sections above and below it are equal. For this reason the days and nights at the equator are always equal. For any place outside the equator the horizon is inclined which accounts for two properties: 1. The division of the Earth into zones where the seasons and the climate are different; 2. The lengths of days and nights in various zones are different.

Let us consider first the climate of a specific zone. The term climate originates from the word *clima* (inclined) that the ancients and the Byzantines used to denote an inclination or a zone on the surface of the Earth. As the Sun moves on its trajectory along the ecliptic (zodiac), its rays warm up the areas directly below it most intensely. All parallels that lie between 24° south and 24° north of the equator intersect the path of the Sun at two points. For example, the Sun crosses the Equator when it moves toward the north at Aries and later on at Libra as it moves toward the south. In addition, in its yearly path, the Sun is on average closer to the equator than at any other place on the surface of the Earth, because its maximum distance from the equator is at most 24°. Consequently, it warms up the regions below the equator for more days during the year making them hot and dry. For this reason Metochites thinks that the areas close to the Equator are uninhabited. The converse happens in zones with latitude greater than 24° because the Sun never crosses their parallel circles. The Sun is never above them, being further away and these locations have a colder climate.

The same geometrical construction explains why the lengths of days and nights are different in various zones. The horizon of a zone is inclined relative to the *sphaera recta* and divides the celestial parallel of a location into two unequal arcs (see Figure 4). For a location in the northern hemisphere the horizon divides the local parallel into two unequal parts with the larger

arc being above (toward the North Pole). When the Sun is in the northern hemisphere it spends more time in the region above the local horizon making the day longer. At that time the reverse occurs for regions in the southern hemisphere where the nights are longer.

A consequence is that we can classify the zones in the north by their longest days in a year. Metochites mentions several locations giving for some of them their latitudes and longest days, where he says that he took the values from the *Handy Tables*. He gives some of the locations and their longest days.

Region	Latitude	Longest day
Equator	0°	12 hours (it is always 12 hours)
Taprobane	4.5°	12 ¼ hours
Avalite Gulf	8.25°	12 ½ hours
Adulitic Gulf	12.5°	12 3/4 hours
Meroë	16° 7′ 27″	13 hours
Little Brittania	61°	19 hours
Thoule	63° 6′	20 hours

Many other locations with values for their latitudes and longest days were recorded by Ptolemy but the Byzantine author does not consider it important to mention more of them explicitly.

4. *Eccentric Circles and their Properties* (Chapters 13-18)

Chapter 13 starts with a maxim: "in our analysis it is necessary to consider carefully for future use everything that follows logically". This prepares the reader for the principles of astronomy and their logical consequences. The basic principles are the celestial spheres and their uniform rotations. The author discussed already the first and highest sphere whose diurnal rotation produces days and nights and the second sphere that produces the precession of the equinoxes. Below them there are seven additional spheres with the next three accounting for the planets Saturn, Jupiter and Mars. After them come three additional spheres which appear to move with the same speed

(ἰσοταχεῖς) and correspond to Sun, Venus and Mercury. Finally is the sphere of the Moon with the largest speed. Each stellar object appears to move with angular velocity which is not constant, but varies along its trajectory being sometimes faster and at other times slower. The models are constructed to account for their apparent rotations in the trajectories.

The Sun is the most regular celestial body always traveling in the middle of the zodiac and having its apogee and perigee always at the same position relative to the fixed stars. The apogee is always at Gemini 5° 30′ and the perigee diametrically opposite at Sagittarius 5° 30′. The five planets and the Moon have trajectories that deviate above and below the path of the Sun, but always inside the zodiac. Their apogees are not fixed but rotate: for the five planets the perigees advance to the next (subsequent) constellations; only for the Moon the perigee moves in the opposite direction toward the preceding constellations. Also their distances from Earth relative to the Sun were considered to be different. Three of them (Saturn, Jupiter and Mars) were assumed to be above and the other three (Venus, Mercury and Moon) below the Sun.

To explain the anomalous rotations the ancients invented two models. The first model introduces a large circle, the deferent, which carries a smaller circle the epicycle. The Sun or a planet is attached on the rim of the epicycle. Both the deferent and the epicycle rotate carrying the Sun or planet with them. We shall return to this model below.

The second model introduces eccentric trajectories and we discuss it with some details. The Earth in this model is located at the centre of the stellar vault and is so small that it can be considered to be a point. The Sun, the Moon and the five planets move on circles with the Earth located outside their centres (eccentric). The centres are different for each stellar object with each circle rotating with its own angular velocity. For us who stand on the Earth, they appear to move with variable speeds (angular velocities). Sometimes are further away from us, at their apogees, and appear to move slower; at other time times they are closer to us and appear to move faster at their perigees.

In chapter 17 Metochites describes the motion of the Sun in terms of an eccentric circle and also the model with one epicycle. Then he makes a comparison between the two and argues that they are equivalent. Ptolemy in

the *Almagest* proved the equivalence of the two theories.[52] Metochites gives a very detailed description of the eccentric model and identifies the angle responsible for the difference between the uniform and anomalous rotation. We repeat the argument with a diagram. In the model the Sun is located on the perimeter of the eccentric κλβζφδ and is carried along. The eccentric rotates around the centre μ and we observe the rotation from point θ which is the centre of the circle homocentric to the zodiac. The difference between the uniform and the true motion is obtained by using the 32 theorem from the *Elements* of Euclid {for any triangle, when a side is extended, the external angle is equal to the sum of the two internal angles on the opposite side}. According to the theorem the angles in Figure 1 are related as follows:

$\pi = \rho + \omega$ therefore

$\omega = \pi - \rho.$

In the equations the anomalous (apparent) rotation is π and the uniform rotation is the angle ρ. Thus ω is the difference of the apparent rotation minus the uniform, i.e. the deviation from uniform. Ptolemy called this angle *prosthaphairesis* and in the medieval times it was renamed the equation.

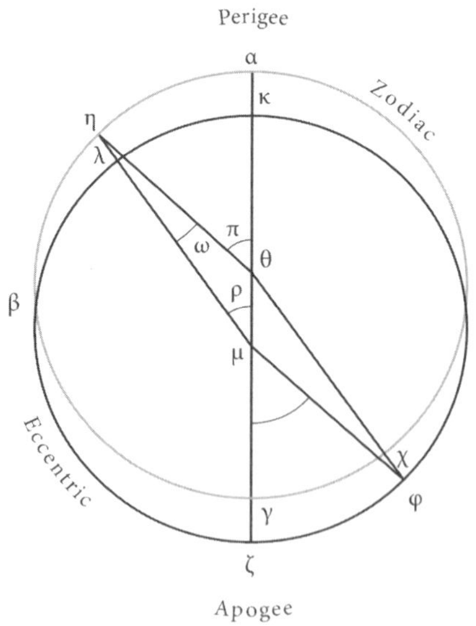

Figure 2. Geometrical demonstration for the *prosthaphairesis*

[52] Ptol. *Alm*. 3.3; see Toomer (1998) 144-45.

As the point λ rotates on the eccentric the angle ω indicates always the equation. It is obvious that at 0° and 180° the equation is zero. In addition he gives a diagrammatic proof that at 90° and 270° the equations are equal. He also discusses a symmetry property that "at equal distances from the perigee the corrections to the uniform motion are equal". In modern terms, it follows from trigonometry that the *prosthaphairesis* is given by $\sin \omega = (r/R) \sin \rho$, where (r/R) is the eccentricity.

The explanation with an epicycle is also interesting because Metochites in fact proposes two models, by varying the sense of the rotations. One of them produces an oval curve and is different from the eccentric model. Perhaps Metocites plans to use it for another planet. The other model has a large circle homocentric to the zodiac (the deferent) with an epicycle attached on its perimeter. As the deferent rotates it carries along the epicycle which rotates with the same angular velocity but in the opposite direction. The equivalence of this model to the eccentric was known as we mentioned earlier. The interested reader can convince himself that the new trajectory is an eccentric circle by drawing a few positions of the Sun on the epicycle.

An unknown reader wanted to confirm the positions of the stars in the models with epicycles and redrew the diagrams introducing several symbols (reproduced photographically in the apparatus of ch. 17.191-217). In the first diagram the epicycle rotates in the same sense as the deferent and there is the remark "toward the subsequent signs of the zodiac". The positions of the stars are indicated by the symbol ⋇ and inside the deferent are drawn the symbols for the constellations Aries ♈, Taurus ♉, and two more which are difficult to identify. Byzantine symbols for the constellations are also found on the top of the table for right ascensions in V (ff. 55v and 56r) and C (ff. 57v and 58r). In the second model the epicycle is rotating in opposite sense and there is the remark "toward the preceding signs of the zodiac". Inside the deferent there are now symbols for the planets: ⊙ for the Sun, ♄ for Saturn, and ♃ for Jupiter. This model is very accurate for the rotation of the Sun, but for the outer planets the first model is more appropriately discussed on chapter 66 of the *Stoicheiosis*. We could not find any deeper meaning for the role of the symbols and we think they were introduced to emphasize a few positions.

The treatment of the models and their equivalence shows that Metochites

had mastered the intricacies of ancient astronomy. We may speculate whether he also calculated numerically some of the values in his tables, for instance the anomalous motion or some other parameters. The impression we obtain from the text is that he uses repeatedly values from the *Almagest* or the *Handy Tables*. In a few cases produces his own results like the calculations for the new calendar (see chapters seven and eight of these comments).

Finally, he considers unphysical that there are two explanations for the motion of the Sun. How could one decide between them? The eccentric method involves only one circle and one angular velocity. By contrast the model with one epicycle involves two circles (the deferent and the epicycle) and two angular velocities in opposite directions. He prefers the eccentric model because "for the philosopher it is preferable for the best theory to select the one with the simplest hypotheses". The simple model for the Sun was introduced by Hipparchos and is so successful, that with the correct parameters its accuracy is better by a factor of twenty than the observational accuracy of that time (the accuracy of measurements at that time was 10′). The reason for this is that it satisfies Kepler's law of equal areas (Horrocks, 1627).[53] Historically, we had to wait 16 centuries, from the time of Hipparchos, for observational accuracy to improve to a level that could detect deviations.

5. *The Uniform Motion of the Sun* (Chapters 19-22)

As described in the previous chapter, the apparent rotation of the Sun is explained in terms of a uniform rotation on a circle whose centre is outside the centre of the zodiac. An observer on Earth is located at the centre of the zodiac and sees a rotation which is non-uniform. The recorded position and angular velocity are expressed as the sum of two terms: a uniform rotation plus an anomaly, described in the previous chapter. After a complete revolution the Sun returns at the initial position and the cycle is repeated over and over again. Thus to find the position of the Sun at any moment we need:

1. the position of the Sun at a specific initial time,
2. the uniform angular velocity,

[53] See, for instance, Aughton (2004) and Dumas – Meyer – Schmidt (1995).

3. a reliable calendar for measuring time, and
4. a table for the anomaly.

In this chapter we discuss the uniform motion and the tables developed for finding how far the Sun rotates in specific intervals of time.

The Sun requires a little less than 365.25 days to complete a revolution and return to the vernal equinox. The correction to this period is very small, but over a period of many years it becomes noticeable. The 365.25 days in a year contain a fraction of a day which complicates the method for measuring time. There are various ways to intercalate the fraction of one-quarter of a day.

The Graeco-Roman calendar includes the excess of one-quarter of a day in the length of every year and after four years a complete day is absorbed into the calendar (now days the extra day is absorbed in a leap year). The Egyptian calendar, on the other hand, has 12 months of 30 days each summing up to 360 days. For the remaining five days they introduce a 13th month of five days. The remaining one-quarter of a day is carried into the next year. In four years they make a full day which is counted as the first day of the fifth year. Thus starting on the first day of the fifth year the Egyptian calendar is ahead of the Graeco-Roman by a whole day. The single days accumulate and in 1460 years the Egyptian calendar is ahead by a whole year.

The astronomers must use long intervals of time in order to refer to old observations. They use the Egyptian calendar for tabulating ancient data and then relate them to recent observations. In fact, the starting instance for counting astronomical events, which was adopted as the initial time by Ptolemy, goes back to the Assyrian king Nabonassar. Another initial time was introduced in the *Handy Tables* as the first year of the reign of Phillip Arrhidaios. For these and other reasons, we need a rule to convert the Graeco-Roman dates to Egyptian. For the conversion, Metochites states that an adjustment was made in the fifth year of the reign of Augustus by adding an entire Egyptian year to the Graeco-Roman calendar and from that date on we count the correction by adding a day for each four-year cycle. To be specific, for the fifth year of Augustus we always add one year and for every four years after Augustus we add a day to the Graeco-Roman calendar.

Metochites devotes a long discussion on the methods of measuring

time because in later chapters he proposes a new starting date for measuring time and also determines the position of the Sun on that date. For the uniform rotation of the Sun he needs the length of the tropical year for which he adopts the value from *Syntaxis* (365.25 - 1/300) = 365.24666 days. Then dividing 360° by the number of days he obtains the rotation in one day to be 360°/365.24666 = 0.98563 degrees or 0° 59′ 8″ 17‴ 13⁗ 12′′′′′ 31′′′′′′ per day. From this we obtain the hourly rotation dividing the above number by 24 and the monthly rotation by multiplying by 30. In addition, the tables give the rotation completed in 18 years and multiples of them up to 360 years. Metochites gives five tables for cycles of 18 years, for single years, months, days and for hours. The tables serve the purpose of an almanac to facilitate the easy calculation for the Sun's rotation. They are the analogue of the tables we use today for trigonometric and other functions. We checked the Tables and they are the same given by Ptolemy in the *Syntaxis*.[54]

We describe one of the tables, for example the one for days. In the first column the days are enumerated, 1, 2, 3 … 30. In the other columns are given the degrees and minutes that the Sun advances in so many days. For two days we multiply the number given above by two, the third day by three and so on. The 30 days form a month. The entry for the 30 days is used as the first row of the table for months which they start increasing again. The many decimals given in the tables are totally unnecessary because the accuracy in observations those days in the best case was 10 minutes of a degree, according to the value given by Al-Tusi.[55] This is confirmed in the marginal note 286 (chapter 26) by Chortasmenos where he uses fewer decimals.

6. *The anomalous motion of the Sun* (Chapters 22-23)

The eccentric trajectory produces an irregular motion whose properties and geometric consequences were explained with a diagram in chapters 17 and 18. The deviation from the uniform motion is a function of the angular distance of the Sun from the apogee, which we denote with the angle θ. We note that the correction must be subtracted for $0 < \theta < 180°$ and must be added to the

54 See Toomer (1998) 142-143.

55 Ragep (1993) 208.

uniform motion for 181° < θ <360°. When we draw the axis from apogee to perigee the symmetric points at either side of the axis have the same correction. This is a consequence of geometric symmetry, as is easily seen and proven in either model with an eccentric or with epicycles. It also follows by inspection that the maximum anomaly occurs at two positions θ = 90° and 270° where it is equal to 2° and 23′. The numerical value depends on the eccentricity and corresponds to r/R = 0.045.[56] At the time of Metochites the eccentricity was improved by Arab/Persian astronomers who measured the modified value for the eccentricity to 0.035, but in our text there is no mention of that. It also follows that the anomaly vanishes at θ = 0° and 180° where the positions of the anomalous and uniform rotations coincide.

The careful reader will notice that the anomaly determines the "momentary position" of the Sun, not its velocity. The reader interested in the velocity must compute how much the Sun rotates per unit of time; in modern terminology we must take a derivative. Still the velocity is discussed qualitatively; it is stated that at the quadrants on each side of the apogee the apparent motion is slower than the uniform motion. The opposite occurs at the quadrants next to perigee where the Sun appears to move faster.

Tables for the anomalous motion were prepared for the *Almagest* and the *Handy Tables*. In the *Almagest* the values are given for every 6° of θ and in the *Handy Tables* for every single degree. Metochites' choice is to present the results from the *Handy Tables* because it is easier and more accurate to use them especially when we need to interpolate between degrees. The accuracy required becomes evident in chapter 26 where an example will be discussed. It is shown there that an error of a few minutes or several seconds accumulates into a visible effect over long periods of time. This will become evident in the example of chapter 26 where the time interval of 1608 years intervenes between the reign of Philip Arrhidaios and the time of Andronikos II.

The table of the anomaly has four columns where the first two columns have the heading common values and the other two the heading "equation". The first column gives the values in degrees from θ = 0° to 180° and the second for 360°- θ. The third column gives the anomaly which is the same for the common values. Ptolemy called the anomaly *prosthaphairesis*

[56] Paschos – Sotiroudis (1998) 111.

(προσθαφαίρεσις) because, as we explained, it is subtractive for the first column and additive for the second.

7. *A New Calendar* (Chapter 25)

In the present and the following chapter Metochites introduces a new calendar and demonstrates how to find the position of the Sun at a date, specifically at the initial date of the new calendar. He argues that starting the calendar at the time of Nabonassar (as done by Ptolemy) or from the first year of Philip Arrhidaios (as done by Theon) is difficult because many years intervened between those dates and his own epoch. Thus he proposes to honor the Emperor Andronikos II Palaiologos, son of Michael, and start the new calendar close to the end of the first year of his reign. There is an interesting precedence to this choice. A new starting date for measuring time also occurs in the Alphonsine Tables which is the eve of the coronation of King Alphonso of Castile, that is the eve of 1st January, 1252. We do not know whether Metochites was aware of the Alphonsine Tables.

There are still two tasks that Metochites must accomplish:

1. He must select the initial time for the new calendar and show how it is correlated to the older initial times either at the epoch of Nabonassar or that of Philip Arrhidaios. He proposes to start the new calendar close to the end of the first year of the reign of Andronikos, which is exactly 1608 Egyptian years after the initial date of the *Handy Tables*. This date is the sixth of October 1283.

2. He must calculate the position of the Sun at this new initial date. Since the intervening time between the starting point in the *Handy Tables* and the starting of the new calendar is exactly 1608 complete years, the calculation for the position of the Sun is easier (see next chapter).

Since we shall use the *Handy Tables* as initial date we must estimate how many Egyptian years intervened from the first year of the reign of Philip Arrhidaios to the new initial time. Metochites uses as reference time the year of creation which we shall denote by the symbol CW (Creation of the World). He states that this date is well documented. In that chronology the first year of the reign of Philip is 5185 years after Creation. The coronation of Andronikos took place in the year 6791 CW. Therefore 1606 Graeco-Roman

years intervened between the two dates which we must convert to Egyptian years that are used in the Tables. For the conversion to Egyptian years there are three corrections:

1) One Egyptian year was reinstated on the sixth year of Caesar Augustus to correct for the Egyptian years before that date. We must add this year because the reign of Augustus is between the times of Philip and Andronikos. Thus the reign of Andronikos began 6972 Egyptian years after the creation of the world.

2) We must correct for the *emvolimous* (intercalary) days starting from the sixth year of Augustus which happens to be the year 5484 CW. This makes (6972 – 5484)/4 = 1308/4 = 327 days.

3) The Greek year begins on the 29th of August. Adopting the new initial date to be the sixth of October brings another 38 days which added to 327 makes a complete Egyptian year of 365 days.

The sum of the three corrections make the intervening years {1606 + 1 year at time of Augustus + 1 year from item (3) =} 1608 Egyptian years. With this number we can use the table of 18-year periods to find where the Sun is located at the moment the new calendar begins.

0	Creation of the world
5185	Beginning of the reign of Philip Arrhidaios
5484	On fifth year of Caesar Augustus add one year
6791	On December begins the reign of Andronikos
6792	Starting of a new calendar on sixth of October

Table with important dates mentioned in the text. For example, the time intervening between Philip and the beginning of the reign of Andronikos is 1606 years.

It is interesting that the discussion in this chapter brings into play the whole machinery developed so far: the various calendars, the regular and anomalous motion of the Sun and the tables. It must have attracted a lot of attention because in manuscript C there are many comments in the margins. In C and on top of folio 50r there is a comment written by the hand of John Chortasmenos, which explains the chronological calculation with an example and follows the points we outlined above. Then there is an extensive remark again by John Chortasmemos that appears in folio 49v and we offer below a translation of the complete comments.

Note in chapter 25.36-45: "Since the beginning of his reign, is on the 11th of October in 6791 from the creation of the world (CW), it is evident that until the 11th of October of 6792 the king ruled for one year. Thus 6792 CW is the first year of his reign and at the end of the 92nd year he completed one year plus a few days".

Note in chapter 25.82ff.: "Taking into account the number of years completed from the creation [of the world] up to this date and subtracting 5185 we obtain the difference, which denotes the years that elapsed since Philip Arrhidaios. To them we add one year to account for the adjustment of Egyptian years. Then we count the years from the sixth year of the reign of Augustus up to now and, by taking one-fourth of this number, not exactly, but including whatever [integer] comes out, we obtain the intercalary (*emvolimous*) days that are derived from the four-year periods. These are the additional days according to the Egyptian time in Alexandria [the Alexandrian method] or according to the Greeks.

Adding them to the days accumulated from the first of September of the aforementioned and present year up to the day in which we carry out the calculation, we count the days from noon of the previous day and with two or three days added from August. In case that the total number exceeds 365 days we take them out and add one year to the recorded number of years. The final years are simple Egyptian years and there may be a remainder of days to be added to the beginning of the month Thoth or to the following months, attributing 30 days to each month. If there is no remainder with less than 30 days, but the resulting number completes whole months, we keep it as the number of months. If there are remaining days after the 12th month of Mesore, we assign them to the next intercalated (*epagomenon*) month and record the remaining one or two up to five days in the second Table.

It is also necessary to know that during a leap year (that is when the number of years from the fifth year of the reign of Caesar Augustus up to the present time is exactly divisible by four, without a remainder), we count the number of days not from the 29th but from the 30th of August, assigning 29 days to February and complete the computation from that time up to now".

The Egyptian year has 12 months of 30 days each (30x12 =360). To complete the 365 days an additional month of 5 days was added (30x12+5 = 365).

> Note in chapter 25.106ff.: “Example: Let us assume that the time we make the calculation is 6792 CW and on the sixth day of the month of October. On the first year of his reign 6791 full years passed plus a remainder. Subtracting from this 5185 for the reign of Philip Arrhidaios we obtain 1606 years.
>
> Add to this one year for the first correction [in the Egyptian calendar] and makes altogether 1607. From the first year of creation up to the completion of the fifth year of the reign of Augustus there are 5484 years, and from the sixth year of the reign of Augustus up to the present year there are 1308 years. One fourth of them, for the four-year periods, gives 327 days without a remainder being a leap year. Add to them 36 days from the beginning of September to the sixth of October and two days from August. Because the year is a leap year, it makes 38 days, which is a complete Egyptian year. Add this to the calculated years from the epoch of Philip and we arrive at the first day of Thoth in the year 1608; thus we calculated that the Roman date sixth of October is the first day of Thoth”.

The interest created by the proposal is attested by the many comments, most of them by John Chortasmenos, in the margins. It is worth noting that another astronomer of this period, Theodore Melitiniotes (died 8 March 1393), included similar computations in his *Three Books on Astronomy* or *Tribiblos*.[57]

8. *Determining the position of the Sun at the beginning of the new calendar—by including Uniform and anomalous rotations* (Chapter 26)

There is still the task of determining the position of the Sun at the beginning of the new calendar. The text gives general instructions arriving at the result 129° 37′ 56″ from the apogee. There is no explicit derivation in the text but the

[57] Ed. by Leurquin (1990).

commentaries in the margin supply the details.

In the first year of Philip the Sun was at 162° 10′ from apogee. To determine the new longitude we estimate the time elapsed from the reference time of the *Handy Tables*. The first year in the Table of Simple Years (chapter 21) corresponds to the second year of the reign of Philip and the second year in the Table corresponds to the third year of his reign and so on. Thus we must calculate the Sun's tragectory for 1607 years (see chapter 26, note on 1-27). This implies that the position of the Sun at 162° 10′ from apogee was at the end of the year 5185 CW.[58] The rotation of the Sun in one year is given in the table of the years as 359° 45.4′. In 18 years it rotates to the new longitude of 355° 37′ where we omitted 17 complete rotations. For 25 years it rotates to 353° 55′ omitting again 24 full rotations. The second value is included in the following comment of chapter 26:

εἰκοσαπενταετηρίδες ἡλίου ἀπ' ἀρχῆς τοῦ ͵ςψϛβ^ου ἔτους

α	ρλα κη´
κστ	ρκε κγ´
να	ριθ ιθ´

ἢ πρόσθες τνγ νε´ ἢ ἄφελε ς δ´ μα´´

1st	131° 28′
26th	125° 23′
51st	119° 19′

either add 353° 55′ or subtract 6° 4′ 41″

We complete the computation using the 25-year cycles, in order to follow the presentation of the commentary. Following the 25-year cycles we can complete it either by adding 353° 55′ for each cycle or subtracting the complement 6° 4′ 41″ as shown above (ἢ πρόσθες τνγ νε´ ἢ ἄφελες στ δ´ μα´´). We shall follow the reduction because the numbers are smaller.

We split the 1607 into 1600 = 64 x 25 plus 7 years. The reduction for 1600 years is (6° 4′ 41″) x 64 = 384° 256′ 2624″ ~ 389° 0′ 0″= 29° 0′ 0″, where again we eliminated complete cycles. We subtract these numbers from the position at Philip's time (162° 10′) – (29° 0′) = 133° 10′.

Finally we must add the rotation from the seven years

[58] For the chronology see Mercier (2011) 63-71.

From the 25-year periods	133°	10′
Single years 7	358°	17′ 53″
Months Beginning of Thoth	0°	0′

491° 27′ 53″ = 131° 28′ approximately (within 7″). This way we calculated the uniform motion of the Sun and found its position at the beginning of the new calendar to be 131° 28′ from its apogee.

This is the calculation presented in the margin as follows.

ψηφοφορία κατὰ τὸν πρόχειρον

		ἀποχῆς ἡλίου	
εἰκοσιπενταετηρίδαι	͵α χα	ρλγ	ι′
ἔτη ἁπλᾶ	ζ	τνη	ιη′
μῆνες	θώθ νεομηνία	~	~

For the last step of the calculation, we must determine the correction from the anomalous rotation and subtract it. For this step we use the table given in chapter 23. Looking at the table for the anomalous rotation we note that the position is between the longitudes

131° where the correction from the anomaly is 51 °1′ and

132° where the correction is 49 °1′.

The longitude of the Sun is less than 180° and the anomaly must be subtracted. To find the numerical value we note that the position is between the two values in the table and we must interpolate. The commentator in the margin makes a linear interpolation which we describe.

In the table we found that a change of 1° = 60′ in longitude produces the change of 2′ in the anomaly (a decrease). By simple proportionality a 28′ increase in longitude corresponds to a decrease of 56″. The position of the Sun is at 131° 28′ from apogee and the corresponding decrease from the anomalous rotation is 1° 50′ 4″. After subtraction we find the exact and true position of the Sun to be (131° 27′ 60″) - (1° 50′ 4″) = 129° 37′ 56″. The interpolation is demonstrated in the text with a diagram which we copy in the original.

ρλα α να΄

ρλα κη΄ πρόσθες α ν΄ δ΄΄
ρλβ α μθ΄

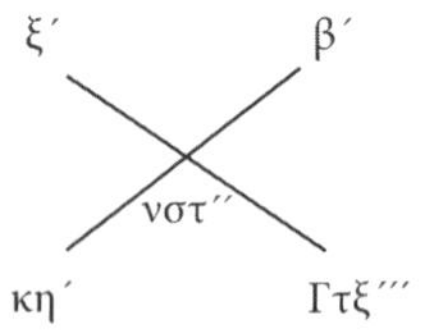

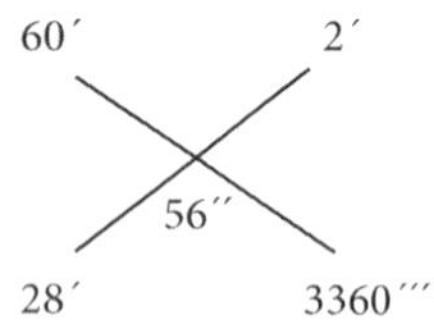

We could read the diagram in the following way: a change of 60′ in longitude brings a change of 2′ in the anomaly. By proportionality a change of 28′ brings a change of 56″ entered in the middle of the diagram. The final correction 56″ = 3360‴ is also written in the lower right-hand side of the diagram. These calculations were standard at that time and many examples appear in the *Tribiblos* by Melitiniotes.[59]

9. *Description of Ascensions and their Tables* (Chapters 27-28)

In these two chapters, Metochites defines the units of time and explains the methods used for correlating events taking place on various locations of the Earth. He divides the presentation into three topics. The first one defines the hours, equinoctial and seasonal and elaborates on the important role played by the meridians. The second topic describes how to use the tables for correlating times and includes an extensive discussion for right and oblique ascensions. The tables were prepared using the mean rotation of the Sun and a correction is necessary for the anomalous rotation on the eccentric. This is included as the third topic describing a method for including the anomaly. The content of these chapters is mathematically advanced relying on spherical geometry. Metochites describes the concepts and methods without providing proofs for which he refers repeatedly to Ptolemy.

At the very beginning we need to define the length of a day plus a night (solar day / *nychthemeron*) and in addition the length of each day from sunrise to sunset. A solar day is defined to be the time interval between two successive passes of the Sun through the same meridian of the Earth. This includes the complete revolution of the first and highest sphere plus the wandering of

[59] Leurquin (1990).

the Sun from West to East on the ecliptic. The mean angular speed for the wandering of the Sun was described earlier to be 0° 59′ per solar day. The definition is still approximate because the Sun rotates on the eccentric and thus the length of each day depends on the date of the year. For the largest part of the chapter, Metochites considers only the uniform rotation and only at the very end of the chapter describes a method for including the correction from the anomalous rotation.

Considering the solar day for a location at the equator, we find an interesting property. The horizon for any location on the equator passes through the north and south poles. Each day of the year, the Sun is located on one of the parallels or the celestial equator. As the first and highest sphere rotates it carries the Sun and all other lower spheres. In this combined rotation, the Sun crosses the horizon of any place on the equator twice, once at sunrise and later at sunset, spending equal times on each side of the horizon. It follows from this that the length of the days and nights for any place on the equator are always equal, which provides a convenient unit of time. For a place on the equator we allocate 12 hours to each day and also 12 hours to each night, called equinoctial hours. We can also define an equinoctial hour in terms of a rotation angle; during one equinoctial hour the celestial equator appears to rotate 15° degrees as it passes over a local meridian of the Earth. We can use the angle of rotation of the celestial equator as a unit of time. The degrees of rotation are called time-degrees and 15° time-degrees correspond to one equinoctial hour.

Finally we also need to define a reference place on the surface of the Earth relative to which we express time. In ancient and Byzantine times the meridian of Alexandria was the reference place. Events happening at other places were given in equinoctial hours of Alexandria. In the ancient world the meridian of Alexandria played the role that the meridian of Greenwich, England plays today. For places to the east of Alexandria, the Sun culminates earlier, and to the west later. We account for the differences by allocating one equinoctial hour to every 15° of geographic longitude. If the meridian of a city is to the east of Alexandria, we add one equinoctial hour for every 15° of terrestrial longitude and for cities to the west we subtract it. Tables for geographic longitudes of famous cities were tabulated in Ptolemy's *Geography*

and also in the *Handy Tables*. We follow the same practice today; when we refer to Central European time we must add one hour to Greenwich-time and for Eastern Europe two hours and so on.

Now that we have the units of time and Alexandria as reference place, we wish to specify the time when the Sun, on a specific day, ascends the horizon of a location on the Earth. Metochites discusses the ascensions at various zones, presents extensive tables and includes explanations on how to use them. The tables were calculated originally by Ptolemy and later annotated by Theon of Alexandria. They are known as the *Handy Tables* and were useful for practical purposes and for this reason survive in numerous manuscripts. Tihon published a critical edition of surviving manuscripts[60] and compared them to each other. Mercier studied their origin, discussed possible methods for their computation and investigated their accuracy.[61] In *Stoicheiosis* the tables for ascensions (right and oblique) are taken from the *Handy Tables*. We compared the tables in *Stoicheiosis* with those published by Tihon and Mercier and found few minor differences which are not worth mentioning.

The time when the Sun rises at the horizon of a location on the equator is called right ascension. A location on the equator of the Earth has a second good property, namely that all stars appear to ascend the local horizon at right angles; hence the name right ascension (ορθή αναφορά) — a term which is still used today. In fact, an observer at the equator observes, together with the rising Sun, the simultaneous rising of all the stars on the same meridian. Metochites elaborates this property and appreciates that for the right ascensions it suffices to give the momentary projection of the Sun onto the equator along its meridian. Since the Sun rotates on the inclined the projections into the equator are not proportional to the longitude of the Sun on the inclined. Metochites expresses this in the following way: the right ascensions are projections of angular intervals of the inclined into the equator; however, equal arcs of the inclined do not have equal projections into the equator. Giving the right ascension for the Sun determines the time of ascend of all stars located on the same meridian. In the *Almagest*, Ptolemy presented tables for right ascensions in intervals of 10° within each constellation and also

[60] Tihon (2011b).
[61] Mercier (2011).

records in the third column the degrees accumulated for that position of the Sun. In the *Handy Tables,* they are tabulated for single degrees and Metochites decides to include those from the *Handy Tables*. In the *Handy Tables* there is a third column which includes the "equation" (*prosthaphairesis*); its purpose was to use the equation in order to correct for the anomalous rotation. Metochites decides to omit this column and accounts for the correction with another method described at the end of the chapter.

In the table, the first column gives the position of the Sun within each constellation which determines the day of the year. The table begins at 1° of Capricorn which corresponds to the 21st of December (winter solstice), and the right ascension is 1° and 6′. Just before is the entry for 30° Sagittarius where the right ascension is 360° 0′. As the days of the year advance, the Sun moves to another position, let us say 30° Capricorn approximately 30 days later. On that day the ascension is not 30° time-degrees, but 32° 16′ as we read from the table. The difference is created by the fact that the Sun moved on the inclined and not on the equator. Sometimes it becomes necessary to find the time when another star ascends a horizon on the equator. In this case we find the position of the star within a constellation. We read in the second column of the table the right ascension of the star and subtract the right ascension of the Sun. The difference expresses in time-degrees the time of its appearance after sunrise.

Finally, the starting point for the table of *sphaera recta* in the *Almagest* is at the beginning of Aries, but in the *Handy Tables* it starts, as we mentioned, at the beginning of Capricorn. Starting with different initial constellations is not crucial because the important thing is to establish how many degrees of the equator correspond to an arc within each constellation of the ecliptic. However, the different starting places require some attention when results from the tables of oblique ascensions are transferred to the table of right ascensions. The different starting times in the tables and the fact that the table for the equation (anomaly) starts at the apogee of the Sun (at Gemini 5° 30′) are the reasons that motivated Metochites to omit the third column in the table of right ascensions and describe the correction from the anomalous motion at the end of this chapter.

We briefly describe how the ancients determined the rising times at cities on the parallels outside the equator. We consider a location on a zone in

the northern hemisphere with its horizon inclined relative to the horizon of the equator. The horizon is illustrated in Figure 3 by the great circle H. We also visualize two meridians: the meridian αβγ for the right ascension (rectangle) and the meridian αδλ defined by the point where the Sun enters the local horizon (point δ). The oblique ascension was computed as the difference of the right ascension of the Sun at the date of the calculation minus a small arc referred to as the ascensional difference. This is the solid angle between the meridian for the right ascension of the Sun and the meridian through the point where the Sun enters the local horizon (point δ in Figure 3). The result of the computation is the solid angle σ in Figure 3 known as oblique ascension. Ptolemy calculated the oblique ascensions for the seven zones (climate) and tabulated them in 10° for each constellation of the ecliptic. The tables for oblique ascensions in both the *Megiste Syntaxis* and for the *Handy Tables* start at the vernal equinox.

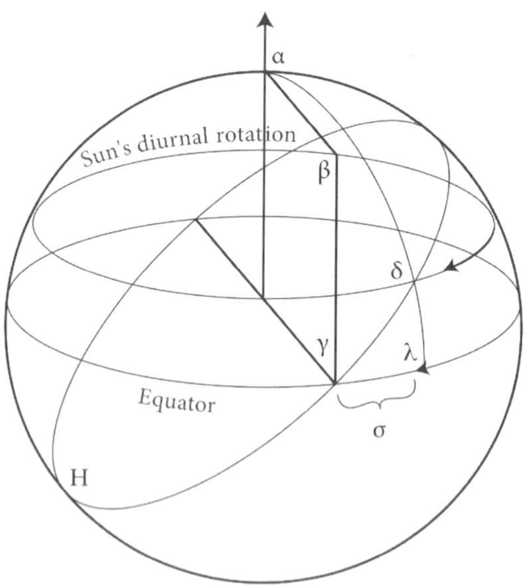

Figure 3. Oblique Ascension

The local horizons for every location on the Earth also define the length of the days. Each horizon intersects and divides the parallel where the Sun is located

on that day into two unequal arcs. In Figure 4 we indicate the local horizon by a plane which intersects two diametrically opposite parallels. As the Sun is located on one of the northern parallels and performs its diurnal rotation it is sometimes above the horizon during the day (solid curve) and at other times below during the night (dashed curve). The opposite happens on the parallel in the southern hemisphere. The time intervals for days and nights are not equal and one can measure their relative lengths by measuring the arcs above and below the horizon.

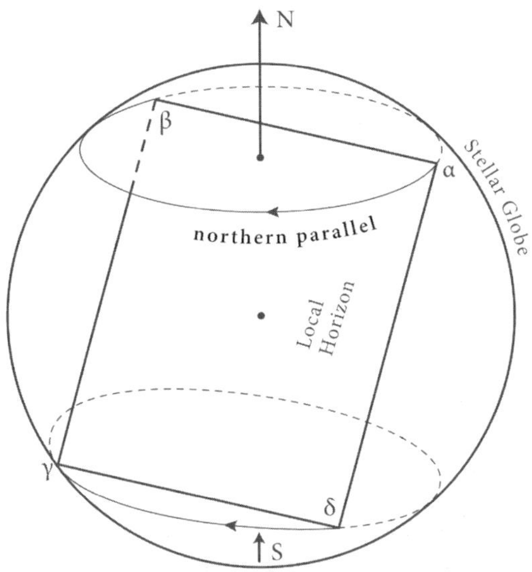

Figure 4. Local horizon of a habitation with two diametrically opposite parallel circles. For the discussion in the text, the Sun rotates in the northern parallel.

In spite of the fact that the lengths of days and nights are different, the ancients and the Byzantines adopted the convention that each day is divided into 12 hours, called seasonal hours. Obviously, the seasonal hours differ from day to day and from place to place. In order to relate the hours between places it became necessary to compute for a specific place and a specific day of the year how many time-degrees correspond to one seasonal hour. This correspondence is called hourly-times (ωριαίοι χρόνοι). The convention is as follows: one equinoctial hour corresponds to 15° degrees of the equator; when

the length of a day is longer than that of a night, an hour of a day can be 16°, 17° or more time-degrees and when shorter to 14°, 13°, ... degrees. As already mentioned, Ptolemy solved this problem geometrically by taking the ratio of the arc of the parallel above the horizon to that below. This property motivated the ancients to describe the latitude of a location not with an angle but instead giving the length of the longest day in time-degrees or the ratio of the length of the longest to that of the shortest day. For the latitude of Alexandria the ratio is 14/10; which means that the longest day contains 14 equinoctial hours and the shortest day 10 hours. Neugebauer mentions that this is a very old tradition going back to the Babylonians.[62]

In *Stoicheiosis* there are also tables which give the rising times of the Sun at a specific day of the year and for a zone on the surface of the Earth. The day of the year is determined by the location of the Sun within a sign of the zodiac and in the *Handy Tables* is given for every degree. Metochites states that for higher accuracy in calculations and especially for long intervals of time it is preferable to use the *Handy Tables*. *Stoicheiosis* contains extensive tables of sunrises on the seven zones (climates) and for the latitude of the Queen City, as they called Byzantium. On the top of each table is the name of the zone, the equinoctial hours for the longest day and also the latitude given in degrees. Then there are three columns for each constellation. The first column gives the position of the Sun within each constellation, which determines the day of the year. The next column gives the rising time of the Sun in time-degrees (ascension). The third column gives how many time-degrees correspond to a seasonal hour (hourly-times).

Once we have the correspondence between seasonal and equinoctial hours, it is easy to convert one to the other. For example, let us assume that for a location (habitation) we are given the hourly-time to be 17° and we wish to find how many equinoctial hours are contained in a day. We multiply 17° by 12 hours in a day to obtain 204°. Then dividing this by 15° we find 13 equinoctial hours, and 9/15~2/3 of an hour is the length of the day in that location. For the conversion of equinoctial hours to hourly-time we reverse the steps. That is, we multiply 13 and 9/15 equinoctial hours by 15° to obtain the total number of equinoctial hours in time-degrees; then we divide it by 12 in order to obtain

[62] See Neugebauer (1969) 159-160 and 183-184.

the hourly time for that moment and the location we consider.

For the conversion of seasonal hours to equinoctial hours we follow another procedure. Let us assume that we wish to convert 3 seasonal hours after midday. We multiply the seasonal hours by the hourly time and divide the product by 15°. Assuming that the hourly-time is 17, we obtain 3x17/15 = 51/15 = 3 and 6/15 equinoctial hours. For similar calculations during the night, we go to the diametrically opposite point of the constellations and use the hourly-times appearing in the table of the habitation under consideration.

Before leaving the second topic, Metochites discusses how we find from the tables the number of equinoctial hours in a day. We demonstrate the method with the help of the geometrical diagram of Figure 4. The rectangle αβγδ represents the horizon for a location on the Earth and the diurnal rotation of the Sun takes place on the northern parallel. Point α defines the meridian for sunrise and point β the sunset. Point γ is diametrically opposite to point α and lies on the meridian that passes through β, i.e. the meridian for sunset. In the tables of oblique ascensions we find in the second column the ascensions for points α and γ and the difference between them is the length of that day expressed in time-degrees.

We now come to the third correction mentioned earlier. The tables were prepared by assuming that the Sun participates in the first and highest rotation and also rotates from West to East on the ecliptic with constant angular velocity of 0° 59′ per day. The two revolutions were included in the definition of the solar day. Since the trajectory on the ecliptic is eccentric, the apparent speed of the Sun varies from day to day which was the reason to introduce the anomaly from the eccentric rotation of the Sun (see chapter 23). Thus the time that the Sun takes to return to a meridian (*nychthemeron*) is not constant but varies from day to day. This requires a third correction. The change brought by the correction over a few days is very small, but over many days becomes noticeable.

For the correction, Metochites proposes to use as initial time the sixth of October 6972 CW when the Sun was located

1. for the uniform motion at Libra 16° 58′, and
2. for the anomalous motion at Libra 15° 6′ and 54″.

These values were determined in chapters 25 and 26 where he presented the new calendar. Starting from the above positions we compute for the date of the new calculation the positions of the Sun for the uniform and the anomalous rotations (see chapters 25, 26 and the table in chapter 23). We enter the initial and the new longitudes in the table of the right ascensions and count the number of ascensions that intervened for the uniform and also for the anomalous rotation. We record the increment in ascensions for each of them. As usually they are given in time-degrees which we divide by 15 in order to convert them to equinoctial hours and fractions thereof. The increments denote the changes for the two types of rotations. Taking their difference gives the "final correction" we need from the two rotations.

When we wish to convert seasonal and anomalous (true) hours to mean, we check if the increment of the anomalous rotation is larger than the one for the mean, then we add the "final correction" to the true value. If the increment of the mean and uniform rotation is larger, then we subtract it from the true time. Sometimes we may wish to convert the result obtained from the uniform rotation to anomalous (true), in that case we reverse the steps. Whenever the increment for the anomalous rotation is larger, then we subtract the "final correction" from the value obtained for the uniform and mean rotation. The third correction is also discussed by Ptolemy in the *Syntaxis*.[63] The discussion in this paragraph of the *Stoicheiosis* and of the *Syntaxis* is complicated because it involves the addition or subtraction of negative numbers which were not invented in those times. The authors were trying to express the results with longer sentences.

After the tables for the seven climates there is a table for the oblique ascensions at Byzantium where the length of the longest day is given as 15 and one-quarter hours and the latitude 43° and 5′. This is not part of the original work of Ptolemy but the work of a later author of the *Handy Tables*. The latitude for Byzantium is half-way between that of Hellespont and the Middle of Pontus; in fact the values in this table are the average values of the two neighbouring zones. Mercier studied possible methods for computing

---

[63] Toomer (1998) 171-72 and 651-52, where he devotes example 8 to the correction.

this table and came to the conclusion that an interpolation was used.[64] He concludes also the values for the table for Byzantium were obtained by averaging the values of the neighbouring zones.

The detailed discussion, the explanations and practical rules included in chapter 27 indicate that Metochites understood the practical content and applications included in the *Almagest*. With *Stoicheiosis* it is much easier for a reader to appreciate and understand the chapters of *Megiste Syntaxis* and also to assess how the *Almagest* was understood in Byzantium when the authors may have had other sources available. Once a reader becomes accustomed to the individualistic writing style of Metochites, he can correlate sections and paragraphs in order to deduce and understand their meaning. These are some of the reasons that astronomers after Metochites studied *Stoicheiosis* and applied its contents to contemporary problems.

10. *Calculation of the Horoscope and of Upper Culmination* (Chapter 30)

The extensive tables of ascensions are useful for solving various problems. Metochites demonstrates their use with a few applications. He begins with two definitions. At a specific seasonal hour of a location on the Earth, we are interested to find the degree of the ecliptic that rises over the eastern horizon of the location. The determination of the degree is called the horoscope. A similar problem asks to find the degree of the ecliptic that culminates at that moment, that is the degree of the ecliptic located at the zenith. The second determination is called upper culmination (μεσουρανεῖν). The seasonal hours we mentioned above may refer to hours after sunrise, after midday or after sunset. Metochites describes a few examples with clarity and we simply repeat his description. The topics are discussed in *Megiste Syntaxis* II.9[65] and Toomer explains the method with several examples in Appendix A.[66] Similar calculations also appear in the book of A. Jones.[67]

Whenever we wish to find the horoscope for a seasonal hour of a day

[64] Mercier (2011) 143-145.

[65] Toomer (1998) 104.

[66] Toomer (1998) 650 (examples 4 and 5).

[67] Jones (1987).

and a specific location of a zone, we use the table of oblique ascensions for the zone. For that day of the year we find in the first column the number of degrees inside the constellation. In the third column we find the hourly-times and multiply the hourly-times with the seasonal hours. We add the product to the ascension appearing in the second column and the resulting number is the "new ascension". Then we search in the table of the same zone and find the number of degrees within a constellation which corresponds to the "new ascension". This is the point of the ecliptic that rises at that moment.

Whenever we wish to find the horoscope for any given seasonal hour of the night we follow the same procedure, but for the seasonal hour of the day we use the hourly times of the day and for the remainder time after 6 P.M. we use the hourly-times for the location diametrically opposite to our location. Metochites, in his paedagogical style, reminds us again that we must subtract 360° (an entire cycle) whenever the result from the multiplication and the subsequent addition exceeds 360°. Furthermore, when the "new ascension" is not found in the tables but lies between two numbers, he suggests to interpolate between them.

Sometimes we are faced with another problem. Let us say that for our zone the Sun rises at specific seasonal hours and we wish to find the point of the ecliptic which culminates at that moment. The culmination takes place on the local meridian which is at right angles to the horizon. This means the local meridian is 90° west of the eastern horizon. Thus we go to the table of the zone under consideration and find in the first column the degrees of the constellation corresponding to that day. We find in the second column the rising time expressed in time-degrees and subtract from them 90°, because the culmination lies to the west. The result is carried over to the table of *sphaera recta*, where we search to find the degrees within a constellation that corresponds to the time established as a result of the subtraction. This is the point of the ecliptic that culminates at that moment. One may ask: why should we use the table of *sphaera recta*? Metochites explains because the time of crossing the meridian of a habitation is equivalent to the culmination in *sphaera recta*.

As we mentioned, the two examples we described are also given in

Chapter 2 of the *Syntaxis*.[68] However, a careful reader will notice that there is a difference of convention. As mentioned in Chapter 27, the table of *sphaera recta* in the *Handy Tables* begins at Capricorn, but in the *Syntaxis* begins at Aries. In addition all the tables for oblique ascensions begin at Aries. Thus we cannot carry over results from tables of the oblique ascensions to the table of right ascensions, i.e. we must make a correction, which is introduced in the following way. We can change the table of right ascensions to begin at Aries by adding to the tabulated ascension 270° and eliminating complete cycles when the resulting number exceeds 360°.

Gregoras noticed the problem and included a comment in the margin of manuscript V, f. 77r, as well as in the other manuscripts. He wrote: "Let it be known that this method of calculation is given by Ptolemy in Chapter II of the *Syntaxis*, where the table of risings in *sphaera recta* begins from Aries, but is different in the *Handy Tables* and in the present treatise, where they begin at Capricorn. Thus in all calculations it is necessary to account for the difference in order for the results to be certain".

Finally, the reader will better appreciate the methods by working in detail two examples using the tables in *Stoicheiosis*. The two exercises are also discussed by Toomer's *Almagest*.[69]

Exercise 1 (horoscope): We consider the location of Alexandria when the Sun is located in Scorpio 13° 17′. For two and one-quarter seasonal hours after midnight we wish to find the degree of the zodiac located at the eastern horizon of Alexandria.

Solution: We use the table for oblique ascensions of Alexandria which is in the third zone (clima). The time is eight and one-quarter hours after 6 P.M. and consequently we must use the values for the location of the Sun diametrically opposite to Scorpio, this is Taurus 13° 17′. We read in the table the hourly-time (ωριαίοι χρόνοι) 16° 36.6′.

The ascension at Taurus 13° 17′ is	31; 3.6
The time-degrees from the additional hours are 8¼ x 16; 36.6 =	137; 2.0

The new ascension is the sum	168; 5.6

[68] Toomer (1998) 104.
[69] Toomer (1998) 650.

The point of the zodiac rising in the eastern horizon of Alexandria is Virgo 20° 4.9′ (Toomer's answer is Virgo 19° 51′).

Exercise 2 (culmination): Given the same data as in the previous exercise find the degree of upper culmination.

Solution: We follow the rules of chapter 30.

The ascension found in the previous exercise is	168; 5.6
From this we subtract 90°	– 90; 0.0

New ascension	78; 5.6

We must now search to find in the table of right ascensions the point of the zodiac corresponding to the new ascension. At this step we must account for the different starting points in the two kinds of tables. Let us denote by y the value in the table of right ascensions. Then

$$y + 270° = 78; 56 + 360°.$$

When the new ascension on right-hand-side is less than 270°, we must add 360°, because otherwise y is negative. This is our case with the solution y = 168; 5.6′. In the table on the right ascensions we search and find the corresponding value within the constellation is Gemini 19°; 5.2′ (Toomer's answer: Gemini 19° 16′).

11. *The obliquity* (Chapter 29)

We mentioned many times that the ecliptic is a great circle inclined relative to the equator. To determine the circle we must show that all constellations lie on the same plane and measure the angle between them. This is accomplished by considering the meridian of the constellation and then measuring along the meridian the angle between the sign and the equator. In modern terms this is the declination for each sign of the zodiac. The determination is somewhat complicated because it occurs on the surface of a sphere. For this reason various values were reported over the centuries. We give a table of the values recorded by Chioniades[70] and add the value of Metochites.

---

[70] Paschos – Sotiroudis (1998) 97.

Table : Values for the obliquity	
Astronomers preceding Ptolemy	24° 5′
Ptolemy (127 – 151 A.D.)	23° 52′
Astronomers after Ptolemy	23° 35′
At the time of Caliph al-Mamun (830 A.D.)	23° 33′
al-Tusi (1260 – 1270 A.D.)	23° 30′
Metochites (1270 – 1332 A.D.)	23° 51′

In his work, Metochites uses the value of 23° 51′ which is the maximal declination measured at solstices occurring at the beginning of Cancer and of Capricorn. In many discussions and calculations it is useful to have, in addition to the longitude of a constellation in the ecliptic, as well the deviation from the great circle of the equator. Since the problem is one of spherical geometry Ptolemy solved it geometrically and gives in the *Syntaxis* a table of the declinations for one quadrant. This is sufficient because the spherical symmetry determines the values at the other quadrants. In the figure below we show the celestial coordinate with a few constellations and the definition of variables.

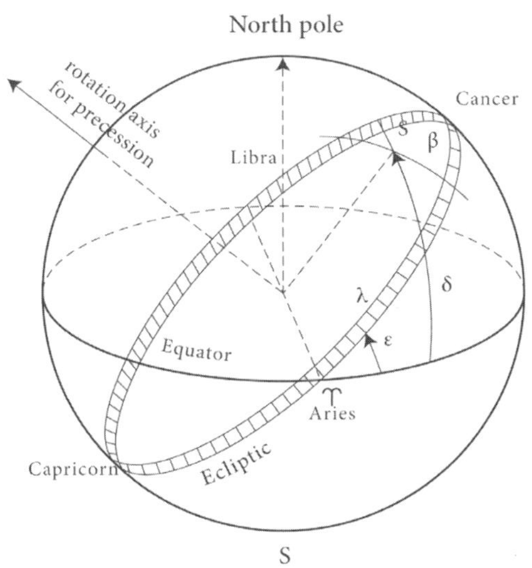

Figure 5. The celestial coordinate system

Metochites presents first the sequence of the signs starting at the spring equinox: Aries, Taurus, Gemini, Cancer, Leo, Virgo, Libra, Scorpio, Sagittarius, Capricorn, Aquarius and Pisces. The equinoctial points are at the beginning of Aries and Libra, and the solstices are at the beginnings of Cancer (in the north) and Capricorn (in the south) as shown in the figure. Then he presents values from the *Handy Tables* for every three degrees.

The table for the obliquity has four columns. The first two columns are labeled "common numbers". The first denotes the longitude for every three degrees, from Cancer up to the beginning of Capricorn, with $\theta = 0°$ to $180°$ and the second column gives the complement $360° - \theta$. The third and fourth columns give the declination in degrees and minutes. As mentioned, the table starts at Cancer and gives the value 23° 51′. The same value holds also for the diametrically opposite point at Capricorn.

We may go a step further and ask how these values were calculated. The problem involves spherical geometry we cannot use a simple proportionality factor as in plane geometry. The correct equation is[71]

$$\sin \delta = \sin \lambda \sin \varepsilon,$$

where δ is the declination,
λ the longitude and
ε the obliquity.

Using this equation we must start the longitude in the standard way at Aries; with Aries having $\lambda=0°$, Cancer with $\lambda = 90°$, Libra $\lambda = 180°$,... We checked a few points and found the agreement with the formula to be very good with differences being at most one minute of a degree.

[71] Mercier (2011) 81.

BIBLIOGRAPHY

Agapitos, P. A. – Hult, K. – Smith, O. L. (1996). *Theodoros Metochites on philosophic irony and Greek history: Miscellanea 8 and 93. Edited with an introduction, translation, and notes* (Nicosia - Göteborg).

Aughton, P. (2004). *The transit of Venus: the brief, brilliant life of Jeremiah Horrocks, father of British astronomy* (London).

Bianconi, D. (2005). "La biblioteca di Cora tra Massimo Planude e Niceforo Gregora. Una questione di mani", *Segno e Testo* 3: 391-438.

Bucossi, A. – Kihlman, E. (eds) (2012). *Ars Edendi*, vol. II (Stockholm).

Bydén, B. (2003). *Theodore Metochites' Stoicheiosis Astronomike and the Study of Natural Philosophy and Mathematics in Early Palaiologan Byzantium* (Göteborg).

Bydén, B. (2011). "Theodore Metochites", in H. Lagerlund (ed.), *Encyclopedia of Medieval Philosophy: Philosophy between 500 and 1500* (Dordrecht; New York), 1266-1269.

Bydén, B. (2012). "Imprimatur? Unconventional Punctuation and Diacritics in Manuscripts of Medieval Greek Philosophical Works", in Bucossi - Kihlman (2012) 155-172.

Christidis, D. A. (1979). "Ἀθηναγόρας, Πρεσβεία περὶ Χριστιανῶν 5,2 (Ὄψις ἀδήλων τὰ φαινόμενα)", *EEThess(philol)* 564-551: 18.

Constantinides, C. N. (1982). *Higher education in Byzantium in the Thirteenth and Early Fourteenth Centuries, 1204 – ca.1310* (Nicosia).

De Groote, M. (2012). *Christophori Mitylenaii Versuum variorum Collectio Cryptensis* (Turnhout).

Dendrinos, Ch. (2011). "Palaeologan scholars at work: Makarios Makres and Joseph Bryennios' autograph", in Giannouli – Schiffer (2011) 23-51.

Dumas, H. S. – Meyer, K. S. – Schmidt, D. S. (eds) (1995). *Hamiltonian Dynamical Systems: History, Theory, and Applications* (New York).

Förstel, C. (2011). "Metochites and his Books between the Chora and the Renaissance", in H.A. Klein, R. Ousterhout, B. Pitarakis (eds.), *The Kariye Camii Reconsidered/Kariye Camii Yeniden* (Istanbul), 241-266.

Fryde, E. (2000). *The Early Palaeologan Renaissance (1261-c. 1360)* (Leiden-Boston).

Giannouli, A. – Schiffer, E. (eds) (2011). *From Manuscripts to Books. Proceedings of the International Workshop on Textual Criticism and Editorial Practice for Byzantine Texts (Vienna, 10-11 December 2009) / Vom Codex zur Edition. Akten des internationalen Arbeitstreffens zu Fragen der Textkritik und Editionspraxis byzantinischer Texte (Wien, 10.-11. Dezember 2009)* (Wien).

Giet, S. (1968). *Basile de Césarée: Homélies sur l'Hexaéméron*, texte grec, introduction et traduction, 2nd ed. (Paris).

Guidetti, F. – Santoni, A. (eds) (2013). *Antiche stelle a Bisanzio: Il codice Vaticano greco 1087* (Pisa).

Hegedus, T. (2007). *Early Christianity and Ancient Astrology* (New York-Oxford).

Hunger, H. (1969). *Johannes Chortasmenos (ca. 1370 – ca. 1436/37): Briefe, Gedichte und kleine Schriften* (Vienna).

Jones, A. (1987). *An Eleventh-Century Manual of Arabo-Byzantine Astronomy* (Amsterdam).

Kunitzsch, P. (1964), "Das Fixsternverzeichnis in der 'Persischen Syntaxis' des Georgios Chrysokokkes", *Byzantinische Zeitschrift* 57: 382-411.

Lamberz, E. (2000)."Das Geschenk des Kaisers Manuel II. an das Kloster Saint-Denis und der 'Metochitesschreiber' Michael Klostomalles," in B. Borkopp – T. Steppan (eds.), *Λιθόστρωτον: Studien zur byzantinischen Kunst und Geschichte. Festschrift für Marcell Restle* (Stuttgart), 155-165.

Lamberz, E. (2006). "Georgios Bullotes, Michael Klostomalles und die byzantinische Kaiserkanzlei unter Andronicos II. und Andronicos III. in den Jahren 1298 – 1329", in B. Mondrain (ed.), *Lire et écrire à Byzance* (Paris), 33-48.

Leurquin, R. (1990). *Théodore Méliténiote: tribiblos astronomique: histoire du texte, édition, traduction et commentaire*, 3 vols (Amsterdam).

Magdalino, P. (2002). "The Byzantine Reception of Classical Astrology", in C. Holmes — J. Waring (eds), *Literacy, Education and Manuscript Transmission in Byzantium and Beyond* (Leiden-Boston-Köln), 33-57.

Markopoulos, A. (2008). "Education", in E. Jeffreys — J. Haldon — R. Cormack (eds), *The Oxford Handbook of Byzantine studies* (New York), 785-795.

Mavroudi, M. (2006). "Exchanges with Arabic Writers during the Late Byzantine Period", in S.T. Brooks (ed.), *Byzantium: Faith and Power (1261–1557). Perspectives on Late Byzantine Art and Culture*. The Metropolitan Museum of Art Symposia (New York; New Haven, CT), 62-75.

Mavroudi, M. (forthcoming). "Scholars and Intellectuals in the work of Ihor Ševčenko", *Dumbarton Oaks Papers*.

Menchelli, M. (2000). "Appunti su manoscritti di Platone, Aristide e Dione di Prusa della prima età dei Paleologi. Tra Teodoro Metochite e Niceforo Gregora", *Studi classici e orientali* 47/2: 141-208.

Menchelli, M. (2013). "Struttura e mani del Vat. gr. 1087 (con osservazioni paleografiche sul copista C e il Marc. gr. 330)", in Guidetti - Santoni (2013) 17-56.

Mercati, I. – Franchi de' Cavalieri, P. (1923). *Codices Vaticani Graeci 1-329* (Rome).

Mercier, R. (2011). *Ptolemaiou Procheiroi Kanones: les Tables Faciles de Ptolémée*, Volume 1b: Tables A1-A2. Transcription and Commentary (Louvain-la-Neuve).

Munitiz, J. A. (1984). *Nicephori Blemmydae autobiographia, sive, Curriculum vitae* (Turnhout).

Neugebauer, O. (1969). *The Exact Sciences in Antiquity* (New York).

Paschos, E. A. – Sotiroudis, P. (1998). *The Schemata of the Stars: Byzantine Astronomy from A.D. 1300* (Singapore; River Edge, N.J.).

Pérez Martín, I. (1997). "La 'escuela de Planudes': notas paleográficas a una publicación reciente sobre los escolios euripideos", *Byzantinische Zeitschrift* 90: 73-96.

Pérez Martín, I. (2008). "El 'estilo Hodegos' y su proyección en las escrituras constantinopolitanas", *Segno e testo* 6: 389-458.

Pérez Martín, I. (2012). Review of Giannouli –Schiffer (2011), *Bryn Mawr Classical Review* 2012.4.29 (http://bmcr.brynmawr.edu/2012/2012-04-29.html)

Pingree, D. (1964). "Gregory Chioniades and Palaeologan Astronomy", *Dumbarton Oaks Papers* 18: 135-160.

Polemis, I. (2015). *Theodori Metochitae Carmina* (Turnhout).

Pontani, F. (2013). "Il Vat. Gr. 1087 e l'astronomia bizantina: cenni introduttivi", in Guidetti – Santoni (2013) 9-15.

Ragep, F. J. (1993). *Nasīr al-Dīn al-Tūsī's Memoir on astronomy* (*al-tadhkira fī 'ilm al-hay'a*), vol. I: Introduction, edition and translation (New York).

Reinsch, D. R. (2012). "What Should an Editor Do with a Text Like the 'Chronographia' of Michael Psellos?," in Bucossi – Kihlman (2012) 131-154.

Sathas, C. N. (1872). *Μεσαιωνική Βιβλιοθήκη*, vol. 1 (Venice).

Ševčenko, I. (1962). *Études sur la polémique entre Théodore Métochite et Nicéphore Choumnos* (Brussels).

Ševčenko, I. (1964). "Some Autographs of Nicephorus Gregoras", *Mélanges Georges Ostrogorsky II. Zbornik radova Vizantološkog instituta* 8: 435-450 (repr. Id., *Society and Intellectual Life in Late Byzantium*, London 1981, item XII).

Ševčenko, I. (1975). "Theodore Metochites, the Chora, and the intellectual trends of his time", in P. Underwood (ed.), *The Kariye Djami*, IV (New York), 19-91.

Tannery, P. (1940). *Quadrivium de Georges Pachymère* (Vatican).

Tihon, A. (1978). *Le Petit commentaire de Théon d'Alexandrie aux tables faciles de Ptolémée: histoire du texte, édition critique, traduction* (Vatican).

Tihon, A. (2000). "Theon of Alexandria and Ptolemy's *Handy Tables*", in N.

M. Swerdlow (ed.), *Ancient Astronomy and Celestial Divination* (Cambridge, Mass.), 357-385.

Tihon A. (2001). "Sous la plume de Jean Chortasmenos: des scolies byzantines sur la trepidation des equinoxes", in M. Cacouros – M.-H. Congourdeau (eds), *Philosophie et sciences à Byzance de 1204 à 1453: les textes, les doctrines et leur transmission: actes de la table ronde organisée au XXe Congrès international d'études Byzantines* (Paris), 157-84.

Tihon, A. (2011). "Barlaam de seminara traité sur la date de pâques", *Byzantion* 81: 362-411.

Tihon, A. (2011b). *Ptolemaiou Procheiroi Kanones: les Tables Faciles de Ptolémée*, Volume 1a: Tables A1-A2: Introduction. Édition critique (Louvain-la-Neuve).

Toomer, G. J. (1998). *Ptolemy's Almagest*, translated and annotated; with a foreword by O. Gingerich (Princeton, N.J.).

Van Dieten, J. L. (1979). *Nikephoros Gregoras, Historia Rhomaïke*, v. 2.1 (Kapitel VIII-XI) (Stuttgart).

Weitzmann, K. (1970). *Illustrations in Roll and Codex: a Study of the Origin and Method of Text Illustration* (Princeton).

Wilson, N. G. (1996). *Scholars of Byzantium* (rev. edn., London).

Index